Springer Series in
SOLID-STATE SCIENCES 137

Springer
Berlin
Heidelberg
New York
Barcelona
Hong Kong
London
Milan
Paris
Tokyo

Springer Series in

SOLID-STATE SCIENCES

The Springer Series in Solid-State Sciences consists of fundamental scientific books prepared by leading researchers in the field. They strive to communicate, in a systematic and comprehensive way, the basic principles as well as new developments in theoretical and experimental solid-state physics.

126 **Physical Properties of Quasicrystals**
Editor: Z.M. Stadnik

127 **Positron Annihilation in Semiconductors**
Defect Studies
By R. Krause-Rehberg and H.S. Leipner

128 **Magneto-Optics**
Editors: S. Sugano and N. Kojima

129 **Computational Materials Science**
From Ab Initio to Monte Carlo Methods
By K. Ohno, K. Esfarjani, and Y. Kawazoe

130 **Contact, Adhesion and Rupture of Elastic Solids**
By D. Maugis

131 **Field Theories for Low-Dimensional Condensed Matter Systems**
Spin Systems and Strongly Correlated Electrons
By G. Morandi, P. Sodano, A. Tagliacozzo, and V. Tognetti

132 **Vortices in Unconventional Superconductors and Superfluids**
Editors: R.P. Huebener, N. Schopohl, and G.E. Volovik

133 **The Quantum Hall Effect**
By D. Yoshioka

134 **Magnetism in the Solid State**
By P. Mohn

135 **Electrodynamics of Magnetoactive Media**
By I. Vagner, B.I. Lembrikov, and P. Wyder

136 **Nanoscale Phase Separation and Colossal Magnetoresistance**
The Physics of Manganites and Related Compounds
By E. Dagotto

137 **Quantum Transport in Submicron Devices**
A Theoretical Introduction
By W. Magnus and W. Schoenmaker

Series homepage – http://www.springer.de/phys/books/sss/

Volumes 1–125 are listed at the end of the book.

Wim Magnus   Wim Schoenmaker

Quantum Transport in Submicron Devices

A Theoretical Introduction

With 40 Figures

Springer

Dr. Wim Magnus
Dr. Wim Schoenmaker
STDI/TCAD, IMEC, Kapeldreef 75, 3001 Leuven, Belgium
e-mail: magnus@imec.be
e-mail: schoen@imec.be

Series Editors:

Professor Dr., Dres. h. c. Manuel Cardona
Professor Dr., Dres. h. c. Peter Fulde*
Professor Dr., Dres. h. c. Klaus von Klitzing
Professor Dr., Dres. h. c. Hans-Joachim Queisser
Max-Planck-Institut für Festkörperforschung, Heisenbergstrasse 1, D-70569 Stuttgart, Germany
* Max-Planck-Institut für Physik komplexer Systeme, Nöthnitzer Strasse 38
D-01187 Dresden, Germany

Professor Dr. Roberto Merlin
Department of Physics, 5000 East University, University of Michigan
Ann Arbor, MI 48109-1120, USA

Professor Dr. Horst Störmer
Dept. Phys. and Dept. Appl. Physics, Columbia University, New York, NY 10027 and
Bell Labs., Lucent Technologies, Murray Hill, NJ 07974, USA

ISSN 0171-1873

ISBN 3-540-43396-1 Springer-Verlag Berlin Heidelberg New York

Library of Congress Cataloging-in-Publication Data.
Magnus, Wim, 1954– . Quantum transport in submicron devices: a theoretical introduction/
Wim Magnus, Wim Schoenmaker. p. cm. – (Springer series in solid-state sciences, ISSN 0171-1873; 137)
Includes bibliographical references and index. ISBN 3-540-43396-1 (alk. paper)
1. Electron transport. 2. Holes (Electron deficiencies) I. Schoenmaker, Wim. II. Title. III. Series.
QC176.8.E4 M34 2002   530.4'16–dc21   2002021663

Springer-Verlag Berlin Heidelberg New York
a member of BertelsmannSpringer Science+Business Media GmbH

http://www.springer.de

Typesetting made by the author
Cover concept: eStudio Calamar Steinen
Cover production: *design & production* GmbH, Heidelberg

Printed on acid-free paper   SPIN: 10859825   57/3141/tr - 5 4 3 2 1 0

To our beloved wives Marie-Paule and Vee
for lending their husbands to science

Preface

In this book the problem of transport of electrons and holes is approached from the point of view that a coherent and consistent physical theory can be constructed for transport phenomena. Along the road we will visit various exciting citadels in theoretical physics. The authors will guide the reader in this tour along the strong and weak aspects of the various theoretical constructions. Moreover, our goal is to make clear the mutual coherence and to put each theoretical model in an appropriate perspective. Merely the fact that so many partial solutions have been proposed to describe transport, be it in condensed matter, fluids, or gases, illustrates that we are entering a world of physics with a rich variety of phenomena. Moreover, the subject is not closed and the final theory of transport has still to be written down. Unavoidably, some aspects are commented by us with remarks that contain our personal views and there is room for discussion. Theoretical physics always aims at a unifying picture, but so far this ultimate goal has not been reached. However, this situation should not withhold us from working towards this goal. By presenting this tour among the many very inventive attempts to build this unifying picture, we hope that the reader is inspired and encouraged to help find the unifying principle behind the many faces of transport.

The present work grew out of a series of lectures that we have presented at IMEC, a research center for microelectronics and populated by a mixture of engineers, physicists, chemists and computer scientists, as far as the scientific staff is concerned. The underlying text reflects this audience. The first part of the book describes a series of physical concepts that can be found in numerous text books in physics libraries. One does not possess an operational knowledge of the Schrödinger equation, if one has no knowledge about the Hamiltonian operator. The latter one can only be properly understood within a realm of analytical mechanics.

Each chapter of Part I describes the essential knowledge that is needed to catch up with the present-day language in device engineering. For example, for engineers, these chapters should trigger interest in further reading physics books. Gradually, the text moves towards engineering problems and the second part of this work should convince physicists that there are numerous fundamental problems to be addressed that originate from microelectronic and mesoscopic physics. Therefore, it is our hope that physicists will find their way to engineering libraries.

Whereas Part I describes well-established result in physics, the second part of this work (Chaps. 11 to 17) contain our research results that can be found in the recent physics literature.

A multidisciplinary text book always runs the risk that a specialist in one field is disappointed by that piece of the text that covers his field of expertise, whereas the areas that are outside his field of interest remain hidden in a layer of fog, because the treatment is too condensed. Being aware of this risk, we have put serious effort into presenting the text in such a way that the reader can follow the line of thoughts. As a consequence, the text is occasionally 'narrative'. The intention of some sections is to introduce the reader to a terminology. The goal of these sections is to encourage further study and to provide access to the modern literature in the field.

Acknowledgment

The writing of this book was triggered by the request of a number of researchers at IMEC with a strong desire to gain a better understanding of what is going on in submicron devices. We have benefited substantially from numerous discussions which have also sharpened our views. In particular we would like to express our gratitude to Serge Biesemans, Stefan Kubicek, Carlos Augusto, Bart Sorée, Brahim Elattari, Marc Van Rossum, Kristin De Meyer, Karl-Heinz Bock, An de Keersgieter and many others scientists who contributed in one way or another to our knowledge of transport theory in the way as it became formulated in this book. We would like to thank our directors Herman Maes and Gilbert Declerck for creating the stimulating environment and support. We also benefited from numerous discussions with co-workers outside IMEC. In particular, we mention Jozef Devreese, Volodya Fomin, Sergei Balaban, Volodya Gladilin, Eugene Pokatilov of the University of Antwerp and the University of Moldova. Their impact can hardly be overrated. We are grateful to many research colleagues for sharing their views. To mention a few: Bob Dutton, Zhiping Yu (Stanford University), Wolfgang Fichtner, Andreas Schenk (ETH Zurich), Andreas Wettstein (ISE-AG), Siegfried Selberherr (TU Vienna), Gilles Le Carval (LETI Grenoble) and Marcel Profirescu (Technical University Bucharest). Some of the results that are presented in this book were obtained in research projects funded by Flemish Institute for Science and Technology (IWT) and the European Commission (IST). Last but not least we would like to express our thanks to Claus Ascheron, Angela Lahee and Elke Sauer (Springer-Verlag) for the smooth collaboration that helped to get this monograph off/on the shelf.

Leuven
April 2002

Wim Magnus
Wim Schoenmaker

List of Symbols

$\boldsymbol{A}$	vector potential
$a_q, a_{\boldsymbol{Q}}$	phonon annihilation operator
$a_q^\dagger, a_{\boldsymbol{Q}}^\dagger$	phonon creation operator
$\boldsymbol{B}$	magnetic induction
$\mathbf{C}$	set of complex numbers
$c_{k\sigma}, c_{\alpha l\boldsymbol{k}\sigma}, c_{n\sigma}$	electron annihilation operator
$c_{k\sigma}^\dagger, c_{\alpha l\boldsymbol{k}\sigma}^\dagger, c_{n\sigma}^\dagger$	electron creation operator
$\mathrm{d}\boldsymbol{r}$	line element
$\mathrm{d}\boldsymbol{S}$	surface element
$\mathrm{d}\tau$	volume element
D_{AC}	acoustic deformation potential
D_{it}	density of interface states
$\boldsymbol{E}$	electric field
$\boldsymbol{E}_{\mathrm{C}}$	conservative electric field
$\boldsymbol{E}_{\mathrm{NC}}$	non-conservative electric field
$E, \mathcal{E}$	energy
E_{F}	Fermi energy
E_{L}	longitudinal electric field
$E_{\alpha\boldsymbol{k}}$	one-electron energy
$H, H_0, H_1, H', H_{\mathrm{int}}$	Hamiltonian
h	Planck's constant
$\hbar$	reduced Planck constant ($h/2\pi$)
I	electric current
$I_{\alpha l}$	leakage current carried by a virtual bound state
$\boldsymbol{J}$	electric current density
$J_{\alpha l}^z$	leakage current density carried by a virtual bound state
$k, k_1, k_2, \ldots$	single-particle quantum numbers
$k_{\mathrm{g},\alpha}, k_{1,\mathrm{ox},\alpha}, k_{2,ox,\alpha}, \ldots, k_{1,\mathrm{w},\alpha}, k_{2,w,\alpha}, \ldots, k_{\mathrm{s},\alpha}$	electron wave number

$\boldsymbol{k}$	electron wave vector
k_{B}	Boltzmann's constant
L	Lagrangian
$\boldsymbol{L}$	total angular momentum
l	subband index
$\boldsymbol{l}$	one-particle angular momentum
m_0	free electron mass
$m_{\mathrm{g}\alpha x}, m_{\mathrm{g}\alpha y}, m_{\mathrm{g}\alpha z}, m_\alpha^{\parallel}, m_\alpha^{\perp}$, $m^{\parallel}, m^{\perp}, m_{1,\mathrm{ox},\alpha}, m_{2,ox,\alpha}, \dots$, $m, m_{\mathrm{n}}, m_{\alpha x}, m_{\alpha y}, m_{\alpha z}$	electron effective masses
m_{p}	hole effective masses
$M(\boldsymbol{Q})$, v_q	electron–phonon coupling strength
N	number of particles, coordinates or modes
N_A	acceptor doping density
n	electron concentration
N_{ox}	number of gate stack layers
N_{w}	number of sublayers in the inversion layer
Φ	wave function, magnetic flux
$\lvert\Phi\rangle$	state vector
$\phi_{\alpha l}(z)$	subband wave function
$p_1, p_2, \dots$	generalized momenta
$\boldsymbol{p}$	one-particle momentum
$\boldsymbol{P}$	total momentum
p	hole concentration
q_{D}	Debye wave number
$q_1, q_2, \dots$	generalized coordinates
$\boldsymbol{q}, \boldsymbol{Q}$	phonon or photon wave vector
$\mathbf{R}$	set of real numbers
$\boldsymbol{r}, \boldsymbol{r}_1, \boldsymbol{r}_2, \dots$	position vector
$\boldsymbol{S}$	spin vector operator
S	entropy
s	spin index
T	lattice temperature
T_{e}	electron temperature
$T_{\mathrm{g},\alpha} T_{1,\mathrm{ox},\alpha}, T_{1,\mathrm{ox},\alpha}, \dots$, $T_{1,\mathrm{w},\alpha}, T_{2,\mathrm{w},\alpha}, \cdots$	transfer matrices
$t_{\mathrm{g}}, t_{\mathrm{ox}}, t_{1,\mathrm{w}}, t_{2,\mathrm{w}}, \dots$	layer thicknesses
U	potential energy
V	scalar electric potential
$\boldsymbol{v}$	velocity, one-particle velocity operator
v_{D}	electron drift velocity
v_{S}	velocity of sound
V_ε	electromotive force
w	electron energy density

$W_{\alpha l}$	subband energy
α	valley index
$\partial\Omega$	boundary surface of Ω
ε_{S}	silicon permittivity
ε_k	one-electron energy
ε_0	vacuum permittivity
$\Gamma_{\alpha l}$	resonance width of virtual bound state
μ	chemical potential
μ_0	vacuum permeability
$\phi_k, \psi_{\alpha \boldsymbol{k}}$	electron wave function
ψ	field operator
Ω	connected subset of $\mathbf{R}^3$
ω_{D}	Debye frequency
ω_q	phonon, photon frequency
ϱ	density matrix, statistical operator, charge density
ϱ_{Si}	silicon mass density
σ	spin index
$\tau_{\alpha l}$	lifetime of a virtual bound state

Contents

Part I. General Formalism

Part I

General Formalism

e

1. The Many Faces of Transport

Nature exhibits a bewildering variety of different transport phenomena. We all are aware that walking from one place to another demands the use of muscular strength to realize the activity. It requires an effort to perform the motion. This simple example shows that transport is related to energy conversion. We must convert potential energy into kinetic energy. However, this is not the whole story. We must also overcome friction during the walk in order to sustain the motion. These two aspects will play a crucial role in identifying motion as transport. In general we could say that motion in a physical system is a transport phenomenon, if the motion is correlated with dissipation. Dissipation is conversion of energy into chaotic motion or heat. The motion of the planets around the sun is usually not considered as transport, because there is no accompanying, or negligible, friction. Other examples of transport are the flow of water through a narrow tube, where friction is caused by the wall of the tube on the water flow: the transport of electrons in semiconductors, where dissipation or friction is caused by collisions of electrons on impurities, lattice defects and lattice vibrations, and possible other mechanisms. The latter example is already somewhat more subtle, because in a perfect lattice there will be no frictional forces, at least at zero temperature. Very complicated transport phenomena are for instance the rising of a smoke ring from a cigarette, whose motion appears very regular for about one meter but then becomes rather abruptly very chaotic. The airflow after the wing of an airplane or a car usually has very turbulent motion. A complete physical theory of turbulence does not yet exist, but it is generally accepted that this motion is characterized by extreme non-linear behavior, i.e. a perturbative approach is not adequate for an accurate description. Perturbation theory will appear over and over again in the course of these lectures. The reason for this is quite simple: we have nothing better to present. Perturbation theories have all in common that some approximate model can be solved and that the part that has been left out, can be incorporated by a limited number of corrective terms. The success or failure of a perturbative model depends on the level of sophistication of the approximation or the zeroth-order model. In the last decade, considerable success has been achieved in finding a good model for the fractional quantum Hall effects. However, a good zeroth order model for quantum transport in mesoscopic systems is still lacking. We do not yet fully understand in a satisfactory way, how the von Klitzing resistance ($R_{\rm K} = \frac{h}{e^2} = 25812.807\Omega$) arises in point contacts, but we believe that a major breakthrough will be achieved if a good zeroth order model is found. What makes perturbation theory so special? Apart from having nothing better to present, it should be noticed that once we have found a solvable zeroth order model, we also have at our disposal an artillery of standard techniques to improve the theoretical description. In the following we will narrow the scope of our tour along the various theories of perturbative approaches. Therefore, we will not touch upon non-perturbative domains of transport theory, such as non-linear dynamics, turbulence models, the theories of Prigogine and coworkers, for long range order in dissipative systems. We will see that despite these restrictions there are left over many exciting stops for being impressed by the beauty of the methods. In particular, realizing that perturbative models may be very ingenious constructions that go far beyond simple non-interactive systems, there are good reasons to hope that transport in semiconductor sub-micron devices may be successfully described with the methods that are presented in the following chapters. However, a well-organized tour starts with filling our rucksack with the appropriate supplies. The collection of the outfit starts with the basics of classical and quantum mechanics as well as equilibrium statistical physics.

The development of non-equilibrium statistical physics or transport theory has been hampered by debates dealing with very fundamental questions concerning the origin of irreversibility. The key question that was clearly posed for the first time by Loschmidt was the following: How can the laws of mechanics that are time-reversal invariant give rise to the irreversible behavior of nature. Now there are at least two answers to this question. (1) Chaos: since real systems are in general highly non-linear, the solution of the equations of motion result into chaotic behavior. Arbitrary small perturbations on the initial conditions lead to fully different results after a finite time lap. Since initial conditions are only known up to finite accuracy, the solutions are deemed to be irreversible. (2) Infinite number of degrees of freedom: real systems consists of a huge number of particles. In the limit of an infinite number of particles, irreversible dynamics is recovered. Simulation of model systems demonstrate that 'infinite' is already reached for ~ 30 degrees of freedom. Chapter 14 contains an explicit demonstration of this claim for a phonon–electron system.

The success of finally getting the Loschmidt paradox under control made researcher rather careless concerning a new problem that is deeming: the unification of irreversibility with quantum mechanics. The problem is at least pertinent as the problem of irreversibility in classical physics and finding the solution becomes urgent due to the fact that experiments penetrate into the realm of transport dominated by quantum-mechanical behavior. Why do we claim that there really is a problem? Transport is pictured as preferred motion from one region in space to another region. The transport is fueled by the presence of different chemical potentials (diffusive transport), different temperatures, or some other intrinsic variable. However, these variables only have a sound physical meaning in equilibrium conditions. Turning these variables into local ones is possible but the price tag is that one has to assume in a rather ad-hoc way that spatial cells exist for which the local values are valid. The (hydrodynamical) coarse graining makes sense provided that sufficient thermalization or interaction occurs. Such a description mimics the classical reasoning but now applied to a quantum-mechanical system. For real mesoscopic systems, above condition is not fulfilled and therefore one should refrain from using these local hydrodynamical variables. Of course, this has profound consequences. Just to mention an example: in order to describe a sustained current in a quantum wire one usually attached reservoirs at different chemical potential to the wire. Next the problem is subdivided into three subsystems (two reservoirs and the wire). Quantum mechanically one is not allowed to subdivide the system, since the position operator and the momentum operator (needed for constructing the Hamiltonian) do not commute. As a consequence, the full system should be treated as a single entity when applying quantum mechanics. This will be elaborated in Chap. 5.

Based on the experiences gathered in the Chaps. 1 to 10, we will discuss a series of applications in Chaps. 11 to 17. The examples were develop by us (occasionally together with a number of colleagues) in recent years and have been published in physics journals. The examples illustrate our approach to quantum transport problems. In particular, the calculation of gate leakage currents is inspired on our desire to respect the rules of the quantum-computational game. Although we realize that alternative formulations e.g. ones that are based on Bardeen's perturbative approach are valuable, it is reassuring that the gate leakage can be obtained using conventional many-particle quantum physics.

The intention of the final Chap. 18 is primarily to trigger active participation of the reader in the fascinating research area that falls under the umbrella named 'quantum transport'. Some lines of thoughts are presented that either may be falsified or verified. At the time of write this monograph we have no hint what will be the outcome. If we manage to leave the reader with a puzzled mind, then we have achieved our goal in the final chapter.

2. Classical Mechanics

2.1 Generalized Coordinates and Constraints

After Newton formulated his principles of mechanics, a more sophisticated formulation of classical mechanics was set up which is currently known as analytical mechanics. Although the principles remain untouched, the mathematical formulation is more general and more powerful and as such, more suitable to attack more complicated problems. In particular, one complication that arises is the inclusion of more than one, i.e. many particles. An analytical description starts with the identification of the generalized coordinates and the constraints [1, 2]. The position vector of a particle of the system becomes a function of the generalized coordinates

$$\boldsymbol{r}_i = \boldsymbol{r}_i(q_1, q_2, \ldots, q_n, t) . \tag{2.1}$$

Let us illustrate this with an example. Consider the configuration of two mass points connected by a bar of length L_2 and mass m_1 is connected to a fixed point by a bar of length L_1 as drawn in Fig. 2.1. There are 6 generalized coordinates, which are in a Euclidean coordinate system $(x_1, y_1, z_1, x_2, y_2, z_2)$. However, there are two constraints, which are determined by the bars of length L_1 and L_2.

$$\begin{aligned} x_1^2 + y_1^2 + z_1^2 &= L_1^2 \\ (x_2 - x_1)^2 + (y_2 - y_1)^2 + (z_2 - z_1)^2 &= L_2^2 . \end{aligned}$$

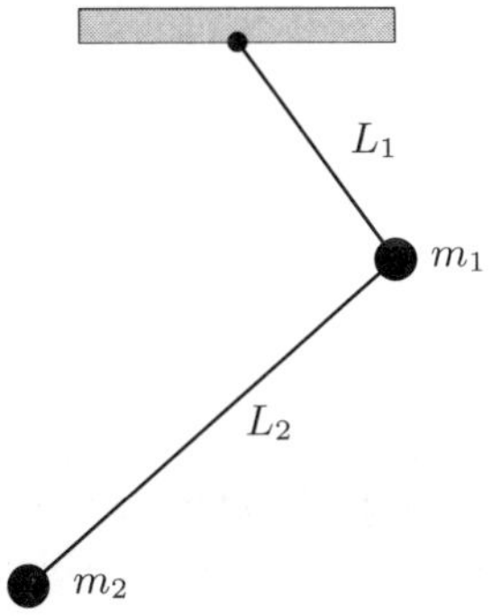

Fig. 2.1. Constrained two-particle system

These constraints limit the real motion, so that effectively there are only four degrees of freedom. Using spherical coordinates we may identify the generalized coordinates as (see Fig. 2.2)

$$
\begin{aligned}
q_1 &= \phi_1 \\
q_2 &= \theta_1 \\
q_3 &= \phi_2 \\
q_4 &= \theta_2
\end{aligned}
$$

and

$$
\begin{aligned}
\boldsymbol{r}_1 &= \boldsymbol{r}_1(\phi_1, \theta_1, t) \\
\boldsymbol{r}_2 &= \boldsymbol{r}_2(\phi_1, \theta_1, \phi_2, \theta_2, t) \ .
\end{aligned}
$$

Analytical mechanics describes this system using a four-dimensional configuration space. The whole topic of analytical mechanics is founded on the ingenious idea of the French mathematician d'Alembert, who invented the concepts of virtual displacements.

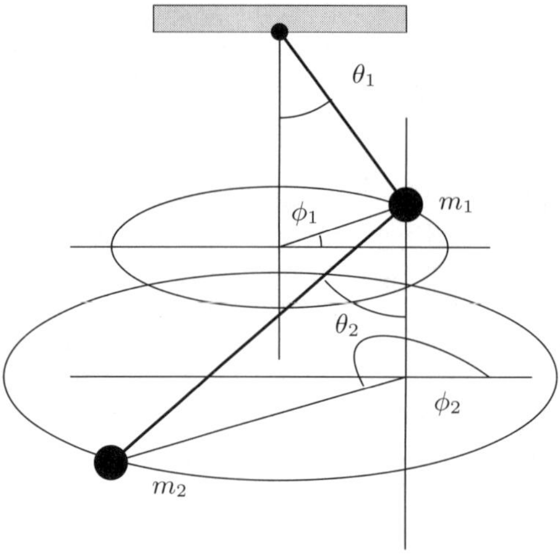

Fig. 2.2. Generalized coordinates for the system of Fig. 2.1

2.2 d'Alembert's Principle

A virtual displacement is a displacement which does not take place in real time but corresponds to a displacement which is imaginary: it takes place in the mind of the scientist. It corresponds to a what–if question. The scientist imagines a 'virtual reality'. Let us illustrate these remarks by an example. Consider a marble being at rest on the bottom of a bowl as is shown in

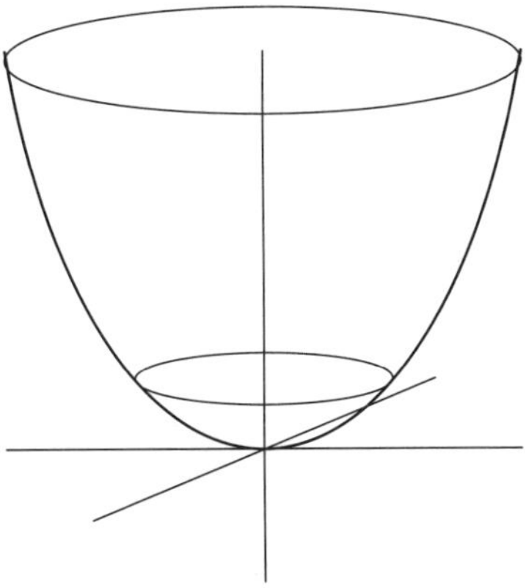

Fig. 2.3. Shape of the potential of a particle in a bowl

Fig. 2.3. We could now ask the following what–if question: 'What would happen to the potential energy, if the position the position of the marble was not $\boldsymbol{r}$ but $\boldsymbol{r}' = \boldsymbol{r} + \delta\boldsymbol{r}$?'
The displacement $\delta\boldsymbol{r}$ has nothing to do with a real motion in time: it is imaginary. If we further imagine that the displacement is compatible with the constraint that the marble is on the surface of the bowl, then the answer of the what–if question is that for all compatible virtual displacements the potential energy increases, i.e.

$$\delta V = V(\boldsymbol{r}') - V(\boldsymbol{r}) > 0 . \tag{2.2}$$

The fact that the marble is at rest or in static equilibrium corresponds to a position $\boldsymbol{r}$ for which $V(\boldsymbol{r})$ is a (local) minimum. This minimum can be found by demanding that for all *infinitesimal* compatible virtual displacements that are compatible with the constraints, we have

$$\delta V = \frac{\partial V}{\partial \boldsymbol{r}} \cdot \delta\boldsymbol{r} = 0 . \tag{2.3}$$

Using the generalized coordinates this equation becomes

$$\delta V = \sum_i \frac{\partial V}{\partial q_i}\, \delta q_i = 0 . \tag{2.4}$$

The potential V is understood to be associated to the *applied* force, $\boldsymbol{F}^{\,\mathrm{a}}$, which is in our example the force of gravity pulling the marble downward, i.e.

$$\boldsymbol{F}^{\,\mathrm{a}} = -\frac{\partial V}{\partial \boldsymbol{r}} \equiv -\nabla V(\boldsymbol{r}) . \tag{2.5}$$

There is also a compensating force, the force of constraint, or the reaction force $\boldsymbol{F}^{\,\mathrm{c}}$, which guarantees that the marble does not fall freely to the center of the earth. Since the marble is at rest, the total force acting on the marble vanishes, i.e.

$$\boldsymbol{F}^{\text{tot}} = \boldsymbol{F}^{\text{a}} + \boldsymbol{F}^{\text{c}} = \boldsymbol{0} \,. \tag{2.6}$$

Since $\boldsymbol{F}^{\text{tot}} = \boldsymbol{0}$, the virtual work done in any virtual displacement amounts to

$$\delta W = \boldsymbol{F}^{\text{tot}} \cdot \delta \boldsymbol{r} = 0 \,. \tag{2.7}$$

Using (2.3) and (2.5), we obtain

$$\delta W = -\delta V + \boldsymbol{F}^{\text{c}} \cdot \delta \boldsymbol{r} \tag{2.8}$$

and therefore

$$\boldsymbol{F}^{\text{c}} \cdot \delta \boldsymbol{r} = 0 \,. \tag{2.9}$$

d'Alembert has elevated this equation to the corner stone or key working principle of analytical mechanics. In words: *The work performed by the forces of constraints is zero for all virtual displacements.*

2.3 Reduction of Dynamics to Statics

So far we considered only static or equilibrium situations. The real genius of d'Alembert shows up in the way how he uses this principle in dynamical applications. Starting from Newton's force law

$$\boldsymbol{F}^{\text{tot}} = m\boldsymbol{a} = m\ddot{\boldsymbol{r}} \,, \tag{2.10}$$

d'Alembert defines a force of inertia as

$$\boldsymbol{F}^{\text{i}} = -m\boldsymbol{a} \,. \tag{2.11}$$

Then Newton's law is

$$\boldsymbol{F}^{\text{TOT}} = \boldsymbol{F}^{\text{tot}} + \boldsymbol{F}^{\text{i}} = \boldsymbol{0} \,. \tag{2.12}$$

This looks like a simple rewriting but there is a deep motivation hidden behind it. Generalizing (2.7), d'Alembert achieved to reduce any *dynamical* problem to a *static* problem. Considering all particles of the system, we obtain

$$\sum_{j=1}^{n} \left(\boldsymbol{F}_j^{\text{a}} + \boldsymbol{F}_j^{\text{c}} + \boldsymbol{F}_j^{\text{i}}\right) \cdot \delta \boldsymbol{r}_j = 0 \,. \tag{2.13}$$

Using d'Alembert's principle, (2.9), as well as a many-particle version of (2.5)

$$\boldsymbol{F}_j^{\text{a}} \cdot \delta \boldsymbol{r}_j = -\sum_{\nu} \frac{\partial V}{\partial \boldsymbol{r}_j} \frac{\partial \boldsymbol{r}_j}{\partial q_\nu} \delta q_\nu = -\sum_{\nu} \frac{\partial V}{\partial q_\nu} \delta q_\nu \,, \tag{2.14}$$

where the sum is taken over all generalized coordinates, we derive:

$$m_j \ddot{\boldsymbol{r}}_j \cdot \delta \boldsymbol{r}_j = -\sum_{\nu} \frac{\partial V}{\partial q_\nu} \delta q_\nu \,. \tag{2.15}$$

2.4 The Lagrangian and Euler–Lagrange Equations

The non-relativistic kinetic energy T is defined in terms of the velocities $\boldsymbol{v}_j = \dot{\boldsymbol{r}}_j$ through

$$T = \sum_j \frac{1}{2} m_j v_j^2 \tag{2.16}$$

from which one can easily derive

$$m_j \ddot{\boldsymbol{r}}_j \cdot \delta \boldsymbol{r}_j = \left[\frac{\mathrm{d}}{\mathrm{d}t} \left(\frac{\partial T}{\partial \dot{q}_\nu} \right) - \frac{\partial T}{\partial q_\nu} \right] \delta q_\nu \; . \tag{2.17}$$

We encourage the reader to prove (2.17), since it will help to obtain a deeper understanding of virtual displacements. Putting everything together we obtain

$$\frac{\mathrm{d}}{\mathrm{d}t} \left(\frac{\partial T}{\partial \dot{q}_\nu} \right) - \frac{\partial T}{\partial q_\nu} = - \frac{\partial V}{\partial q_\nu} \; . \tag{2.18}$$

If the potential has no velocity dependence, then

$$\frac{\mathrm{d}}{\mathrm{d}t} \left(\frac{\partial L}{\partial \dot{q}_\nu} \right) - \frac{\partial L}{\partial q_\nu} = 0 \; , \qquad \nu = 1, \ldots, n \tag{2.19}$$

where $L = T - V$ is the *Lagrangian.* The equations (2.19) are known as the *Euler–Lagrange equations.*

2.5 Maupertuis and Maserati

Is this where the action is?

Can we develop some physical intuition for the mathematical construction $L = T - V$? At first sight it looks as if we only tried to write (2.17) in a more compact notation. There is however more behind it [3]. Define the *action functional* S as

$$S = \int_{t_0}^{t_1} L\,(q, \dot{q}, t)\, \mathrm{d}t \; . \tag{2.20}$$

The Euler–Lagrange equations correspond to the condition that S is minimal, i.e. $\delta S = 0$. Of all space-time paths between $(\boldsymbol{r}_0, t_0)$ and $(\boldsymbol{r}_1, t_1)$, nature realizes that path for which S is minimal. This is illustrated in Fig. 2.4.

In order to show the equivalence, it suffices to prove that for any virtual deviation from the the physical realized path, the action is stationary, i.e. $\delta S = 0$, if the Euler–Lagrange equations are valid and vice versa. If you ride your bicycle from A to B, you apply the principle of least action, provided you avoid using your brakes and avoid quick accelerations. *Porsches, Ferraris and Maseratis are not made for applying the principle of least action!*

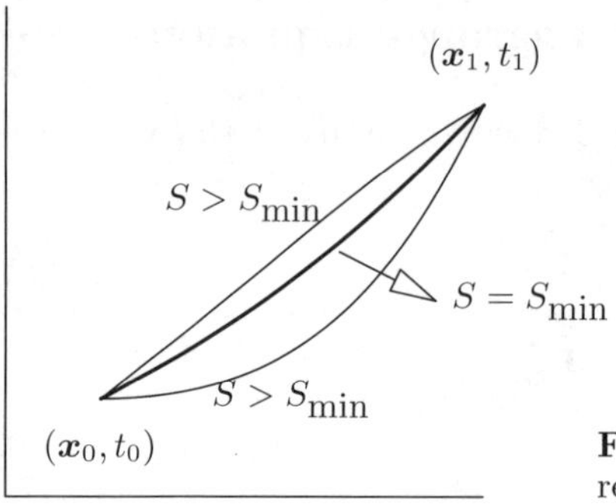

Fig. 2.4. The path with minimal action corresponds to the real motion

2.6 From Lagrangian to Hamiltonian

The Euler–Lagrange equations contain second-order partial derivatives with respect to time. The Lagrangian is a function defined on a $(2n + 1)$-dimensional space with coordinates $q_\nu, \dot{q}_\nu, t$ and $\nu = 1, ..n$. Hamilton managed to find a formulation in which only *first-order* partial derivatives with respect to time occur. The Hamilton function or Hamiltonian is defined as

$$H = \sum_{\nu=1}^{n} p_\nu \, \dot{q}_\nu - L(q, \dot{q}, t) \; , \tag{2.21}$$

where $\{p_\nu\}$ denote the canonical momenta defined as

$$p_\nu = \frac{\partial L}{\partial \dot{q}_\nu} \; . \tag{2.22}$$

The Hamiltonian does not depend on $\dot{q}_\nu$, i.e. $H = H(p, q, t)$, since

$$\frac{\partial H}{\partial \dot{q}_\nu} = p_\nu - \frac{\partial L}{\partial \dot{q}_\nu} = 0 \tag{2.23}$$

and therefore it is also defined on a $(2n + 1)$-dimensional space with coordinates (q_ν, p_ν, t).

For various physical systems the kinetic energy takes the following form:

$$T = \frac{1}{2} \sum_{\nu=1}^{n} \dot{q}_\nu^2 \; . \tag{2.24}$$

Hence, the Hamiltonian reduces to the total energy of the system

$$H = T + V \; . \tag{2.25}$$

The Hamilton equations of motion, i.e. the first-order partial differential equations, are

$$\begin{aligned} \frac{\partial H}{\partial q_\nu} &= -\frac{\partial L}{\partial q_\nu} = -\frac{\mathrm{d}}{\mathrm{d}t}\left(\frac{\partial L}{\partial \dot{q}_\nu}\right) = -\dot{p}_\nu \\ \frac{\partial H}{\partial p_\nu} &= \dot{q}_\nu \end{aligned} \tag{2.26}$$

or

$$\dot{p}_\nu = -\frac{\partial H}{\partial q_\nu} \qquad \text{and} \qquad \dot{q}_\nu = \frac{\partial H}{\partial p_\nu} \ . \tag{2.27}$$

The Hamilton equations can also be derived from an action principle in the (p, q, t)-space by defining the action integral

$$A = \int_{t_0}^{t_1} (p_\nu q_\nu - H(q_\nu, p_\nu, t))\, \mathrm{d}t \ . \tag{2.28}$$

In order to evaluate the integral it should be noted that $\dot{p}_\nu$ and $\dot{q}_\nu$ are dependent quantities for each trajectory in the (q, p, t)-space as is illustrated in Fig. 2.5. The time evolution of the Hamiltonian $H = H(q, p, t)$ is given by

$$\frac{\mathrm{d}H}{\mathrm{d}t} = \sum_\nu \left(\frac{\partial H}{\partial q_\nu} \dot{q}_\nu + \frac{\partial H}{\partial p_\nu} \dot{p}_\nu \right) + \frac{\partial H}{\partial t} = \frac{\partial H}{\partial t} \ . \tag{2.29}$$

Using Hamilton's equations, we obtain for any observable defined on q, p and t that

$$\begin{aligned} \frac{\mathrm{d}A}{\mathrm{d}t} &= \sum_\nu \left(\frac{\partial A}{\partial q_\nu} \frac{\partial H}{\partial p_\nu} - \frac{\partial A}{\partial p_\nu} \frac{\partial H}{\partial q_\nu} \right) + \frac{\partial A}{\partial t} \\ &= [A, H] + \frac{\partial A}{\partial t} \ . \end{aligned} \tag{2.30}$$

The Poisson bracket is defined as

$$[u, v] = \sum_\nu \left(\frac{\partial u}{\partial q_\nu} \frac{\partial v}{\partial p_\nu} - \frac{\partial u}{\partial p_\nu} \frac{\partial v}{\partial q_\nu} \right) \ . \tag{2.31}$$

An observable is time independent if the following conditions are satisfied

- There is no explicit time dependence: $\partial A / \partial t = 0$.
- The Poisson bracket with H vanishes: $[A, H] = 0$.

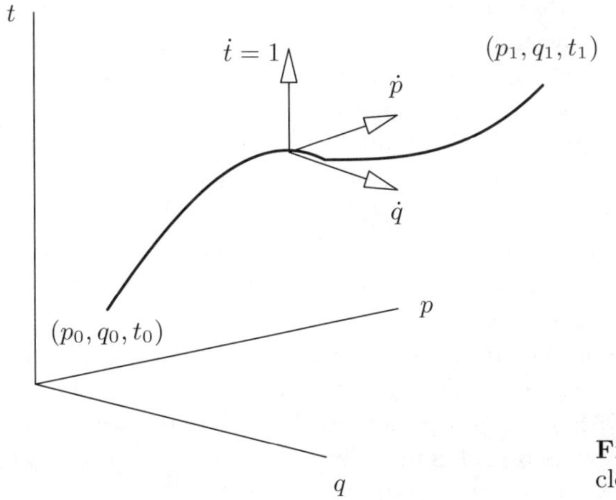

Fig. 2.5. Trajectory of a particle in the (q, p, t)-space

2.7 Phase Space and Configuration Space

We have collected here the key ingredients of analytical mechanics. There is much more to say but for the applications that we have in mind, we restrict ourselves to the following remarks. A dynamical system of N particles in three-dimensional space has $n = 3N$ degrees of freedom, provided that there are no constraints. The time evolution of the system can be described as a *single point* moving in a $6N$-dimensional space. This space with the $6N$ coordinates q_ν, p_ν is called phase space, Gibbs space or Γ-space. The space, on which H is defined and which is spanned by q_ν, p_ν, t is called *extended* phase space.

If there are $k < 6N$ constraints then the total number of degrees of freedom is $n = 3N - k$ and the system moves as a single point in a $(6N - 2k)$-dimensional space.

The choice of the coordinate system is not unique. One may choose a new coordinate system

$$Q_\nu = Q_\nu(q_1, q_2, \ldots, q_n; p_1, p_2, \ldots, p_n, t) \tag{2.32}$$

$$P_\nu = P_\nu(q_1, q_2, \ldots, q_n; p_1, p_2, \ldots, p_n, t)\ . \tag{2.33}$$

A canonical transformation corresponds to the condition that the Hamilton equations in the new coordinate system are identical in appearance as in the old coordinate system, i.e. there exists a Hamilton function $K = K(P, Q, t)$ such that

$$\dot{P}_\nu = -\frac{\partial K}{\partial Q_\nu} \tag{2.34}$$

$$\dot{Q}_\nu = \frac{\partial K}{\partial P_\nu}\ . \tag{2.35}$$

In applying the variational method to the action integrals it was required to consider path variations, such that the end points remain fixed. Schwinger has constructed a more general action principle in which this limitation is not included. The canonical momenta, p_ν i.e. (2.22) as well as the Hamiltonian can be *derived* from Schwinger's action principle [4].

We have presented the fundamentals of analytical mechanics with the goal the acquaint the reader with the concept of phase space. Another goal has been to give the reader a good notion of virtual moves. We have seen virtual displacements and virtual paths. *Virtual reality* traces back to the 18th century, when sublime minds as those of Lagrange, d'Alembert, Euler invented analytical mechanics.

Is it possible to formulate non-equilibrium statistical mechanics by reducing the problem of transport to an equilibrium statistical physics problem, and mimicing in this way d'Alembert's ingenious step? We do not know the answer to this question, but it is an intriguing line of thought. In the final chapter we will speculate how such a formulation might look.

Viewing the dynamics of a collection of many particles as a *single point* traversing in a high-dimensional space ($6n$ coordinates), is one possible description. We may also annotate the $\boldsymbol{r}$ and $\boldsymbol{p}$ value of each particle at each instant in time. Each particle can be represented as a point in a six-dimensional space spanned by the coordinates (x, y, z, p_x, p_y, p_z). The full collection of particles are described by N trajectories in this space. The scattering of these points is described by the distribution function $f(\boldsymbol{r}, \boldsymbol{p}, t)$. The six-dimensional space is referred to as the Boltzmann space or μ-space.

Space	Synonym	Dimension	System representation
Γ-space	Gibbs space phase space	$6N - 2k$	single point
μ-space	Boltzmann space configuration space	6	N points

2.8 The Liouville Equation

Statistical mechanics does not aim at solving the equations of motion for all the particles in a system in full microscopic detail [5, 6, 7]. Instead, it aims at identifying the collective variables which are of real interest for understanding the dynamical behavior and which are useful in engineering the system. The introduction of probability in a macroscopic description using collective variables, is possible because there are many *different* microscopic states, represented by points in phase space which correspond to equivalent macroscopic states. A macroscopic state can be characterized by the the total energy, the total number of particles, the total charge, the volume of the system, but also other variables may be relevant. In general it is advised to start with observables which are conserved in time. The wide variety of different micro-states is mimiced by producing a large number of virtual copies ('clones' or 'replicas') of the same physical system. The copies have the same energy, the same number of particles, the same total charge, same volume, etc. The collection of all virtual copies is called the *ensemble* and we want to determine the probability distribution $\varrho(q, p, t)$ for the ensemble over the phase space. The function ϱ is named the *density function.*

Since the total number of ensembles N is fixed, we have at any time that $\int_{\text{phase space}} \mathrm{d}\Omega \, \varrho(q, p, t) = N(t)$ is a constant. Just as with charge conservation in ordinary space, where $\int_{\text{volume}} \mathrm{d}V \, \varrho(\boldsymbol{r}, t) = Q(t)$ is constant. We may transpose this result to the phase space and obtain a local conservation law analogous to

$$\begin{aligned} \frac{\partial \varrho}{\partial t} &= -\nabla \cdot \boldsymbol{J} \\ \boldsymbol{J} &= \varrho \boldsymbol{v} \,, \end{aligned} \tag{2.36}$$

which in this case becomes

$$\begin{aligned}\frac{\partial \varrho}{\partial t} &= -\sum_{\nu}\left(\frac{\partial}{\partial q_\nu}(\varrho \dot{q}_\nu) + \frac{\partial}{\partial p_\nu}(\varrho \dot{p}_\nu)\right) \\ &= -\sum_{\nu}\left(\frac{\partial \varrho}{\partial q_\nu}\dot{q}_\nu + \varrho\frac{\partial \dot{q}_\nu}{\partial q_\nu} + \frac{\partial \varrho}{\partial p_\nu}\dot{p}_\nu + \varrho\frac{\partial \dot{p}_\nu}{\partial p_\nu}\right) . \end{aligned} \tag{2.37}$$

Since $\frac{\partial \dot{q}_\nu}{\partial q_\nu} = \frac{\partial \dot{p}_\nu}{\partial p_\nu} = 0$, we derive from Hamilton's equations that

$$\frac{\partial \varrho}{\partial t} = [H, \varrho] = \mathcal{L}\, \varrho . \tag{2.38}$$

The above equation is the so-called Liouville equation and $\mathcal{L}$ is the Liouville operator.

2.9 The Micro-Canonical Ensemble

In equilibrium statistical physics, we expect that ϱ does not depend on time explicitly. As a consequence $[H, \varrho] = 0$ and this is guaranteed if ϱ is merely a function of H. What else can we say about ϱ? Of course, each virtual copy should be consistent with the boundary conditions that are provided by the macroscopic state. If the macroscopic state is characterized by a fixed amount of energy, then the least-biased guess for the distribution function ϱ is the one that treats all micro states that are compatible with the energy condition on a equal footing.

The micro-canonical ensemble is defined by the condition [6]

$$\varrho(q,p) = \begin{cases} 1 \text{ if } E < H(q,p) < E + \Delta \\ 0 \text{ otherwise} . \end{cases} \tag{2.39}$$

To this density function corresponds a volume in phase space, being

$$\Gamma(E) = \int \varrho \,\mathrm{d}\Omega = \int_{E<H<E+\Delta} \mathrm{d}^{3N}q \,\mathrm{d}^{3N}p . \tag{2.40}$$

Furthermore, let us define

$$\Sigma(E) = \int_{H<E} \mathrm{d}^{3N}q \,\mathrm{d}^{3N}p . \tag{2.41}$$

Then, the volume in phase space is given by

$$\Gamma = \frac{\partial \Sigma(E)}{\partial E}\, \Delta . \tag{2.42}$$

The entropy S is proportional to the logarithm of the volume occupied in phase space, i.e.

$$S(E) = k_{\mathrm{B}} \log \Gamma(E) \,, \tag{2.43}$$

where $k_{\mathrm{B}} = 1.38062 \times 10^{-23}\mathrm{J/K}$ is the Boltzmann constant.

A remarkable property of a high-dimensional object is that the major part of its volume is concentrated near its surface. Therefore, we may just as well define the entropy through the relation

$$S(E) = k_{\mathrm{B}} \log \Sigma(E) \,. \tag{2.44}$$

An example will illustrate these remarks. Suppose there are N atoms of some gas contained in a real-space volume of size V and let us assume that the total energy is E. For a gas of non-interacting atoms, this means that

$$E = \sum_{i=1}^{N} \frac{p_i^2}{2m} \tag{2.45}$$

and

$$\Sigma(E) = V^N \int_{\sum_i^N p_i^2 < 2mE} \mathrm{d}^{3N} p = V^N \, \Omega_{3N}(\sqrt{2mE}) \,, \tag{2.46}$$

where $\Omega_k(r)$ is the volume of a kth-dimensional sphere with radius r, i.e.

$$\Omega_k(r) = \frac{\pi^{\frac{k}{2}} r^k}{(\frac{k}{2})!} \,. \tag{2.47}$$

From this formula we easily derive that indeed for $k \to \infty$ all volume is found in the surface, i.e.

$$\frac{\mathrm{d}\Omega/\mathrm{d}r}{\Omega} = \frac{k}{r} \to \infty \,. \tag{2.48}$$

The phase-space volume is therefore

$$\Sigma(E) = \frac{\pi^{\frac{3N}{2}} (2mE)^{\frac{3N}{2}}}{(\frac{3N}{2})!} V^N \,. \tag{2.49}$$

With the use of Stirling's formula, $\log n! \simeq n \log n - n$, we find that the entropy is

$$S = k_{\mathrm{B}} N \log \left(\left[\frac{4m\pi E}{3N} \right]^{\frac{3}{2}} V \right) + \frac{3}{2} k_{\mathrm{B}} N \,. \tag{2.50}$$

The ideal gas is fully described by this entropy formula. The internal energy is found by solving this expression for E and identify the result as $U(N, V)$. The temperature and pressure are

$$\frac{1}{T} = \left(\frac{\partial S}{\partial E} \right)_V \quad \text{and} \quad p = T \left(\frac{\partial S}{\partial V} \right)_E \tag{2.51}$$

and the equation of state $pV = N k_{\mathrm{B}} T$ is obtained.

The micro-canonical ensemble is well-suited for explanatory reasons. It gives a clear picture how statistical considerations enter a macroscopic modeling of systems. However, there are a number of serious drawbacks.

1. For practical calculations the constraint of the system, moving exactly in an energy shell or on an energy surface, is difficult to implement.
2. It is also not satisfactory from a the point of view that it pretends to know exactly how many particles participate as well as the exact value of the total energy that is contained in the system.

In order to relax these constraints we will be satisfied with obtaining the *expectation values* for these quantities rather than the exact values. This leads to the canonical and grand-canonical ensembles that will be discussed in Chap. 7.

2.10 The Boltzmann Equation

In this section we will consider the dynamical behavior of a system as a collection of points moving in the configuration space. Each particle is represented by a point in a 6-dimensional space. As time progresses, each particle sweeps out a trajectory in this space. We may introduce a 'macroscopic' view on this space, and make a subdivision of of cells which look large compared to the size of one particle but are small compared to the outer dimensions of the system. By counting the number of particles in each cell, we obtain a distribution functions, which from the outside looks as a smooth continuous function of its arguments, i.e. the representing coordinates $\boldsymbol{q}$, $\boldsymbol{p}$ and the time instant t. A connection to the microscopic world is provided by the following relation

$$f(\boldsymbol{q},\boldsymbol{p},t) = \int_{\Delta\omega(\boldsymbol{q},\boldsymbol{p})} \mathrm{d}^3 q\, \mathrm{d}^3 p \sum_{i=1}^{N} \delta\left(\boldsymbol{q}_i(t) - \boldsymbol{q}\right) \delta\left(\boldsymbol{p}_i(t) - \boldsymbol{p}\right) . \tag{2.52}$$

The process of dividing the μ-space into cells of size $\Delta\omega = \delta q^3 \delta p^3$ is known as 'coarse graining'. We can illustrate the role of the coarse-graining scaling parameter $\Delta\omega$. If we take the cell size arbitrary small then we will find occasionally a single particle in the cell. Such a choice corresponds to a fully microscopic description of the mechanical system and no reduction of the number of degrees of freedom is obtained. On the other hand, if we choose $\Delta\omega$ arbitrary large, then all degrees of freedom are represented by one point f, and we lost all knowledge of the system. Therefore, an more appropriate choice is to set $\Delta\omega$ such that it acts as the 'communicator' between the microscopic and macroscopic worlds. The integration variables $\mathrm{d}^3 q\, \mathrm{d}^3 p$ belong to the microscopic world, whereas the arguments in $f(\boldsymbol{q},\boldsymbol{p},t)$ belong to the macroscopic world. The Boltzmann equation describes the time evolution of the distribution function f.

In order to derive the Boltzmann equation it is advisable to consider an hierarchy of interactions which are switched on one after the other. The simplest system consists of N particle which are not submitted to external forces nor are there mutual interactions. In this case it is easy to predict the locations of the particles at $t+dt$ provided that one knows the location at time t, since the equations of motion for the particles in the cell $\Delta\omega(\boldsymbol{q},\boldsymbol{p})$ give

$$\boldsymbol{q}(t+\mathrm{d}t) = \boldsymbol{q}(t) + \frac{\boldsymbol{p}(t)}{m}\mathrm{d}t \,. \tag{2.53}$$

Since there are neither external nor internal forces, we have

$$\boldsymbol{p}(t+\mathrm{d}t) = \boldsymbol{p}(t) \,. \tag{2.54}$$

The distribution function at $t + dt$ and the distribution function at t are connected by the following relation

$$f(\boldsymbol{q}+\mathbf{d}\boldsymbol{q}, \boldsymbol{p}+\mathbf{d}\boldsymbol{p}, t+\mathrm{d}t) = f(\boldsymbol{q},\boldsymbol{p},t) \,. \tag{2.55}$$

Expanding the left-hand side to first order in dt, we obtain

$$\frac{\partial f}{\partial t} + \nabla_{\boldsymbol{q}} f \cdot \frac{\boldsymbol{p}(t)}{m} = 0 \,. \tag{2.56}$$

This is the Boltzmann equation of a non-interacting system of particles without external forces. In the presence of external forces (2.55) is modified as follows:

$$\begin{aligned}\boldsymbol{p}(t+\mathrm{d}t) &= \boldsymbol{p}(t) + \boldsymbol{F}^{\mathrm{ext}}\,\mathrm{d}t \\ &= \boldsymbol{p}(t) - \nabla_{\boldsymbol{q}} V(\boldsymbol{q})\,\mathrm{d}t \,.\end{aligned} \tag{2.57}$$

Consequently, the Boltzmann equation becomes

$$\frac{\partial f}{\partial t} + \nabla_{\boldsymbol{q}} f \cdot \frac{\boldsymbol{p}}{m} - \nabla_{\boldsymbol{p}} f \cdot \nabla_{\boldsymbol{q}} V(\boldsymbol{q}) = 0 \,. \tag{2.58}$$

Finally, we include all other interactions in a scattering term or collision term. This term may include inter-particle scattering, the interaction with radiation, or any other type of interaction we may think of. For microelectronic devices, a number of possible scattering mechanisms naturally arise. Inter-particle scattering generally refers to electron–electron scattering, electron–hole scattering including recombination and annihilation phenomena, electron–phonon scattering, electron–impurity scattering and hole–impurity scattering. For optical devices and photo-voltaic devices, electron–hole–photon interactions may become very important.

In any case, the detailed structure of the interaction potentials needs to be known in order to perform calculations with the Boltzmann equation that now takes the form

$$\frac{\partial f}{\partial t} + \nabla_{\boldsymbol{q}} f \cdot \frac{\boldsymbol{p}}{m} - \nabla_{\boldsymbol{p}} f \cdot \nabla_{\boldsymbol{q}} V(\boldsymbol{q}) = \left(\frac{\partial f}{\partial t}\right)_{\mathrm{coll}} \tag{2.59}$$

and

$$\left(\frac{\partial f}{\partial t}\right)_{\text{coll}} = \left(\frac{\partial f}{\partial t}\right)_{\text{electron–electron}} + \left(\frac{\partial f}{\partial t}\right)_{\text{electron–phonon}} + \dots \quad (2.60)$$

The collision term is often written as an *operator* acting on the distribution function according to

$$\left(\frac{\partial f}{\partial t}\right)_{\text{coll}} = S_{\text{op}} f(\boldsymbol{q}, \boldsymbol{p}, t) \ . \quad (2.61)$$

As an operator it produces a new distribution function from a given one.

In a semi-quantum mechanical treatment one replaces the momentum by $\hbar$ times the wave vector, $\boldsymbol{p} = \hbar \boldsymbol{k}$, thereby obtaining a Boltzmann equation of the following type for electrons

$$\frac{\partial f}{\partial t} + \nabla_{\boldsymbol{q}} f \cdot \frac{\hbar \boldsymbol{k}}{m} - \frac{e}{\hbar} \nabla_{\boldsymbol{k}} f \cdot \boldsymbol{E}(\boldsymbol{q}) = S_{\text{op}} f \ , \quad (2.62)$$

where e is the elementary charge and m the effective mass of the electron in some energy band. So far, the dissipative features of the transport problem has not been addressed. All we did was reformulating the equations of motion in another space. Since the motion is microscopically reversible, no increase of entropy can be expected from the results that we have obtained up to here. In order to describe transport, i.e. irreversible or dissipative behavior, the following approximation for the collision term is introduced

$$S_{\text{op}} = \textstyle\sum_{\mathbf{k}'} S(\boldsymbol{k}', \boldsymbol{k})\, f(\boldsymbol{q}, \boldsymbol{k}', t)\, [1 - f(\boldsymbol{q}, \boldsymbol{k}, t)]$$
$$- \textstyle\sum_{\mathbf{k}} S(\boldsymbol{k}, \boldsymbol{k}')\, f(\boldsymbol{q}, \boldsymbol{k}, t)\, [1 - f(\boldsymbol{q}, \boldsymbol{k}', t)] \ . \quad (2.63)$$

The first sum describes the increase of $f(\boldsymbol{q}, \boldsymbol{k}, t)$ due to scattering of any initial state $\boldsymbol{k}'$ into the final state $\boldsymbol{k}$, for which the transition probability is given by $S(\boldsymbol{k}', \boldsymbol{k})$. This transition is assumed to be proportional to the probability that the particle is in the state $\boldsymbol{k}'$, which is given by $f(\boldsymbol{q}, \boldsymbol{k}', t)$.
Similarly, the second term describes the decrease of $f(\boldsymbol{q}, \boldsymbol{k}, t)$ due to scattering of the initial state $\boldsymbol{k}$ to any other state $\boldsymbol{k}'$. The Pauli exclusion principle has been accounted for by the factors: $[1 - f(\boldsymbol{q}, \boldsymbol{k}, t)]$ and $[1 - f(\boldsymbol{q}, \boldsymbol{k}', t)]$.

Although above assumptions may look quite reasonable, they are crucial for the further applicability of this formalism. Strictly speaking the collision term is a higher-order correlation function describing the scattering of incoming particles to outgoing particles. The above approximation neglects any correlation or memory effects in the collision process. In fact, the approximation is sufficiently crude such that the entropy increases, where 'entropy' is defined as 'information entropy'

$$S = -k_{\text{B}} \int \mathrm{d}^3 q \, \mathrm{d}^3 k \; f(\boldsymbol{q}, \boldsymbol{k}, t) \log f(\boldsymbol{q}, \boldsymbol{k}, t) \ . \quad (2.64)$$

In Chap. 7, further justification will be given for identifying expressions such as $\int f \log f$ as a measure for entropy.

For a semi-quantum mechanical treatment the transition probabilities can be evaluated using perturbative methods. A popular method exploits Fermi's Golden rule. However, this rule must be used with care. Equilibrium considerations imply that

$$\frac{S(\boldsymbol{k}',\boldsymbol{k})}{S(\boldsymbol{k},\boldsymbol{k}')} = \frac{f(\boldsymbol{q},\boldsymbol{k},t)/\left[1-f(\boldsymbol{q},\boldsymbol{k},t)\right]}{f(\boldsymbol{q},\boldsymbol{k}',t)/\left[1-f(\boldsymbol{q},\boldsymbol{k}',t)\right]} \tag{2.65}$$

and, bearing in mind that in equilibrium $f_{\boldsymbol{k}} = f_E$, we obtain

$$S(\boldsymbol{k}',\boldsymbol{k}) = S(\boldsymbol{k},\boldsymbol{k}')\exp\left(-\beta(E_{\mathbf{k}} - E_{\mathbf{k}'})\right) \tag{2.66}$$

with $\beta = 1/k_{\mathrm{B}}T$.

The use of Fermi's Golden rule for some scattering process, calculated from some microscopic process gives $S(\boldsymbol{k}',\boldsymbol{k}) = S(\boldsymbol{k},\boldsymbol{k}')$, as one expects from microscopic reversibility. The factor $\exp(-\beta(E_{\mathbf{k}} - E_{\mathbf{k}'}))$ arises due to the many-particle character of the statistical environment. For example, the elementary processes of one phonon absorption and one phonon emission, have equal transition probabilities

$$S^{(\mathrm{a})}(\boldsymbol{k}',\boldsymbol{k}) = S^{(\mathrm{e})}(\boldsymbol{k},\boldsymbol{k}') = \frac{\pi K^2}{\varrho\,\Omega\omega}\delta\left(E_{\mathbf{k}} - E_{\mathbf{k}'} + \hbar\omega\right) , \tag{2.67}$$

where K is the strength of the interaction, ϱ the mass density of the material, Ω the volume and $\hbar\omega$ the phonon energy. In a phonon bath the absorption and emission scattering rates differ by the number of phonons which are available for the processes:

$$\begin{aligned} S^{(\mathrm{a})}(\boldsymbol{k}',\boldsymbol{k}) &= \frac{\pi K^2}{\varrho\,\Omega\omega}\, n(\beta)\,\delta\left(E_{\mathbf{k}} - E_{\mathbf{k}'} + \hbar\omega\right) \\ S^{(\mathrm{e})}(\boldsymbol{k},\boldsymbol{k}') &= \frac{\pi K^2}{\varrho\,\Omega\omega}\left[n(\beta)+1\right]\delta\left(E_{\mathbf{k}} - E_{\mathbf{k}'} + \hbar\omega\right) , \end{aligned} \tag{2.68}$$

where $n(\beta)$ is the Bose–Einstein factor

$$n(\beta) = \frac{1}{\mathrm{e}^{\beta\hbar\omega} - 1} . \tag{2.69}$$

Therefore we obtain

$$\frac{S^{(\mathrm{e})}(\boldsymbol{k},\boldsymbol{k}')}{S^{(\mathrm{a})}(\boldsymbol{k}',\boldsymbol{k})} = \frac{n(\beta)+1}{n(\beta)} = \exp\left(\beta(E_{\mathbf{k}} - E_{\mathbf{k}'})\right) \tag{2.70}$$

which agrees with above considerations.

2.11 Drude's Model of Carrier Transport

The first simple model for electrons in a conducting material was given by Drude [8, 9]. In many materials, the conduction current that flows due to the presence of an electric field, $\mathbf{E}$, is proportional to $\mathbf{E}$, so that

$$J = \sigma E \ , \tag{2.71}$$

where the electrical conductivity σ is a material parameter. In metallic materials, Ohm's law, (2.71) is accurate. In Drude's model, the electrons move as independent particles in the metallic region suffering from scattering during their travel from the cathode to the anode. The distribution function is assumed to be of the following form:

$$f(\boldsymbol{q}, \boldsymbol{p}, t) = f_{\mathrm{O}}(\boldsymbol{q}, \boldsymbol{p}, t) + f_{\mathrm{A}}(\boldsymbol{q}, \boldsymbol{p}, t) \ , \tag{2.72}$$

where f_{O} is the equilibrium distribution function, being symmetric in the momentum variable $\boldsymbol{p}$, and f_A is a perturbation due to an external field that is anti-symmetric in the momentum variable. The collision term in Drude's model is crudely approximated by the following assumptions: (1) only kick-out, (2) all $S(\boldsymbol{p}, \boldsymbol{p'})$ are equal, (3) no Pauli exclusion principle and (4) no carrier heating, i.e. low-field transitions. The last assumption implies that only the anti-symmetric part participates in the collision term [10].

Defining a characteristic time $\tau_{\mathbf{p}}$, the momentum-relaxation time, we find that

$$\left(\frac{\partial f}{\partial t}\right)_c = -\frac{f_{\mathrm{A}}}{\tau_{\mathbf{p}}} \quad \text{and} \quad \tau_{\mathbf{p}} = \sum_{\mathbf{p}'} S(\boldsymbol{p}, \boldsymbol{p'}) \ . \tag{2.73}$$

Furthermore, assuming a constant electric field $\boldsymbol{E}$ and a spatially uniform charge electron distribution, the Boltzmann transport equation becomes

$$-q\, \boldsymbol{E} \cdot \nabla (f_{\mathrm{O}} + f_{\mathrm{A}}) = -\frac{f_{\mathrm{A}}}{\tau_{\mathbf{p}}} \ . \tag{2.74}$$

Finally, if we assume that $f \simeq f_{\mathrm{O}} \propto \exp\left(-p^2/2mkT\right)$ then

$$f_{\mathrm{A}} = q\tau_{\mathbf{p}}\, \boldsymbol{E} \cdot \nabla_{\mathbf{p}} f_{\mathrm{O}} = \frac{q\tau_{\mathbf{p}}}{mkT}\, \boldsymbol{E} \cdot \boldsymbol{v} f_{\mathrm{O}} \ . \tag{2.75}$$

Another way of looking at this result is to consider $f = f_{\mathrm{O}} + f_{\mathrm{A}}$ as a Taylor series for f_{O}:

$$f(\mathbf{p}) = f_{\mathrm{O}}(\mathbf{p}) + (q\tau_{\mathbf{p}}\, \boldsymbol{E} \cdot \nabla_{\mathbf{p}})\, f_{\mathrm{O}}(\mathbf{p}) + \ldots = f_{\mathrm{O}}(\mathbf{p} + q\tau_{\mathbf{p}}\boldsymbol{E}) \ . \tag{2.76}$$

This is a *displaced* Maxwellian distribution function in the direction opposite to the applied field $\boldsymbol{E}$. The current density $\boldsymbol{J} = qn\boldsymbol{v}$ follows from the averaged velocity

$$\boldsymbol{J} = qn \frac{\int \mathrm{d}^3p\, (\boldsymbol{p}/m)\, f(\boldsymbol{p})}{\int \mathrm{d}^3p\ f(\boldsymbol{p})} = \frac{q^2\tau_{\mathbf{p}}}{m}\, n\boldsymbol{E} \ . \tag{2.77}$$

The electron mobility, μ_n, is defined as the proportionality constant in the constitutive relation $\boldsymbol{J} = q\mu_n n\boldsymbol{E}$, such that

$$\mu_n = \frac{q\tau_{\mathbf{p}}}{m} \ . \tag{2.78}$$

So we have been able to 'derive' Ohm's law from the Boltzmann transport equation.

It is a remarkable fact that Drude's model is quite accurate, given the fact that no reference was made to Pauli's exclusion principle and the electron waves do not scatter while traveling in a perfect crystal lattice. Indeed, it was recognized by Sommerfeld that ignoring these effects will give rise to errors in the calculation of the order of 10^2, but both these errors cancel. Whereas Drude's model explains the existence of resistance, more advanced models are needed to accommodate for the non-linear current–voltage characteristics, the frequency dependence and the anisotropy of the conductance for some materials. A 'modern' approach to derive conductance properties was initiated by Kubo [11]. His theory naturally leads to the inclusion of anisotropy, non-linearity and frequency dependence.

2.12 Currents in Semiconductors

In metals the high conductivity prevents local charge accumulation to occur at a detectable time scale. The situation in semiconductors is quite different. In uniformly doped semiconductors, the decay of an excess charge spot occurs by a diffusion process, that takes place on much longer time scale. In non-uniformly doped semiconductors, there are depletion layers, or accumulation layers of charges that permanently exists even in thermal equilibrium.

The charge and current densities in semiconductors follow also from the general Boltzmann transport theory, but this theory needs to be completed with the specific details such as the band gap, dopant distribution, and the properties related to the interfaces with other materials.

Starting from the Boltzmann transport equation, the *moment expansion* considers variables that are averaged quantities as far as the momentum dependence is concerned. The generic expression for the moment expansion is

$$\frac{1}{\Omega}\sum_{\mathbf{p}} \mathcal{Q}(\boldsymbol{p})\left(\frac{\partial}{\partial t}+\frac{\boldsymbol{p}}{m}\cdot\nabla_{\mathbf{q}}+\boldsymbol{F}\cdot\nabla_{\mathbf{p}}\right) f(\boldsymbol{p},\boldsymbol{q},t)=\frac{1}{\Omega}\sum_{\mathbf{p}} \mathcal{Q}(\boldsymbol{p})\left(\frac{\partial f}{\partial t}\right)_{\mathrm{coll}} \tag{2.79}$$

where $\mathcal{Q}(\boldsymbol{p})$ is an polynomial in the components of $\boldsymbol{p}$ and the normalization $1/\Omega$ allows for a smooth transition to integrate over all momentum states in the Brillouin zone

$$\frac{1}{\Omega}\sum_{\mathbf{p}} \rightarrow \frac{1}{4\pi^3}\int_{\mathrm{BZ}} \mathrm{d}^3k \,. \tag{2.80}$$

The zeroth order expansion gives [10]

$$\frac{\partial n}{\partial t} - \frac{1}{q}\nabla \cdot \boldsymbol{J}_{\mathrm{n}} = -U$$

$$\frac{\partial p}{\partial t} + \frac{1}{q}\nabla \cdot \boldsymbol{J}_{\mathrm{p}} = U \,, \tag{2.81}$$

and where the various variables for the electrons and holes are

electrons	holes
$n(\boldsymbol{r},t) = \frac{1}{\Omega}\sum_{\mathbf{p}} f_{\mathrm{n}}(\boldsymbol{p},\boldsymbol{r},t)$	$p(\boldsymbol{r},t) = \frac{1}{\Omega}\sum_{\mathbf{p}} f_{\mathrm{p}}(\boldsymbol{p},\boldsymbol{r},t)$
$\boldsymbol{J}_{\mathrm{n}}(\boldsymbol{r},t) = -q\, n(\boldsymbol{r},t)\boldsymbol{v}_{\mathrm{n}}(\boldsymbol{r},t)$	$\boldsymbol{J}_{\mathrm{p}}(\boldsymbol{r},t) = q\, p(\boldsymbol{r},t)\boldsymbol{v}_{\mathrm{p}}(\boldsymbol{r},t)$
$\boldsymbol{v}_{\mathrm{n}}(\boldsymbol{r},t) = \frac{1}{\Omega}\sum_{\mathbf{p}}(\boldsymbol{p}/m) f_{\mathrm{n}}(\boldsymbol{p},\boldsymbol{r},t)$	$\boldsymbol{v}_{\mathrm{p}}(\boldsymbol{r},t) = \frac{1}{\Omega}\sum_{\mathbf{p}}(\boldsymbol{p}/m) f_{\mathrm{p}}(\boldsymbol{p},\boldsymbol{r},t)$

and U is the recombination/generation rate

$$U = \frac{1}{\Omega}\sum_{\mathbf{p}} \left(\frac{\partial f}{\partial t}\right)_{\mathrm{coll}} = R - G \,. \tag{2.82}$$

The particle velocities provide an expression for the current densities but by choosing $\mathcal{Q}(\boldsymbol{p}) = \boldsymbol{p}$, we obtain the first moment of the expansion that can be further approximated to give alternative expressions for the current densities. Defining the momentum relaxation time $\tau_{\mathbf{p}}$ as a characteristic time for the momentum to reach thermal equilibrium from a non-equilibrium state and the electron and hole temperature tensors [12]

$$\begin{aligned} \frac{1}{2} n k_{\mathrm{B}} T_{ij}(\boldsymbol{r},t) &= \frac{1}{\Omega}\sum_{\mathbf{p}} \frac{1}{2m}(p_i - m v_i)(p_j - m v_j) f(\boldsymbol{p},\boldsymbol{r},t) \\ &= T(\boldsymbol{r},t)\delta_{ij} \,, \end{aligned} \tag{2.83}$$

where the last equality follows from assuming an isotropic behavior, then one arrives at the following constitutive equation for the currents in semiconducting materials

$$\boldsymbol{J}_{\mathrm{n}} + n\,\tau_{\mathbf{p}\mathrm{n}} \frac{\mathrm{d}}{\mathrm{d}t}\left(\frac{\boldsymbol{J}_{\mathrm{n}}}{n}\right) = q\,\mu_{\mathrm{n}}\, n \left(\boldsymbol{E} + \frac{k_{\mathrm{B}}}{q}\nabla T_{\mathrm{n}}\right) + q D_{\mathrm{n}} \nabla n$$

$$\boldsymbol{J}_{\mathrm{p}} + p\,\tau_{\mathbf{p}\mathrm{p}} \frac{\mathrm{d}}{\mathrm{d}t}\left(\frac{\boldsymbol{J}_{\mathrm{p}}}{p}\right) = q\,\mu_{\mathrm{p}}\, p \left(\boldsymbol{E} - \frac{k_{\mathrm{B}}}{q}\nabla T_{\mathrm{p}}\right) - q D_{\mathrm{p}} \nabla p \,. \tag{2.84}$$

The momentum relaxation times, the electron and hole mobilities and the electron and hole diffusivities are related through the Einstein relation

$$D = \frac{k_{\mathrm{B}}T}{q}\mu = \frac{k_{\mathrm{B}}T}{m}\tau \,. \tag{2.85}$$

The second terms on the left-hand sides of (2.84) are the *convective currents*. The procedure of taking moments of the Boltzmann transport equation always involves a truncation, i.e. the nth order equation in the expansion demands information of the $(n+1)$th order moment to be supplied. For the second-order moment, one thus needs to provide information on the third moment

$$\frac{1}{\Omega}\sum_{\mathbf{p}} p_i p_j p_k f(\boldsymbol{p}, \boldsymbol{r}, t) . \tag{2.86}$$

In the above scheme the second-order expansion leads to the *hydrodynamic model* [12]. In this model the carrier temperatures are determined self-consistently with the carrier densities. The closure of the system of equations is achieved by assuming a model for the term (2.86) that only contains lower order variables. The thermal flux $\boldsymbol{Q}$, being the energy that gets transported through thermal conductance can be expressed as

$$\boldsymbol{Q} = \frac{1}{\Omega}\sum_{\mathbf{p}} \frac{1}{2m}|\boldsymbol{p} - m\boldsymbol{v}|^2 \left(\frac{\boldsymbol{p}}{m} - \boldsymbol{v}\right) = -\kappa \nabla T , \tag{2.87}$$

where κ_{n} and κ_{p} are the thermal conductivities.

Besides the momentum flux, a balance equation is obtained for the energy flux

$$\frac{\partial(n w_{\mathrm{n}})}{\partial t} + \nabla \cdot \boldsymbol{S}_{\mathrm{n}} = \boldsymbol{E} \cdot \boldsymbol{J}_{\mathrm{n}} + n\left(\frac{\partial w_{\mathrm{n}}}{\partial t}\right)_{\mathrm{coll}}$$

$$\frac{\partial(p w_{\mathrm{p}})}{\partial t} + \nabla \cdot \boldsymbol{S}_{\mathrm{p}} = \boldsymbol{E} \cdot \boldsymbol{J}_{\mathrm{p}} + p\left(\frac{\partial w_{\mathrm{p}}}{\partial t}\right)_{\mathrm{coll}} . \tag{2.88}$$

The energy flux is denoted as $\mathbf{S}$ and w is the energy density. In the isotropic approximation, the latter reads

$$w_{\mathrm{n}} = \frac{3}{2}k_{\mathrm{B}}T_{\mathrm{n}} + \frac{1}{2}m_{\mathrm{n}}v_{\mathrm{n}}^2$$

$$w_{\mathrm{p}} = \frac{3}{2}k_{\mathrm{B}}T_{\mathrm{p}} + \frac{1}{2}m_{\mathrm{p}}v_{\mathrm{p}}^2 . \tag{2.89}$$

The energy flux can be further specified as

$$\boldsymbol{S}_{\mathrm{n}} = \kappa_{\mathrm{n}}\nabla T_{\mathrm{n}} - (w_{\mathrm{n}} + k_{\mathrm{B}}T_{\mathrm{n}})\frac{\boldsymbol{J}_{\mathrm{n}}}{q}$$

$$\boldsymbol{S}_{\mathrm{p}} = \kappa_{\mathrm{p}}\nabla T_{\mathrm{p}} + (w_{\mathrm{p}} + k_{\mathrm{B}}T_{\mathrm{p}})\frac{\boldsymbol{J}_{\mathrm{p}}}{q} . \tag{2.90}$$

Just as for the momentum, one usually assumes a characteristic time, τ_{e}, for a non-equilibrium energy distribution to relax to equilibrium. Then the collision term in the energy balance equation becomes

$$n\left(\frac{\partial w_{\mathrm{n}}}{\partial t}\right)_{\mathrm{coll}} = -n\frac{w_{\mathrm{n}} - w^*}{\tau_{\mathrm{en}}} - Uw_{\mathrm{n}}$$

$$p\left(\frac{\partial w_{\mathrm{p}}}{\partial t}\right)_{\mathrm{coll}} = -p\frac{w_{\mathrm{p}} - w^*}{\tau_{\mathrm{ep}}} - Uw_{\mathrm{p}} \tag{2.91}$$

and w^* is the carrier mean energy at the lattice temperature. In order to complete the hydrodynamic model the thermal conductivities are given by the Wiedemann–Franz law for thermal conductivity

$$\kappa = \left(\frac{k_{\mathrm{B}}}{q}\right)^2 T\sigma(T)\Delta\,. \tag{2.92}$$

Here Δ is a value obtained from evaluating the steady-state Boltzmann transport equation for uniform electric fields. The electrical conductivity is $\sigma(T) = q\mu c$, where $c = n, p$ is the carrier concentration. If a power-law dependence for the energy relaxation times can be assumed, i.e.

$$\tau_{\mathrm{e}} = \tau_0 \left(\frac{w}{k_{\mathrm{B}}T^*}\right)^{\nu} \tag{2.93}$$

then $\Delta_T = 5/2 + \nu$. Occasionally, ν is considered to be a constant: $\nu \simeq 1/2$. However, this results into a too-restrictive expression for the $\tau_{\mathrm{e}}(w)$. Therefore Δ_T is often tuned towards Monte-Carlo data.

Comparing the present elaboration on deriving constitutive equations from the Boltzmann transport equation with the derivation of the currents in metals we note that we did not refer to a displaced Maxwellian distribution. Such a derivation is also possible for semiconductor currents. The method was used by Stratton [13]. A difference pops up in the diffusion term of the carrier current. In the results above we obtained: $\boldsymbol{J}$ (diffusive part) $\propto \mu\nabla T$. In Stratton's model one obtains: $\boldsymbol{J}$ (diffusive part) $\propto \mu\nabla\left(\mu T\right)$, the difference being a term $\xi = \partial \log \mu(T)/\partial \log(T)$. Stratton's model is usually referred to as the *energy-transport* model.

3. Mathematical Interlude

Before collecting the fundamentals of quantum mechanics, it may be worth to take a closer look at some key ingredients of the toolbox of theoretical physics which provide us with some elementary comfort during our study of quantum mechanics: the so-called generalized functions and more specifically, the step function and the delta function as well as the space of functionals [14].

3.1 Generalized Functions

Globally speaking, generalized functions can be considered appropriate generalizations of ordinary functions, exhibiting striking singularities or discontinuities and being capable of performing real odd tasks that are beyond the scope of ordinary functions.

In practice each generalized function may be constructed as a well-defined limit of a generating sequence of ordinary functions and in many cases several generating sequences are converging to the same limit.

3.1.1 The Step Function $\theta(x)$

Defining Properties.

$$\theta(x) = \begin{cases} 1 & \text{if } x \geq 0\,, \\ 0 & \text{if } x < 0\,. \end{cases} \tag{3.1}$$

Generating Sequences $\{\theta_\varepsilon(x)\}$.

- A bunch of Fermi–Dirac like functions

$$\theta_\varepsilon(x) = \frac{1}{1 + \mathrm{e}^{-x/\varepsilon}}\,. \tag{3.2}$$

- A tougher one:

$$\theta_\varepsilon(x) = \frac{1}{2\pi\mathrm{i}} \int_{-\infty}^{+\infty} \mathrm{d}k \frac{\mathrm{e}^{\mathrm{i}kx}}{k - \mathrm{i}\varepsilon}\,. \tag{3.3}$$

One recovers the θ-function from the last expression by considering $\varepsilon \downarrow 0$, i.e. $\theta(x) = \lim_{\varepsilon\downarrow 0} \theta_\varepsilon(x)$.

3.1.2 The Delta Function $\delta(x)$

Defining Properties.

$$\delta(x) = \begin{cases} \infty & \text{if } x = 0 \,, \\ 0 & \text{if } x \neq 0 \,. \end{cases}$$

$$\int_{-\infty}^{+\infty} \delta(x) f(x)\, \mathrm{d}x = f(0) \quad \text{if } f \text{ is continuous in } x = 0 \,.$$

Generating Sequences $\{\delta_\varepsilon(x)\}$.

- A square pulse:

$$\delta_\varepsilon(x) = \begin{cases} 1/2\varepsilon & \text{if } \; |x| \leq \varepsilon \\ 0 & \text{elsewhere.} \end{cases} \tag{3.4}$$

- A Gauss pulse:

$$\delta_\varepsilon(x) = \frac{1}{\sqrt{\pi\varepsilon}} \exp\left[-\left(\frac{x^2}{\varepsilon}\right)\right] \,. \tag{3.5}$$

- The Lorentzian family

$$\delta_\varepsilon(x) = \frac{1}{\pi} \frac{\varepsilon}{\varepsilon^2 + x^2} \,. \tag{3.6}$$

- Decaying sines

$$\delta_\varepsilon(x) = \frac{1}{\pi} \frac{\sin x/\varepsilon}{x} \,. \tag{3.7}$$

  Written in a dirty but useful way, it also reads

$$\delta_\varepsilon(x) = -\frac{\mathrm{i}}{\pi} \frac{\mathrm{e}^{\mathrm{i}x/\varepsilon} - 1}{x} \,. \tag{3.8}$$

One recovers the δ-function from the last expression by considering $\varepsilon \downarrow 0$, i.e. $\delta(x) = \lim_{\varepsilon \downarrow 0} \delta_\varepsilon(x)$.

3.1.3 Some Useful Relations

$$\theta'(x) = \delta(x) \tag{3.9}$$

$$\int_{-\infty}^{x} \mathrm{d}y\, \delta(y) = \theta(x) \tag{3.10}$$

$$\int_{-\infty}^{+\infty} \mathrm{d}x\, f(x)\, \delta'(x) = -f'(0) \tag{3.11}$$

$$\delta(x) = \frac{1}{2\pi}\int_{-\infty}^{+\infty} \mathrm{d}k\, \mathrm{e}^{ikx} \tag{3.12}$$

$$\delta'(x) = \frac{-\,\delta(x)}{x} \tag{3.13}$$

$$\delta[F(x)] = \sum_n \frac{\delta(x-x_n)}{|F'(x_n)|} \tag{3.14}$$

where $\{x_n\}$ are the solutions of $F(x) = 0$.

3.2 Functions of Functions – Functionals

Suppose $x(t)$ is a function of t on some domain $D \subset \mathbf{R}$. A mapping which associates to each $x(t)$ a real or complex number is called a *functional.* An elementary example is the integral of $x(t)$, i.e.

$$F[x(t)] = \int_D \mathrm{d}s\, x(s)\ . \tag{3.15}$$

Furthermore suppose that $\eta(t)$ is a function of t which remains small for all t, i.e. $|\eta(t)| < \varepsilon$. To first order in η the number $F[x(t)+\eta(t)]$ is

$$F[x(t)+\eta(t)] = F[x(t)] + \int_D \mathrm{d}s\, K(s)\,\eta(s)\ . \tag{3.16}$$

The integral kernel $K(s)$ is the *functional derivative*, i.e.

$$\frac{\delta F}{\delta x(s)} = K(s)\ . \tag{3.17}$$

It is very illustrative to divide the t-domain into small intervals of length ε such that $t_{j+1} = t_j + \varepsilon$ or $t_j = j\varepsilon$. The function $x(t)$ can be approximated by giving the values $x_j \equiv x(t_j)$ for each t_j. The functional F then becomes an ordinary function of all x_j

$$\begin{aligned} &F[x(t)] \to F(\ldots, x_i, x_{i+1}, \ldots) \\ &\frac{\delta F}{\delta x(s)} \to \frac{1}{\varepsilon}\frac{\partial F}{\partial x_j} = K_j\ . \end{aligned} \tag{3.18}$$

Since by changing the path $x(t)$ to $x(t)+\eta(t)$, the value x_j becomes $x_j + \eta_j$ and to first order

$$\begin{aligned} F(\ldots, x_i+\eta_i, x_{i+1}+\eta_{i+1}, \ldots) &= F(\ldots, x_i, x_{i+1}, \ldots) \\ &\quad + \sum_j \frac{\partial F}{\partial x_j}\eta_j + \ldots \end{aligned} \tag{3.19}$$

and

$$\sum_j \left(\frac{1}{\varepsilon}\frac{\partial F}{\partial x_j}\right)\eta_j\varepsilon \to \int K(s)\,\eta(s)\,\mathrm{d}s\,. \qquad (3.20)$$

3.3 Functional Integration – Path Integrals

Consider the functional $F[x(t)]$ and suppose we want to make an average over all possible paths $x(t)$, with some weight factor $W[x(t)]$. Thus we need to make sense of the following expression

$$I = \sum_{\text{all paths}} F[x(t)]\,W[x(t)]\,. \qquad (3.21)$$

In order to interpret this sum, we subdivide the t-domain in small intervals of size ε and the functionals F and W become ordinary functions of x_i again. The sum over all paths is obtained by the multiple integral of all x_i. So the following limiting procedure provides this sum

$$\begin{aligned}\sum_{\text{all paths}} F[x(t)]\,W[x(t)] &= \lim_{N\to\infty}\int \mathrm{d}x_1 \int \mathrm{d}x_2 \ldots \int \mathrm{d}x_N \\ &\quad\times F(\ldots x_i, x_{i+1},\ldots)\,W(\ldots x_i, x_{i+1},\ldots) \\ &= \int \mathcal{D}x\, F[x(t)]\,W[x(t)]\,. \end{aligned} \qquad (3.22)$$

This integral is a *path integral* since it is an integral over all paths [15]. The path integral considered as the limit of an multiple integral is illustrated in Fig. 3.1. For each value of t, all values of $\boldsymbol{x}$ are considered. In modern theoretical physics path integrals are used intensively.

The generalizations to functionals can be pushed quite far. For instance, we can define a functional delta function as

$$\begin{aligned}\delta[x(t)-g(t)] &= \lim_{N\to\infty} \delta(x_1-g_1)\,\delta(x_2-g_2)\,\ldots\,\delta(x_N-g_N) \\ &= \int \mathcal{D}c\,\exp\left[\mathrm{i}\int \mathrm{d}t\; c(t)\;(x(t)-g(t))\right]\,. \end{aligned} \qquad (3.23)$$

Strictly speaking, we should include a factor $(\frac{1}{2\pi})^\infty$, however, such normalization factors are usually omitted because path integrals are mostly used for determining averages, and therefore a simular factor shows up in the denominator, i.e. an average is given by

$$\langle F\rangle = \frac{\int \mathcal{D}x\; F[x]\;W[x]}{\int \mathcal{D}x\; W[x]}\,. \qquad (3.24)$$

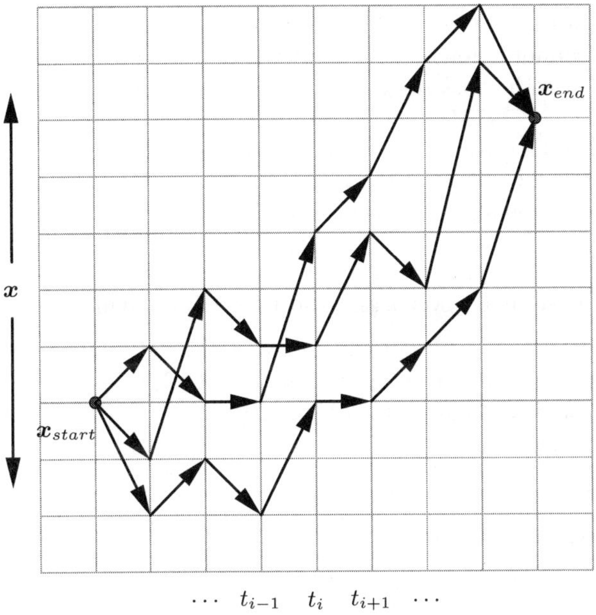

Fig. 3.1. Illustration of the path integral: a sum over all paths is realized by including all values of $\boldsymbol{x}$ at each instant t

3.4 Vector Identities

Below we have summarized a few vector identities that may be helpful to manipulate algebraic expression involving scalars and vector fields. All these fields are supposed to be differentiable on a connected subset Ω of $\mathbf{R}^3$.

$$\nabla \cdot (\nabla \times \boldsymbol{A}) \equiv 0 \tag{3.25}$$

$$\nabla \times (\nabla f) \equiv \boldsymbol{0} \tag{3.26}$$

$$\begin{aligned} \nabla \times (\boldsymbol{A} \times \boldsymbol{B}) = & -\boldsymbol{A}(\nabla \cdot \boldsymbol{B}) + \boldsymbol{B}(\nabla \cdot \boldsymbol{A}) - (\boldsymbol{A}\cdot\nabla)\boldsymbol{B} \\ & +(\boldsymbol{B}\cdot\nabla)\boldsymbol{A} - \boldsymbol{A} \times (\nabla \times \boldsymbol{B}) + \boldsymbol{A} \times (\nabla \times \boldsymbol{B}) \end{aligned} \tag{3.27}$$

$$\nabla \times (f\boldsymbol{A}) = f\nabla \times \boldsymbol{A} + \nabla f \times \boldsymbol{A} \tag{3.28}$$

$$\nabla \cdot (f\boldsymbol{A}) = f\nabla \cdot \boldsymbol{A} + \nabla f \cdot \boldsymbol{A} \tag{3.29}$$

$$\nabla \times (\nabla \times \boldsymbol{A}) = \nabla(\nabla \cdot \boldsymbol{A}) - \nabla^2 \boldsymbol{A} \tag{3.30}$$

Equation (3.30) should generally be considered as a convenient definition of the vectorial Laplace operator, rather than a vector identity, particularly if curvilinear coordinate systems are addressed. Indeed, although the (3.30) expands to

$$[\nabla \times (\nabla \times \boldsymbol{A})]_x = \frac{\partial}{\partial x}\nabla \cdot \boldsymbol{A} - \nabla^2 A_x \tag{3.31}$$

etc. in Cartesian coordinates, thereby justifying the identification $\nabla^2 \boldsymbol{A} = (\nabla^2 A_x, \nabla^2 A_y, \nabla^2 A_z)$ we cannot follow that strategy when dealing with curvilinear coordinates where also the basis vectors are changing with position.

The derivation of the above identities and various others can be greatly simplified by using the following identity relating the summed product of two Levi–Civita symbols to a combination of Kronecker deltas:

$$\varepsilon_{ijk}\,\varepsilon_{irs} = \delta_{jr}\delta_{ks} - \delta_{kr}\delta_{js} \tag{3.32}$$

where the left-hand side is summed over i and the Levi–Civita symbol ε_{ijk} is defined by

$$\varepsilon_{ijk} = \begin{cases} 1 & \text{if } (i,j,k) \text{ is an even permutation of } (1,\,2,\,3) \\ -1 & \text{if } (i,j,k) \text{ is an odd permutation of } (1,\,2,\,3) \\ 0 & \text{in any other case}\ . \end{cases} \tag{3.33}$$

3.5 A Few Theorems

At several stages of transport theory one is relying on the results of differential geometry to invoke the existence of particular vector and scalar fields as well as to manipulate volume, surface and line integrals. In this light, we have listed four theorems that have been employed in the course of this book.

Stokes' theorem. Let Σ be an open, orientable, multiply connected surface in $\mathbf{R}^3$ bounded by an outer, closed curve $\partial\Sigma_0$ and n inner, closed curves $\partial\Sigma_1$, $\ldots$, $\partial\Sigma_n$ defining n holes. If Σ is oriented by a surface element $\mathrm{d}\boldsymbol{S}$ and if $\boldsymbol{A}$ is a differentiable vector field defined on Σ, then

$$\int_\Sigma \nabla \times \boldsymbol{A} \cdot \mathrm{d}\boldsymbol{S} = \oint_{\partial\Sigma_0} \boldsymbol{A} \cdot \mathrm{d}\boldsymbol{r} - \sum_{j=1}^{n} \oint_{\partial\Sigma_j} \boldsymbol{A} \cdot \mathrm{d}\boldsymbol{r} \tag{3.34}$$

where the orientation of all boundary curves is uniquely determined by the orientation of $\mathrm{d}\boldsymbol{S}$.

Gauss' theorem. Let Ω be a closed, orientable, multiply connected subset of $\mathbf{R}^3$ bounded by an outer, closed surface $\partial\Omega_0$ and n inner, closed surfaces defining n holes. If $\boldsymbol{E}$ is a differentiable vector field defined on Ω then

$$\int_\Omega \nabla \cdot \boldsymbol{E}\,\mathrm{d}\tau = \int_{\partial\Omega_0} \boldsymbol{E} \cdot \mathrm{d}\boldsymbol{S} - \sum_{j=1}^{n} \int_{\partial\Omega_j} \boldsymbol{E} \cdot \mathrm{d}\boldsymbol{S} \tag{3.35}$$

and

$$\int_\Omega \nabla \times \boldsymbol{E}\,\mathrm{d}\tau = \int_{\partial\Omega_0} \mathrm{d}\boldsymbol{S} \times \boldsymbol{E} - \sum_{j=1}^{n} \int_{\partial\Omega_j} \mathrm{d}\boldsymbol{S} \times \boldsymbol{E} \tag{3.36}$$

where all boundary surfaces have the same orientation as the outward pointing surface element of the outer surface.

$\boldsymbol{J}\cdot\boldsymbol{E}$ theorem. Let Ω be a closed, multiply connected, bounded subset of $\mathbf{R}^3$ with one hole and boundary surface $\partial\Omega$. If $\boldsymbol{J}$ and $\boldsymbol{E}$ are two differentiable vector fields on Ω, circulating around the hole and satisfying the conditions

$$\nabla \cdot \boldsymbol{J} = 0 \tag{3.37}$$

$$\nabla \times \boldsymbol{E} = \mathbf{0} \tag{3.38}$$

$$\boldsymbol{J} \parallel \partial\Omega \quad \text{or} \quad \boldsymbol{J} = \mathbf{0} \text{ in each point point of } \partial\Omega \tag{3.39}$$

then [52]

$$\int_{\Omega} \boldsymbol{J}\cdot\boldsymbol{E}\,\mathrm{d}\tau = \left(\int_{\Sigma} \boldsymbol{J}\cdot\mathrm{d}\boldsymbol{S}\right)\left(\oint_{\Gamma} \boldsymbol{E}\cdot\mathrm{d}\boldsymbol{r}\right) \tag{3.40}$$

where Σ is an arbitrary cross section, intersecting Ω only once and Γ is a simple closed curve, encircling the hole and lying within Ω but not intersecting $\partial\Omega$. The orientation of Σ is uniquely determined by the positive orientation of Γ.

Helmholtz' theorem. Let Ω be a simply connected, bounded subset of $\mathbf{R}^3$. Then, any finite, continuous vector field $\boldsymbol{F}$ defined on Ω can be derived from a differentiable vector potential $\boldsymbol{A}$ and a differentiable scalar potential χ such that

$$\boldsymbol{F} = \boldsymbol{F}_{\mathrm{L}} + \boldsymbol{F}_{\mathrm{T}} \tag{3.41}$$

$$\boldsymbol{F}_{\mathrm{L}} = \nabla\chi \tag{3.42}$$

$$\boldsymbol{F}_{\mathrm{T}} = \nabla \times \boldsymbol{A}\,. \tag{3.43}$$

Due to the obvious properties

$$\nabla \times \boldsymbol{F}_{\mathrm{L}} = \mathbf{0}$$
$$\nabla \cdot \boldsymbol{F}_{\mathrm{T}} = 0\,. \tag{3.44}$$

$\boldsymbol{F}_{\mathrm{L}}$ and $\boldsymbol{F}_{\mathrm{T}}$ are respectively called the longitudinal and transverse components of $\boldsymbol{F}$.

Exercises

3.1. Assuming that the action reads $S = \int_{t_0}^{t_1} L(x,\dot{x},t)\,\mathrm{d}t$, show that

$$\frac{\delta S}{\delta x(s)} = -\frac{\mathrm{d}}{\mathrm{d}s}\left(\frac{\partial L}{\partial \dot{x}}\right) + \frac{\partial L}{\partial x}\,.$$

3.2. Suppose $F[x(t)] = x(\tau)$ for some fixed value $t = \tau$. Show that

$$\frac{\delta F}{\delta x(s)} = \delta(s-\tau)\,.$$

4. Quantum Mechanics

4.1 Rules of the Game

The theory of relativity has changed our way of thinking about the structure of space–time. Nevertheless it has resulted in a fully coherent and consistent picture. Quantum mechanics on the other hand has revolutionized our thinking about physics. It has degraded physicists back to the level of shamans applying rules that they do not fully understand themselves. It took them a quarter of a century (1900–1925) to discover these rules which will be discussed below [16–22].

4.1.1 States in Fock Space

- For any physical system there exists a state function $|\Phi\rangle$ which contains all information that there is to know about the system. Such a state function is identified as an abstract vector of a complex linear vector space, which is called Fock space.
- The physical state of the system is completely specified by the 'direction' of $|\Phi\rangle$ in Fock space. In other words, if $|\Phi\rangle$ and $\lambda\,|\Phi\rangle$ are specifying the same state regardless the (complex) value of λ.
- As in any other vector space, each state vector $|\Phi\rangle$ may be expanded in a basis $\{|\phi_1\rangle\,, |\phi_2\rangle\,, \ldots, |\phi_n\rangle\,, \ldots\}$:

  $$|\Phi\rangle = \sum_n a_n\,|\phi_n\rangle \ , \tag{4.1}$$

  where the sum in principle runs from 1 to N, the dimension of the vector space, which in most relevant cases however, turns out to be infinite!

Example. Consider a localized electron in a one-dimensional lattice. The lattice is illustrated in Fig 4.1 as a linear chain. The state corresponding to the situation that the electron is localized near the jth atom is given by $\langle\phi\rangle_j$. The collection of all these states build the Fock space of the system.

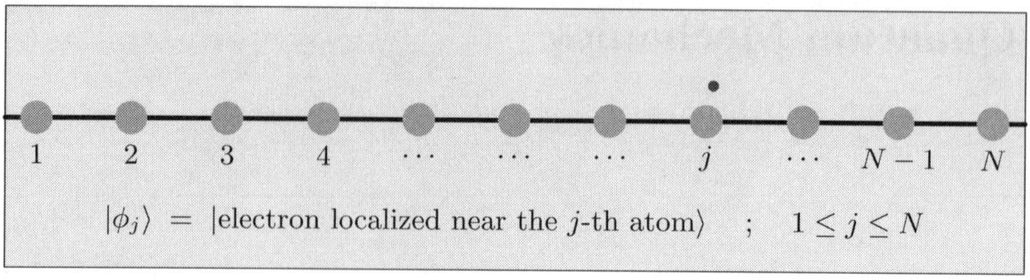

Fig. 4.1. Localized electron states on a linear chain of N atoms

4.1.2 The Superposition Principle

If $|\Phi_1\rangle$ and $|\Phi_2\rangle$ are two possible states of the system then every linear combination

$$a_1 |\Phi_1\rangle + a_2 |\Phi_2\rangle$$

is also a physical realizable state of the system for any couple of complex values (a_1, a_2).

4.1.3 Scalar Products and Probability Amplitudes

- In complete analogy with $\mathbf{R}^3$, the three-dimensional space we are living in, a scalar product is defined in Fock space. The scalar product of two state vectors $|\Phi_1\rangle$ and $|\Phi_2\rangle$ is denoted by $\langle\Phi_2|\Phi_1\rangle$ and satisfies the equality

$$\langle\Phi_2|\Phi_1\rangle = \langle\Phi_1|\Phi_2\rangle^* , \tag{4.2}$$

  where * stands for complex conjugate and the 'bra' and 'ket' notation is due to Dirac [18].
- Since $|\Phi\rangle$ and $\lambda\,|\Phi\rangle$ represent the same state, one usually replaces $|\Phi\rangle$ by its unit vector imposing the normalization

$$\langle\Phi|\Phi\rangle = 1 . \tag{4.3}$$

  Unless stated otherwise, we will mostly be dealing with such normalized state vectors.
- Physically, $\langle\Phi_2|\Phi_1\rangle$ represents the *probability amplitude* of the system to be found in a state $|\Phi_2\rangle$ when it is already in the state $|\Phi_1\rangle$. Unlike conventional probability theory, the probability P of such an event is found to be derived from the probability *amplitude* simply as

$$P = |\,\langle\Phi_2|\Phi_1\rangle\,|^2 . \tag{4.4}$$

4.1.4 Observable Quantities and Operators

To each physical observable quantity or 'observable' $\mathcal{A}$ corresponds a linear and Hermitian (or self-adjoint) operator A acting upon the state vectors in Fock space.

Examples. The position, velocity, momentum, angular momentum, energy, force, torque, etc. are examples of physical observable quantities.

- Mathematically, the hermiticity of A is most easily expressed by representing A by its matrix in an arbitrary basis $\{|\phi_n\rangle\}$:

$$A^*_{nm} = A_{mn} \qquad \text{for all } n,\, m \tag{4.5}$$

  with $A_{nm} = \langle\phi_n|A|\phi_m\rangle$.
- The general matrix element $\langle\phi_n|A|\phi_m\rangle$ is to be interpreted as the *probability amplitude* for the *transition* from $|\Phi_1\rangle$ to $|\Phi_2\rangle$ due to the action of A.
- In many cases one is interested in the eigenvalues of the operator A since they provide a well-defined set of 'allowed' values the observable $\mathcal{A}$ can take in a measurement. In this light is goes without saying that real observable quantities should give rise to real eigenvalues and that is precisely what is guaranteed by the hermiticity of the operators associated with the observables. Moreover, the real eigenvalues correspond to an orthogonal set of eigenvectors which can often be used as a convenient set of basis vectors for Fock space.

There is however also a huge class of non-Hermitian operators corresponding to auxiliary physical quantities that, albeit non-measurable may be quite important, such as creation and annihilation operators which will be considered later. The operator A associated with such a quantity has then a so-called Hermitian conjugate counterpart which is denoted by $A^\dagger$ and which can be unequivocally extracted from A by taking the complex conjugate of the transposed matrix elements of A

$$(A^\dagger)_{nm} = A^*_{mn} \qquad \text{for all } n,\, m\ . \tag{4.6}$$

4.1.5 Measurements and Expectation Values

Measuring an observable A in a system which is in a state $|\Phi\rangle$, one obtains the *expectation value*

$$\frac{\langle\Phi|A|\Phi\rangle}{\langle\Phi|\Phi\rangle}\ .$$

Two remarks are in order:

- It immediately follows that an expectation value is independent of the normalization of $|\Phi\rangle$, as it should be. However, since we intent to consider mostly normalized states, we may skip the denominator $\langle\Phi|\Phi\rangle$.
- In general, the system state $|\Phi\rangle$ need not be an eigenstate of the observable $\mathcal{A}$ one is trying to measure and consequently, the expectation value will generally be different from any of the eigenvalues of A.

It is instructive to try the 'bra' and 'ket' tagliatelle and to discover the elegance with which several crucial concepts like 'expectation value', 'probability amplitude', 'eigenvalues and eigenvectors', ... can be related.

Even if $|\Phi\rangle$ is not an eigenstate of A, we can always *expand* $|\Phi\rangle$ in the set of eigenstates since the latter provide a suitable basis

$$|\Phi\rangle = \sum_n c_n |a_n\rangle , \tag{4.7}$$

where $|a_n\rangle$ is the normalized eigenvector of A corresponding to the eigenvalue a_n

$$A|a_n\rangle = a_n |a_n\rangle . \tag{4.8}$$

Bearing in mind that $\{|a_n\rangle\}$ is an orthogonal basis, one easily finds

$$c_n = \langle a_n|\Phi\rangle . \tag{4.9}$$

Therefore, c_n is nothing but the probability amplitude to find the system in the nth eigenvector $|a_n\rangle$ whereas it is known to be in the state $|\Phi\rangle$.
The normalization condition $\langle\Phi|\Phi\rangle = 1$ further imposes the constraint

$$\sum_n |c_n|^2 = 1 , \tag{4.10}$$

which automatically meets the requirement that the sum of all probabilities add up to 1. Using once more the orthogonality of the eigenvectors, one finally arrives at the expression

$$\langle\Phi|A|\Phi\rangle = \sum_n |c_n|^2 a_n \tag{4.11}$$

relating the probability amplitudes, the eigenvalue spectrum and the expectation value of a system state.

4.1.6 Dynamics – The Schrödinger Equation

In order to investigate the time dependence of the system state vector $\langle\Phi\rangle$ one needs to have a dynamical equation.
It is now postulated that the rate at which the state vector evolves in time is proportional to the action of the Hamilton operator on the state vector. So, written as a concise formula, the requested 'deus ex machina' reads

$$i\hbar \frac{\partial |\Phi(t)\rangle}{\partial t} = H|\Phi(t)\rangle . \tag{4.12}$$

This is the most general form of the *time-dependent* Schrödinger equation.
Being the Hamilton operator or Hamiltonian of the system, H is the operator in Fock space that is associated with the total energy of the system and, as such it is does not explicitly depend on time (except for the case of time dependent perturbations, which won't be discussed in this premature phase).

Before entering the quantum mechanical kitchen and grasping the book of recipes that is supposed to tell us how to construct all these fancy operators, we can already see a way of making life a bit easier; as in the case of ordinary vector calculus, when we have to differentiate a vector we feel much more comfortable if we can manage to express that vector in a basis consisting of 'constant' basis vectors. Then we just need to differentiate the vector components and mostly this is a lot easier to do. In our case 'constant basis vectors' means 'time independent basis vectors' and since the Hamiltonian H is governing the Schrödinger equation, it is obvious to expand $|\Phi(t)\rangle$ in the set of eigenvectors of H. In other words, our first task would be to derive the energy spectrum of H, i.e. to extract the energy eigenvalues $\{E_n\}$ and the energy eigenvectors or *eigenstates* $\{|\phi_n\rangle\}$ from the *time independent Schrödinger equation*

$$H\,|\phi_n\rangle \;=\; E_n\,|\phi_n\rangle \;, \tag{4.13}$$

for all possible n.

Then, writing $|\Phi(t)\rangle$ as a linear combination of the system eigenstates with time-dependent expansion coefficients $\{c_n(t)\}$,

$$|\Phi(t)\rangle \;=\; \sum_n c_n(t)\;|\phi_n\rangle \tag{4.14}$$

we obtain from direct substitution into the time-dependent Schrödinger equation

$$\mathrm{i}\hbar\,\frac{\mathrm{d}c_n(t)}{\mathrm{d}t} \;=\; E_n\;c_n(t)\;. \tag{4.15}$$

Clearly, the solution is given by

$$c_n(t) \;= c_n(0)\;\mathrm{e}^{-\mathrm{i}E_n t/\hbar}\;, \tag{4.16}$$

so that

$$|\Phi(t)\rangle \;=\; \sum_n\;|\phi_n\rangle\,\langle\phi_n|\Phi(0)\rangle\;\;\mathrm{e}^{-\mathrm{i}E_n t/\hbar}\;. \tag{4.17}$$

In principle the dynamical problem is now reduced to an elementary initial-value problem being determined entirely by the energy spectrum of the physical system.

4.2 The Book of Recipes

Before we can entertain ourselves by making attempts toward the solution of Schrödinger equation, we still need recipes showing how to construct quantum mechanical operators in Fock space.

4.2.1 First Recipe – Correspondence Principle

If the physical system has a classical counterpart, go and get the expression for the classical Hamiltonian, i.e. the total energy functional. Next, quantize all generalized coordinates and momenta $\{q_1, q_2, q_3 \ \ldots\}$,$\{p_1, p_2, p_3 \ \ldots\}$ by replacing them by coordinate and momentum operators $\{Q_1, Q_2, Q_3 \ \ldots\}$, $\{P_1, P_2, P_3 \ \ldots\}$.

4.2.2 Second Recipe – Canonical Commutation Rules

Construct the coordinate and momentum operators by invoking the following set of 'canonical commutation rules':

$$[Q_\mu, P_\nu] = \mathrm{i}\hbar\, \delta_{\mu\nu} \tag{4.18}$$

$$[Q_\mu, Q_\nu] = 0 \tag{4.19}$$

$$[P_\mu, P_\nu] = 0 \,, \tag{4.20}$$

where $[A, B] = AB - BA$ is the commutator of A and B and $\delta_{\mu\nu}$ represents a Kronecker delta

$$\delta_{\mu\nu} = \begin{cases} 1 & \text{if } \mu = \nu \,, \\ 0 & \text{if } \mu \neq \nu \,. \end{cases} \tag{4.21}$$

4.2.3 Third Recipe – Use Your Imagination

If the physical system has no classical counterpart, use your imagination And try for instance to rely on analogies or generalizations of classical pictures. For instance, the analogy with orbital angular momentum which is first encountered in classical physics has been proved helpful for introducing the concept of spin angular momentum which has an exclusively quantum mechanical nature. If everything else fails, ... go to a real magician!

4.2.4 Fourth Recipe – Choose an Appropriate Basis

Finally, in order to start real cooking, we need to know how to do the operator calculus, how to solve Schrödinger's equation, how to concretize the above mentioned commutation relations etc. Therefore we need a suitable basis for our Fock space, so that all formal stuff looks somewhat less abstract. As was suggested before, we will choose the eigenvectors of an operator which is perhaps the most familiar one: the position operator. For the sake of simplicity we will confine ourselves in the following to one degree of freedom (for example one particle moving on a straight line) but the generalization to more degrees of freedom is straightforward and will be extensively discussed in the chapter on statistical physics. So, let us start with a position operator Q and a momentum operator P, being guided merely by the commutation rules

$$[Q, P] = \mathrm{i}\hbar \tag{4.22}$$
$$[Q, Q] = 0 \tag{4.23}$$
$$[P, P] = 0\,. \tag{4.24}$$

What about the eigenvectors of the position operator Q? Of course, it's quite simple to write down the formal eigenvalue equation

$$Q\,|q\rangle \;=\; q\,|q\rangle\;, \tag{4.25}$$

where $|q\rangle$ is an eigenvector of Q and q its corresponding (real) eigenvalue, but what does it mean? The answer is just a matter of simple geometrical intuition. If $|q\rangle$ is an eigenvector of the *position* operator, we feel that it should describe the state of a particle located at a sharply defined *position* q. This is a simple and correct conclusion but it rises a new problem: since the eigenvalues q represent real positions, they may generally take any real value, at least within the boundaries of a particular domain Ω to which the particle's motion is confined. Hence the spectrum of Q is found to be continuous which, in turn, makes the basis $\{|q\rangle\,; q \in \mathbf{R}\}$ not only infinite, but even uncountable. As an immediate consequence we must now treat the expansion of any state vector $|\Phi\rangle$ in the 'coordinate basis' as an integral over Ω

$$|\Phi\rangle \;=\; \int_\Omega \mathrm{d}q\;|q\rangle\,\langle q|\Phi\rangle \qquad \text{for any } |\Phi\rangle\,. \tag{4.26}$$

In particular, we may set $|\Phi\rangle = |\,|q'\rangle$ for some $q' \in \Omega$ and find

$$|q'\rangle = \int_\Omega \mathrm{d}q\;|q\rangle\,\langle q|q'\rangle\;. \tag{4.27}$$

In other words, we are left to conclude that the scalar product $\langle q|q'\rangle$ appears to be delta function instead of the expected Kronecker delta:

$$\langle q|q'\rangle \;=\; \delta(q - q')\,, \tag{4.28}$$

which further means that we cannot normalize the eigenvectors $|q\rangle$ to one. Now, this is certainly not the end of the world since we can handle most of the mathematics by carefully exploiting the properties of the delta function, but it illustrates the mathematical subtleties to be dealt with when continuous spectra are being investigated.

On the other hand, the good news is that we now have all necessary information to make the action of an arbitrary operator explicit in the coordinate representation. But it should be noticed first that (4.26) leads naturally to the definition of the concept wave function. Any state vector $|\Phi\rangle$ can be characterized by a wave function $\Phi(q)$ which is proportional to the probability amplitude that a particle is located at a position q, when it is known to be in the state $|\Phi\rangle$.

Mathematically, the definition is

$$\Phi(q) \;=\; \langle q|\Phi\rangle\;. \tag{4.29}$$

'Geometrically' speaking, the wave function $\Phi(q)$ is the projection of the state vector $|\Phi\rangle$ onto the basis vector $|q\rangle$.

Let us now translate the action of P on the abstract state vectors into the coordinate representation. Sandwiching the elementary commutator $[Q, P] = \mathrm{i}\hbar$ between $\langle q|$ and $|q'\rangle$, we obtain

$$\langle q|\,[Q,P]\,|q'\rangle = \mathrm{i}\hbar\,\langle q|q'\rangle \tag{4.30}$$

$$\langle q|\,QP\,|q'\rangle - \langle q|PQ|q'\rangle = \mathrm{i}\hbar\,\delta(q-q')\,. \tag{4.31}$$

Inserting the eigenvalues equations $\langle q|\,Q = q\,\langle q|$ and $Q\,|q'\rangle = q'\,|q'\rangle$ the left-hand side of (4.31) simplifies to

$$(q-q')\,\langle q|P|q'\rangle \;=\; \mathrm{i}\hbar\;\delta(q-q') \tag{4.32}$$

or

$$\begin{aligned}\langle q|P|q'\rangle &= \mathrm{i}\hbar\;\frac{\delta(q-q')}{q-q'}\\ &= -\mathrm{i}\hbar\;\frac{\delta(q'-q)}{q'-q}\\ &= \mathrm{i}\hbar\;\frac{\partial}{\partial q'}\,\delta(q'-q)\,.\end{aligned} \tag{4.33}$$

Having constructed all matrix elements of P in the coordinate representation, we are now in a position to evaluate its action upon $|\Phi\rangle$ straightaway using (4.33)

$$P\,|\Phi\rangle \;=\; \int_{\Omega} \mathrm{d}q'\;P\,|q'\rangle\,\langle q'|\Phi\rangle \tag{4.34}$$

$$\;=\; \int_{\Omega} \mathrm{d}q'\;P\,|q'\rangle\,\Phi(q')\,. \tag{4.35}$$

Projecting $P\,|\Phi\rangle$ onto $\langle q|$, we obtain for $q \in \Omega$:

$$\begin{aligned}\langle q|P|\Phi\rangle &= \int_{\Omega} \mathrm{d}q'\;\langle q|P|q'\rangle\,\Phi(q')\\ &= \mathrm{i}\hbar\int_{\Omega} \mathrm{d}q'\left[\frac{\partial}{\partial q'}\,\delta(q'-q)\right]\Phi(q')\\ &= -\mathrm{i}\hbar\;\frac{\mathrm{d}\Phi(q)}{\mathrm{d}q}\,.\end{aligned} \tag{4.36}$$

Since the left-hand side is just the wave function of the state $P\,|\Phi\rangle$, we conclude that in the momentum operator is represented by the differential operator $-\mathrm{i}\hbar\;\mathrm{d}/\mathrm{d}q$ in the coordinate representation

$$\boxed{\mathcal{P}\Phi(q) = -\mathrm{i}\hbar\;\frac{\mathrm{d}\Phi(q)}{\mathrm{d}q}} \qquad \text{for all } \Phi(q).$$

The generalization to several degrees of freedom is trivial: the wave function of a state $|\Phi\rangle$ describing a system with N generalized coordinates and momenta is now defined as

$$\Phi(q_1, q_2, \ldots, q_N) = \langle q_1, q_2, \ldots, q_N|\Phi\rangle , \tag{4.37}$$

whereas

$$P_j \to \mathcal{P}_j = -\mathrm{i}\hbar \frac{\partial}{\partial q_j} \qquad \text{for all } j = 1, \ldots, N . \tag{4.38}$$

4.3 The Position and Momentum Representation

The consequences of the fourth recipe are far reaching in the sense that they allow to work out the correspondence principle explicitly and to set up convenient representations for the operators.

4.3.1 Position Representation

The example that was elaborated in the previous section was particularly addressed the *position representation* which means that we have chosen the eigenvectors of the position operators as a basis for the Fock space of state vectors and, consequently that we are bound to express all relevant operators, including the Hamiltonian, in terms of their matrix elements in the 'position basis', which has led to the prescriptions (4.38). So let us be guided by these prescriptions and construct the Hamiltonian of 1 particle with mass m moving around in the presence of some potential (energy) $U(\boldsymbol{r})$. Here the generalized coordinates are simply the Cartesian coordinates of the particle position and momentum vectors $\boldsymbol{r}$ and $\boldsymbol{p}$.

Classically, the Hamiltonian represents the total energy of the particle:

$$E = \frac{p^2}{2m} + U(\boldsymbol{r}) . \tag{4.39}$$

In the position representation its quantum mechanical counterpart is obtained by the substitution

$$\boldsymbol{p} \to -\mathrm{i}\hbar\nabla$$
$$p^2 = \boldsymbol{p}\cdot\boldsymbol{p} \to -\hbar^2\nabla^2 ,$$

so that

$$H = -\frac{\hbar^2}{2m}\nabla^2 + U(\boldsymbol{r}) . \tag{4.40}$$

In a similar way the above formulation can be extended to a many-particle system and to the representation of other operators. For instance, the angular momentum vector $\boldsymbol{L} = \boldsymbol{r} \times \boldsymbol{p}$ is represented in the position representation by the differential operator

$$\boldsymbol{L} = -\mathrm{i}\hbar\, \boldsymbol{r} \times \nabla . \tag{4.41}$$

4.3.2 Momentum Representation

Instead of starting from the basis generated by the position eigenvectors, we may as well choose the momentum eigenstates and from thereon construct the operators in the momentum representation. In complete analogy with the case of the position representation, it may be shown that this time the prescriptions are given by:

$$Q_j \to \mathrm{i}\hbar \frac{\partial}{\partial p_j} \qquad \text{for all } j = 1, \ldots, N \ , \tag{4.42}$$

where the wave functions are now defined by the scalar products

$$\Phi(p_1, p_2, \ldots, p_N) = \langle p_1, p_2, \ldots, p_N | \Phi \rangle \ . \tag{4.43}$$

4.3.3 Stationary Schrödinger Equation

As a final remark on representations, we summarize below the stationary Schrödinger equation yielding the one-particle energy spectrum.

While in Fock space the Schrödinger equation appears to be a formal eigenvalue equation

$$H\,|\Phi_n\rangle = E_n\,|\Phi_n\rangle \ , \tag{4.44}$$

it may take the form of a partial differential equation in the position and momentum representation

$$\frac{-\hbar^2}{2m}\nabla^2 \Phi_n(\boldsymbol{r}) + U(\boldsymbol{r})\,\Phi_n(\boldsymbol{r}) = E_n \Phi_n(\boldsymbol{r}) \tag{4.45}$$

$$\frac{p^2}{2m}\,\tilde{\Phi}_n(\boldsymbol{p}) + U(\mathrm{i}\hbar\nabla_{\boldsymbol{p}})\,\tilde{\Phi}_n(\boldsymbol{p}) = E_n \tilde{\Phi}_n(\boldsymbol{p}) \ , \tag{4.46}$$

where $\Phi_n(\boldsymbol{r}) = \langle \boldsymbol{r} | \Phi_n \rangle$ and $\tilde{\Phi}_n(\boldsymbol{p}) = \langle \boldsymbol{p} | \Phi_n \rangle$ denote the wave functions of the state $|\Phi_n\rangle$ in the position and momentum representation respectively.

4.4 Commuting Operators, 'Good' Quantum Numbers

In many cases the sequential action of two operators A and B in Fock space depends on the order and in general we expect that $AB \neq BA$ or, equivalently,

$$[A, B] \neq 0 \ . \tag{4.47}$$

However, if one is dealing with commuting operators, a special situation occurs: a theorem from the theory of Hilbert spaces states that if $[A, B] = 0$, one can always find a complete set of eigenvectors that will diagonalize *both* A and B:

$$\exists\{|\phi_n>\} \quad \forall n : \begin{array}{l} A\,|\phi_n\rangle = \alpha_n\,|\phi_n\rangle \\ B\,|\phi_n\rangle = \beta_n\,|\phi_n\rangle \end{array} . \tag{4.48}$$

This statement has a very pronounced physical meaning: if A and B commute, one may in principle obtain well defined measurement values for both observables simultaneously. For instance, if we were to extract sharp values of a particle's x-coordinate, we would have to face a rather blurred value of its momentum p_x due to the Heisenberg uncertainty relation

$$\Delta p_x \geq \frac{\hbar}{2\Delta x} . \tag{4.49}$$

However, since $[x, p_y] = 0$ we might determine the y-component of the momentum with absolute certainty and complete uncertainty regarding the y-coordinate.

There is also another application of this rule that is frequently used when one of the operators is the Hamiltonian of the system and the other represents a symmetry operation that leaves the physical system invariant in real space (configuration space). In general, the action of a symmetry operation S on the physical system will induce a similarity transformation of the Hamiltonian:

$$H \to H' \;=\; S^{-1}HS . \tag{4.50}$$

For instance, if S is an inversion with respect to the origin of the coordinate axes changing every $\boldsymbol{r}$ into $-\boldsymbol{r}$, the one-particle Hamiltonian of equation (4.40) would transform into

$$H \;=\; -\frac{\hbar^2}{2m}\nabla^2 \;+\; U(-\boldsymbol{r}) . \tag{4.51}$$

On the other hand, if S is a real symmetry operation of the system, it must leave the Hamiltonian invariant:

$$H \;=\; H' \;=\; S^{-1}HS \tag{4.52}$$

or

$$[H, S] = 0 . \tag{4.53}$$

We therefore conclude that every symmetry operation S of a physical system commutes with the system's Hamiltonian H so there exists a common set of eigenvectors simultaneously diagonalizing H and S. This is a very helpful tool in many practical cases since it may be easier to find the eigenvectors of S first and then to seek the energy eigenvalues that correspond to a given eigenvalue of S. These eigenvalues of S can thus efficiently be used to label the energy eigenstates and therefore they are called 'good quantum numbers'.

4.5 Path Integrals

R. P. Feynman [23] has formulated the theory of quantum mechanics in an entirely different way. Starting from the action integral S (see (2.20)) he pointed out that the probability amplitude for a particle to traverse from an initial space point (q_0, t_0) to a final one (q_1, t_1) along the path $q(t)$ can be expressed in terms of the action integral as follows:

$$A[q(t)] = \exp\left(\frac{\mathrm{i}}{\hbar} S[q(t)]\right) \tag{4.54}$$

with

$$S[q(t)] = \int_{t_0}^{t_1} \mathrm{d}t\, L(q, \dot{q}, t) \,. \tag{4.55}$$

The global probability amplitude for finding the particle at (q_1, t_1) provided it started in (q_0, t_0) is then obtained by the superposition principle, i.e. by summing the contributions of all possible space paths $q(t)$

$$K(q_1, t_1; q_0, t_0) = \sum_{\{q(t)\}} A[q(t)] \tag{4.56}$$

$$= \int \mathcal{D}q \, \exp\left(\frac{\mathrm{i}}{\hbar} S[q(t)]\right) \tag{4.57}$$

Clearly, although all paths are contributing equally in magnitude, their relative phase factors are different which may give rise to drastic interference effects. The probability of the above mentioned event is given by

$$P = |K(q_1, t_1; q_0, t_0)|^2 \,. \tag{4.58}$$

The function $K(q_1, t_1; q_0, t_0)$ is called *the particle propagator* and it takes the form of a Green function.

Feynman's rules are not complementary to the ones that we have given so far, but can be derived from the earlier given ones. Alternatively, one may start with Feynman's rules as postulates and derive the others (as Feynman did himself). In order to see the equivalence, we will derive the formula for the propagator $K(q_1, t_1; q_0, t_0)$.
The Schrödinger equation is a first order differential equation is time. We may ask the following question:

Given the wave function $\psi(q, t_0)$ at time t_0, what will be the wave function at some later time t_1?

In general, the amplitude at position q_1 at time t_1 is a superposition of all contributions from all points q where each contribution is proportional to the amplitude at t_0:

$$\psi(q_1, t_1) = \int \mathrm{d}q\, K(q_1, t_1; q, t_0)\, \psi(q, t_0) \quad \text{with} \quad t_1 > t_0\ , \tag{4.59}$$

or equivalently,

$$\theta(t_1 - t_0)\psi(q_1, t_1) \;=\; \int \mathrm{d}q\, K(q_1, t_1; q, t_0)\, \psi(q, t_0)\ . \tag{4.60}$$

Here we have explicitly applied Huygens' principle or the superposition principle to the waves in quantum mechanics. It is a remarkable fact that Huygens' principle is exact in quantum mechanics, whereas it is only approximate in optics. The correct formula in optics was given by Kirchhoff a few centuries after Huygens, starting from the second order Maxwell equations. In particular, let $\psi(q, t_0)$ correspond to the wave function of a particle being at time t_0 in the eigenstate $|q_0\rangle$ of the position operator

$$\psi(q, t_0) = \delta(q - q_0)\ . \tag{4.61}$$

Then we find

$$\psi(q_1, t_1) = K(q_1, t_1; q_0, t_0)\ . \tag{4.62}$$

This wave function describes the probability amplitude of finding the particle at q_1 at time t_1, so the kernel K is the probability amplitude for a particle being in q_1 at t_1, provided that it was in q_0 at t_0.

Another way of writing K is obtained by the following three steps:

- put the particle in a position eigenstate $|q_0\rangle$ at t_0
- perform a time evolution of this state over the period $t_1 - t_0$
- project the resulting state onto the eigenstate $|q_1\rangle$

In summary, the result is

$$K(q_1, t_1; q_0, t_0) \;=\; \left\langle q_1 \right| \exp\left(-\frac{\mathrm{i}}{\hbar} H (t_1 - t_0)\right) \left| q_0 \right\rangle \tag{4.63}$$

which may be obtained by formally integrating of the time dependent Schrödinger equation between t_0 and t_1:

$$\begin{aligned} \mathrm{i}\hbar\, \frac{\mathrm{d}}{\mathrm{d}t}\, |q(t)\rangle \;&=\; H\ |q(t)\rangle \\ &\Downarrow \\ |q(t_1)\rangle \;&=\; \exp\left(-\frac{\mathrm{i}}{\hbar} H\ (t_1 - t_0)\right) |q(t_0)\rangle\ . \end{aligned} \tag{4.64}$$

The Green function of the Schrödinger equation constitutes an answer to the following question:

Which wave is generated by a point source in the Schrödinger equation?

The mathematical formulation of this problem reads:

$$\mathrm{i}\hbar\frac{\partial}{\partial t}\psi(q,t) - H\psi(q,t) = \delta(q)\delta(t) \qquad t > 0 \,. \tag{4.65}$$

The corresponding solution is the Green function $\psi(q,t) = G(q,t)$ because, if we now consider an arbitrary source term $J(q,t)$ in the Schrödinger equation

$$\mathrm{i}\hbar\frac{\partial}{\partial t}\psi(q,t) - H\psi(q,t) = J(q,t) \,, \tag{4.66}$$

then the solution reads

$$\psi(q,t) = \int \mathrm{d}q_0 \int_{-\infty}^{t} \mathrm{d}t_0\, G(q,t;q_0,t_0)\, J(q_0,t_0) \,. \tag{4.67}$$

Of course, one can not add extra terms to the Schrödinger equation in an arbitrary way. However, it is possible to split the Hamiltonian in a lowest order part and a perturbative part:

$$H = H_0 + \Delta H \,. \tag{4.68}$$

We may identify the source term as

$$J(q,t) = \Delta H\, \psi(q,t) \tag{4.69}$$

such that the solution becomes to first order in J:

$$\psi(q,t) = \int \mathrm{d}q_0 \int_{-\infty}^{t} \mathrm{d}t_0\, G_0(q,t;q_0,t_0)\, \Delta H\, \psi(q_0,t_0) \tag{4.70}$$

where G_0 corresponds to H_0, the zeroth order part of the Hamiltonian.

The kernel K and the Green function G are actually the same object. This can be demonstrated by acting with the operator $\left[\mathrm{i}\hbar\frac{\partial}{\partial t} - H\right]$ on the kernel equation

$$\begin{aligned}\left[\mathrm{i}\hbar\frac{\partial}{\partial t} - H\right]\theta(t-t_0)\psi(q,t) &= \mathrm{i}\hbar\delta(t-t_0)\psi(q,t)\\ &= \int \mathrm{d}q_0 \left[\mathrm{i}\hbar\frac{\partial}{\partial t} - H\right] K(q,t;q_0,t_0)\,\psi(q_0,t_0) \,.\end{aligned} \tag{4.71}$$

Since this should be generally true for all ψ we obtain

$$\left[\mathrm{i}\hbar\frac{\partial}{\partial t} - H\right] K(q,t;q_0,t_0) \;=\; \mathrm{i}\hbar\delta(q-q_0)\delta(t-t_0) \tag{4.72}$$

which shows that K and G satisfy the same equation. The boundary condition follows from causality, which implies that $G = K/\mathrm{i}\hbar = 0$ for $t < t_0$.

We will now continue to derive Feynman's formula for K. For this purpose we subdivide the time interval $t' - t_0$ into N parts of size ϵ. By plugging in the identity operator over and over again, we obtain

$$K(q',t';q_0,t_0) = \int \mathrm{d}q_1 \int \mathrm{d}q_2 \ldots \int \mathrm{d}q_N \left\langle q'|\exp\left(-\frac{\mathrm{i}}{\hbar}H\epsilon\right)|q_N\right\rangle$$
$$\left\langle q_N|\exp\left(-\frac{\mathrm{i}}{\hbar}H\epsilon\right)|q_{N-1}\right\rangle \ldots \left\langle q_1|\exp\left(-\frac{\mathrm{i}}{\hbar}H\epsilon\right)|q_0\right\rangle .$$

Each transition amplitude can now be further evaluated by exploiting the fact that ϵ is infinitesimally small:

$$\begin{aligned}\left\langle q_{i+1}|\exp\left(-\frac{\mathrm{i}}{\hbar}H\epsilon\right)|q_i\right\rangle &= \left\langle q_{i+1}|\left(1-\frac{\mathrm{i}}{\hbar}H\epsilon\right)|q_i\right\rangle \\ &= \delta(q_{i+1}-q_i) - \frac{\mathrm{i}}{\hbar}\epsilon\,\langle q_{i+1}|H|q_i\rangle \\ &= \int \frac{\mathrm{d}p_i}{2\pi\hbar}\mathrm{e}^{\frac{\mathrm{i}}{\hbar}p_i\,(q_{i+1}-q_i)} \\ &\quad -\frac{\mathrm{i}}{\hbar}\epsilon\left(\left\langle q_{i+1}|\frac{p^2}{2m}|q_i\right\rangle + \langle q_{i+1}|V(q)|q_i\rangle\right) .\end{aligned}$$

The third term can be evaluated in the representation in which the momentum is diagonal, i.e. using the fact that the momentum eigenstate $|p\rangle$ in the q-representation is

$$\langle q|p\rangle = \frac{1}{\sqrt{2\pi\hbar}}\mathrm{e}^{\frac{\mathrm{i}}{\hbar}p\cdot q} \tag{4.73}$$

we obtain

$$\begin{aligned}\left\langle q_{i+1}|\frac{p^2}{2m}|q_i\right\rangle &= \int \frac{\mathrm{d}p_{i+1}}{\sqrt{2\pi\hbar}}\int\frac{\mathrm{d}p_i}{\sqrt{2\pi\hbar}}\,\mathrm{e}^{\frac{\mathrm{i}}{\hbar}p_{i+1}q_{i+1}}\,\frac{p^2}{2m}\delta(p_{i+1}-p_i)\mathrm{e}^{-\frac{\mathrm{i}}{\hbar}p_i q_i} \\ &= \int\frac{\mathrm{d}p_i}{2\pi\hbar}\mathrm{e}^{\frac{\mathrm{i}}{\hbar}p_i(q_{i+1}-q_i)}\,\frac{p_i^2}{2m} .\end{aligned}$$

Therefore the matrix elements becomes

$$\begin{aligned}\left\langle q_{i+1}|\exp\left(-\frac{\mathrm{i}}{\hbar}H\epsilon\right)|q_i\right\rangle &= \int\frac{\mathrm{d}p_i}{2\pi\hbar}\mathrm{e}^{\frac{\mathrm{i}}{\hbar}p_i(q_{i+1}-q_i)}\left(1-\frac{\mathrm{i}}{\hbar}\epsilon\frac{p_i^2}{2m}-\frac{\mathrm{i}}{\hbar}\epsilon V(q_i)\right) \\ &= \int\frac{\mathrm{d}p_i}{2\pi\hbar}\mathrm{e}^{\frac{\mathrm{i}}{\hbar}p_i(q_{i+1}-q_i)-\frac{\mathrm{i}}{\hbar}\epsilon\left(\frac{p_i^2}{2m}+V(q_i)\right)} .\end{aligned}$$

Putting all elements together results into

$$\begin{aligned}K(q',t';q_0,t_0) &= \lim_{N\to\infty}\int\left[\frac{\mathrm{d}p_i\,\mathrm{d}q_i}{2\pi\hbar}\right] \\ &\quad\times\exp\left[\frac{\mathrm{i}}{\hbar}\sum_{i=1}^{N}\epsilon\left(p_i\frac{q_{i+1}-q_i}{\epsilon}-\left(\frac{p_i^2}{2m}+V(q_i)\right)\right)\right] \\ &= \int\mathcal{D}p\mathcal{D}q\,\exp\left(\frac{\mathrm{i}}{\hbar}\int\mathrm{d}t\,(p\,\dot{q}-H(p,q)\,)\right) .\end{aligned}$$

We recognize the Lagrangian in the integrand of the exponent. By performing all momentum integrations we arrive at the result we wanted to prove, i.e.

$$K(q_1, t_1, q_0, t_0) = \int \mathcal{D}q \exp\left(\frac{\mathrm{i}}{\hbar}\int_{t_0}^{t_1} \mathrm{d}t\, L(q\,\dot{q})\right) . \tag{4.74}$$

As was mentioned in Sect. 3.3 the dealing with an overall normalization factor in front of the path integral is usually done rather carelessly, because the practical use of the path integrals involves the determination of averages. Indeed, doing the momentum integration very carefully generates a factor A^{-1}

$$A = \sqrt{\frac{2\pi\epsilon\hbar\mathrm{i}}{m}} \tag{4.75}$$

and

$$\int \mathrm{d}p_i \exp\left[\frac{\mathrm{i}}{\hbar}\epsilon\left(p_i\frac{q_{i+1}-q_i}{\epsilon} - \frac{p_i^2}{2m}\right)\right]$$
$$= \sqrt{\frac{m}{2\pi\epsilon\hbar\mathrm{i}}} \exp\left[\frac{\mathrm{i}}{\hbar}\,\epsilon\,\frac{1}{2}m\left(\frac{q_{i+1}-q_i}{\epsilon}\right)^2\right] . \tag{4.76}$$

Therefore, the following interpretation should be given to (4.74):

$$\mathcal{D}q = \lim_{N\to\infty} \frac{dq_1}{A}\frac{dq_2}{A}\cdots\frac{dq_N}{A} . \tag{4.77}$$

The path-integral approach is very powerful because it helps avoiding operator ordering ambiguities. The p and q variables are ordinary numbers and not operators!

Furthermore, continuing along the path-integral approach, many aspects of quantum mechanics get an interpretation which could not be obtained otherwise, e.g. the wave function $\psi(q, t)$ is the arrival amplitude for arriving in q at time t coming from *anywhere* in the past. So the wave function contains the full history of the particle. In a simular way, the future of the particle is fully determined by all paths starting at time t. All this information is contained in the complex conjugate of a wave function $\phi^*(q, t)$. We could name it the departure amplitude. If the future implies that the particle is submitted to a measurement, the outcome is fully determined by this wave function. A system, whose history up to time t results into a wave function $\psi(q, t)$, will have a probability of being found in a state $\phi^*(q, t)$ (e.g. the eigenstate of some measurement device), equal to

$$\langle\phi|\psi\rangle = \int dq\, \phi^*(q, t)\psi(q, t) .$$

Exercises

4.1. Using the prescriptions (4.38) calculate the eigenfunctions of the momentum operator in the position representation for a single particle moving on a one-dimensional interval $[0, L]$ when periodic boundary conditions are imposed.

4.2. Prove that the plane-wave states given by

$$\psi_{\boldsymbol{k}}(\boldsymbol{r}) = \exp(\mathrm{i}\boldsymbol{k} \cdot \boldsymbol{r})$$

are eigenfunctions of the translation operator $T(\boldsymbol{a})$, generating a translation over $\boldsymbol{a}$ in configuration space.

4.3. Consider a particle described by cylindrical coordinates (r, ϕ, z) Prove that the z-component of the angular momentum operator is represented by $-\mathrm{i}\hbar\partial/\partial z$ in the coordinate representation.

4.4. Check out by explicit calculation in the coordinate representation whether the commutation rule $[q, p] = \mathrm{i}\hbar$ is consistently fulfilled. *Hint*: let $[q, -\mathrm{i}\hbar\partial/\partial q]$ operate on an arbitrary wave function.

5. Single-Particle Quantum Mechanics

This work is not aiming at a general tutorial on quantum physics nor does it intend to emulate the $(n+1)$-th book about quantum mechanics after mankind was overwhelmed by the huge reservoir of n preceding good, bad and average editions. Therefore we will not systematically repeat all crucial topics and concepts and we will restrict our exposure to those features of quantum mechanics that are essential to the understanding of (quantum) transport. In particular, we refer to the standard text books and courses for more elaborate discussions on single particle quantum mechanics.

Although transport is definitely a many-particle phenomenon, its description may be formulated entirely in terms of one-particle eigenstates which are thus playing a central role. This is the main motivation of the present chapter.

5.1 Charge Density, Current and Single Particle Wave Functions

Probing the charge density, predicting the current flowing through a device and estimating the unavoidable noise, ... these are typical skills one may expect to result from a decent transport theory. Although 'charge density' and 'current' are typical many-particle quantities, it is instructive to have them in mind already when solving the single-particle Schrödinger equation. After all, the macroscopic charge and current densities are nothing but statistical weighted sums of microscopic charge and current densities carried by single particle states or arising from transitions between them. It therefore makes sense to look at the single particle eigenstates from that point of view.

It is not very hard to associate a charge density with a single-particle wave function. Generally, if $\Phi(\boldsymbol{r},t)$ is the wave function of some single-particle state $|\Phi(t)\rangle$, then $|\Phi(\boldsymbol{r},t)|^2$ is the probability density of finding the particle at a position $\boldsymbol{r}$ when it is known to be in the time dependent state $|\Phi(t)\rangle$. It is then intuitively clear that the charge density of the corresponding state $|\Phi(t)\rangle$ is given by

$$\varrho(\boldsymbol{r},t) = q|\Phi(\boldsymbol{r},t)|^2 \qquad \text{for a particle with charge } q \ . \tag{5.1}$$

It is further possible to assign a current density $\boldsymbol{J}(\boldsymbol{r},t)$ to the state $|\Phi(t)\rangle$ so as to satisfy the continuity equation

$$\nabla \cdot \boldsymbol{J}(\boldsymbol{r},t) + \frac{\partial \varrho(\boldsymbol{r},t)}{\partial t} = 0 \,. \tag{5.2}$$

Starting from the time-dependent Schrödinger equation in the position representation, it is a good exercise to show that

$$\begin{aligned} \boldsymbol{J}(\boldsymbol{r},t) &= \frac{-iq\hbar}{2m}\left[\Phi^*(\boldsymbol{r},t)\nabla\Phi(\boldsymbol{r},t) - \left(\nabla\Phi^*(\boldsymbol{r},t)\right)\Phi(\boldsymbol{r},t)\right] \\ &= \frac{-q\hbar}{m}\Im\left[\left(\nabla\Phi^*(\boldsymbol{r},t)\right)\Phi(\boldsymbol{r},t)\right] \end{aligned} \tag{5.3}$$

is the right candidate. From (5.3) we may conclude that single states can only carry current if their wave functions have a non-vanishing imaginary part.

Remark. If $|\Phi\rangle$ is an energy eigenstate, i.e. if $|\Phi\rangle$ is an eigenvector of the one-particle Hamiltonian, then clearly the time variable t may be suppressed in (5.1) and (5.3). In that case the continuity equation reduces to

$$\nabla \cdot \boldsymbol{J}(\boldsymbol{r}) = 0 \tag{5.4}$$

and the current is 'incompressible'.

5.2 Constant Potential, Energy Bands and Energy Subbands

The solution of Schrödinger's equation for a particle in a constant potential is interesting in many respects. First, the solution requires the least mathematical effort. Denoting the constant potential energy by U_0, we may rewrite the Schrödinger's equation in the form

$$\nabla^2\phi(\boldsymbol{r}) + \alpha\phi(\boldsymbol{r}) = 0 \,, \tag{5.5}$$

where α is not dependent on the position vector $\boldsymbol{r}$:

$$\alpha = \frac{2m}{\hbar^2}(E - U_0) \,. \tag{5.6}$$

If the domain Ω has a relatively simple geometry (such as rectangular boxes, cylinders, spheres, etc.) such that the coordinate surfaces[1] are coinciding with the domain boundaries, one may solve (5.5) by separation of variables. More elaborate treatments and examples of this well-known technique can be found in standard textbooks on mathematical physics and quantum mechanics such as Morse and Feshbach, Flügge, and have been omitted here [24, 25].

Next, the constant potential represents an idealization of the periodic crystal potential which governs all perfect metals and semiconductors, at least if they are considered infinite in all directions. In that case the translational

[1] A coordinate surface is a surface along which one particular coordinate remains constant.

invariance of the crystal lattice gives rise to the well-known Bloch theorem stating that the eigenfunctions have the form of a modulated plane wave

$$\phi_{n\mathbf{k}}(\boldsymbol{r}) = \exp(\mathrm{i}\boldsymbol{k}\cdot\boldsymbol{r})\, u_{n\mathbf{k}}(\boldsymbol{r}) \ , \tag{5.7}$$

where n, $\boldsymbol{k}$ and $u_{n\mathbf{k}}(\boldsymbol{r})$ are representing the band index, the crystal momentum of the particle and an 'envelope function' reflecting the periodicity of the crystal lattice respectively. The allowed energy eigenvalues are grouped in bands, separated by 'forbidden' regions. Since the forbidden regions are located near the edge of the so-called Brillouin zones, one usually adopts the 'reduced zone scheme' to get a better overview of the band structure: all crystal momenta $\boldsymbol{k}$ exceeding the first Brillouin zone (1BZ) are confined to the latter by adding an appropriate lattice vector $\boldsymbol{G}$ of the reciprocal lattice. This can always be done since the energies are known to be periodic functions as well:

$$E_{n\mathbf{k}} = E_{n,\boldsymbol{k}+\boldsymbol{G}} \qquad \text{for all } n \text{ and } \boldsymbol{G}. \tag{5.8}$$

In a second step the bands themselves are horizontally shifted to the first Brillouin zone so that $E_{n\mathbf{k}}$ becomes a multivalued function of $E_{n\mathbf{k}}$ the branches of which are identified as the energy bands. The band index n labels the subsequent bands whereas the crystal momentum $\boldsymbol{k}$ determine the allowed energies within the individual bands, as is illustrated in Fig. 5.1.

In semiconductors the band index n typically runs over the valence band(s) and the relevant conduction band valleys (X in silicon, Γ, L and X in GaAs etc.). Our constant potential model makes life much easier (sometimes too easy) and treats the crystal as a continuum thereby replacing the restricted lattice periodicity by full translational invariance, smearing out all crystal potential fluctuations and reducing the set of multiple bands to one single parabolic band. Correspondingly, the envelope functions are just constants $\{u_{\mathbf{k}}\}$ and the wave functions are

$$\phi_{\mathbf{k}}(\boldsymbol{r}) = u_{\mathbf{k}} \exp(\mathrm{i}\boldsymbol{k}\cdot\boldsymbol{r}) \ , \tag{5.9}$$

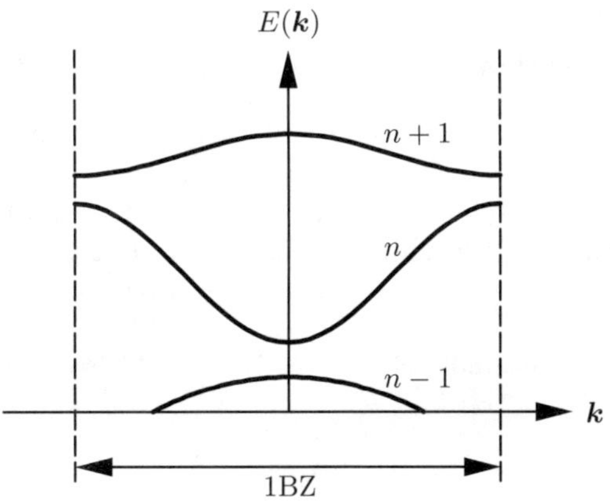

Fig. 5.1. Band structure in the reduced zone scheme

with energy eigenvalues

$$E_{n\mathbf{k}} = U_0 + \frac{\hbar^2 k^2}{2m} . \tag{5.10}$$

The mass m in (5.10) is actually the effective mass as extracted from more realistic band structure calculations. Usually it is denoted by m^*, but in order not to overload the notation, we will drop the asterisk and further represent the effective mass by m.

If we assume that the particle is moving in a rectangular box with edges L_x, L_y and L_z on which periodic boundary conditions are imposed, the components of the allowed $\boldsymbol{k}$-vectors are discretized according to

$$k_x = \frac{2\pi}{L_x} n_x, \quad k_y = \dots , \tag{5.11}$$

where

$$n_x,\ n_y,\ n_z = 0,\ \pm 1,\ \pm 2,\ \dots . \tag{5.12}$$

The discretization should not be taken too seriously in this case since the box is here assumed to be extremely large (bulk material!) and then of course $2\pi/L_x, \dots \to 0$ whence $\boldsymbol{k}$ becomes quasi-continuous. On the other hand it helps us normalize the wave functions in an elegant manner by choosing

$$u_{\mathbf{k}} = \frac{1}{\sqrt{\Omega}} , \tag{5.13}$$

with

$$\Omega = L_x L_y L_z .$$

Finally, it is nicely illustrated that one should pay attention to the choice of appropriate boundary conditions. If we had imposed infinite potential walls at the edges of the box, the eigenfunctions would be bound to vanish at the walls and would take the form

$$\phi_{\mathbf{k}}(\boldsymbol{r}) = \sqrt{\frac{8}{\Omega}} \sin k_x x \ \sin k_y y \ \sin k_z z . \tag{5.14}$$

In other words, being real, the wave functions would not carry any current and would not normally provide a convenient basis for a transport study. On the contrary, if we adopt periodic boundary conditions in the transport direction, say the x-direction and keep the infinite walls at the remaining edges, we can imagine the box as a torus or a snake biting its own tail, a picture that at least visualizes an electric circuit. The wave functions would then still vanish at the edges in the y- and z-directions but would become periodic in the x-direction:

$$\begin{aligned} \phi_{\mathbf{k}}(0,y,z) &= \phi_{\mathbf{k}}(L_x,y,z) && \text{for all } y,\ z \\ \phi_{\mathbf{k}}(x,0,z) &= \phi_{\mathbf{k}}(x,L_y,z) = 0 && \text{for all } x,\ z \\ \phi_{\mathbf{k}}(x,y,0) &= \phi_{\mathbf{k}}(x,y,L_z) = 0 && \text{for all } x,\ y\ . \end{aligned} \tag{5.15}$$

The eigenfunctions and the corresponding energy eigenvalues are now

$$\begin{aligned} \phi_{\mathbf{k}}(\boldsymbol{r}) &= \frac{2}{\sqrt{L_x L_y L_z}} \exp\left(\mathrm{i}k_x x\right) \sin k_y y \, \sin k_z z \\ E_{\mathbf{k}} &= U_0 \ + \ \frac{\hbar^2 k^2}{2m} \\ k_x &= \frac{2\pi}{L_x}\, n_x \quad ; \ n_x = 0 \ \pm 1, \ \pm 2 \ \ldots \\ k_y &= \frac{\pi}{L_y}\, n_y \quad ; \ n_y = 1, \ 2 \ \ldots \\ k_z &= \frac{\pi}{L_z}\, n_z \quad ; \ n_z = 1, \ 2 \ \ldots \end{aligned} \tag{5.16}$$

and the eigenstates are also eigenvectors of the momentum operator p_x:

$$p_x \left|\phi_{\mathbf{k}}\right\rangle \ = \ \hbar k_x \left|\phi_{\mathbf{k}}\right\rangle \ . \tag{5.17}$$

As expected, the current density carried by $|\phi_{\boldsymbol{k}}\rangle$ reads:

$$J_x(y,z) = \frac{4q\hbar k_x}{\Omega m} \ \sin^2 k_y y \ \sin^2 k_z z \tag{5.18}$$

$$J_y(y,z) = 0 \tag{5.19}$$

$$J_z(y,z) = 0 \ , \tag{5.20}$$

whereas the charge density

$$\varrho(y,z) \ = \ \frac{4q\hbar k_x}{\Omega} \ \sin^2 k_y y \ \sin^2 k_z z \ . \tag{5.21}$$

is oscillating in the y- and z-directions and does not vary at all with x. The latter is an immediate consequence of the Heisenberg uncertainty relations: since the state $|\phi_{\mathbf{k}}\rangle$ is an eigenvector of p_x, the x-component of the momentum is sharply defined which must be compensated by an infinite uncertainty about the particle's x-coordinate. Therefore the probability of finding a particle at x should be the same for all x-values.

As one might imagine from the above considerations the boundary conditions (5.15) are very suitable for the description of a very long quantum wire for which the transverse dimensions L_y and L_z are very small (50–200 nm) compared to the dimension L_x along the transport direction (several microns). In such a case, discretization in the transverse directions is real and for each quasi-continuous k_x there are several allowed energies E_{k_x,k_y,k_z} corresponding to the discrete values of k_y and k_z. In other words, instead of having a three-dimensional single band traversed by continuous values of all components of $\boldsymbol{k}$, we are now dealing with 2 ladders of one-dimensional subbands, which are labeled by the discrete values of k_y and k_z.

5.3 Potential Wells

The infinite potential barriers arising in the quantum wire model briefly discussed in the last section originated from the particular boundary condition that the wave functions be vanishing at the wire edges. As such this model provides the simplest example of a (square) potential well:

$$U(\boldsymbol{r}) = \begin{cases} \text{constant inside the domain } \Omega, \\ \infty \qquad \text{elsewhere.} \end{cases} \tag{5.22}$$

However, in a more profound treatment of the quantum wire and many other more or less sophisticated semiconductor structures the potential acting on the charge carriers may considerably fluctuate and give rise to more complicated potential wells. In most cases of interest, the solutions of Schrödinger's equation cannot be cast in closed-form analytical expressions and one has to rely on numerical methods or elaborate perturbation methods. Here we will only illustrate the notion of electric subbands and consider one particular potential well that provides a very crude but exactly solvable model. For simplicity we further assume that the potential exists only in one dimension which may yield a very acceptable picture in many layered structures where the thickness of the subsequent layers is small compared to the size of the structure in the other directions of space.

The potential energy describing the potential well we have in mind looks like a one-dimensional attractive Coulomb potential proportional to $-1/x$ and for the sake of notational simplicity it is expressed as follows:

$$U(x) = \begin{cases} -\dfrac{\hbar^2}{2mbx} & \text{for } x > 0 \\ \infty & \text{for } x < 0 \,, \end{cases} \tag{5.23}$$

where b is a geometrical parameter measuring the steepness of the well. Since we are focusing on the wave functions of this well we will not worry about how to extract – from a self-consistent calculation – a relevant value of b. Qualitatively, the model could serve as a zeroth order approximation for describing the potential well in MOS inversion layers for which the region $x < 0$ would correspond to the oxide and $x > 0$ would extend from inversion layer to bulk substrate.

Since $U(x)$ depends only on x we can apply the usual trick and reduce the three dimensional Schrödinger equation to a one-dimensional eigenvalue problem. Indeed, separation of variables enables us to write the wave functions as plane waves, modulated by subband wave functions depending only on x:

$$\phi_{l\mathbf{k}}(\boldsymbol{r}, x) = \frac{1}{\sqrt{L_y L_z}} \exp(\mathrm{i}\boldsymbol{k}\cdot\boldsymbol{r})\, \chi_l(x) \,, \tag{5.24}$$

where $\boldsymbol{r} = (y, z)$ and $\boldsymbol{k} = (k_y, k_z)$ are now two-dimensional vectors perpendicular to the x-axis and $\chi_l(x)$ and W_l are representing the wave function and energy of the lth subband. Stated explicitly, $\chi_l(x)$ is the normalized solution of

$$\frac{d^2\chi_l(x)}{dx^2} + \left[\frac{2mW_l}{\hbar^2} + \frac{1}{bx}\right]\chi_l(x) = 0 . \qquad (5.25)$$

Once the subband wave functions and energies are found, we may reconstruct the global wave functions from (5.24) and the total energy eigenvalues from

$$E_{l\mathbf{k}} = \frac{\hbar^2(k_y^2 + k_z^2)}{2m} + W_l . \qquad (5.26)$$

The solutions to (5.25) turn out to be a subset of the well-known Laguerre functions – actually the subset that describes the s orbitals of the hydrogen atom – and may be written as follows:

$$\chi_l(x) = \frac{l!\exp(-x/2lb)}{\sqrt{2bl^3}} \sum_{r=1}^{l} \frac{(-1)^{r-1}}{(r-1)!r!(l-r)!}\left[\frac{x}{lb}\right]^r \qquad (5.27)$$

$$W_l = -\frac{\hbar^2}{8mb^2l^2} \quad l = 1, 2, 3, \ldots . \qquad (5.28)$$

In Fig. 5.2 the squares of the first three Laguerre functions are plotted as a function of x, together with the potential energy. From (5.28) it becomes clear that the subband spectrum is discrete and consists of an infinite but countable number of subband states accumulating at $E = 0$. It should also denoted that the ground-state wave function coincides with the variational wave function

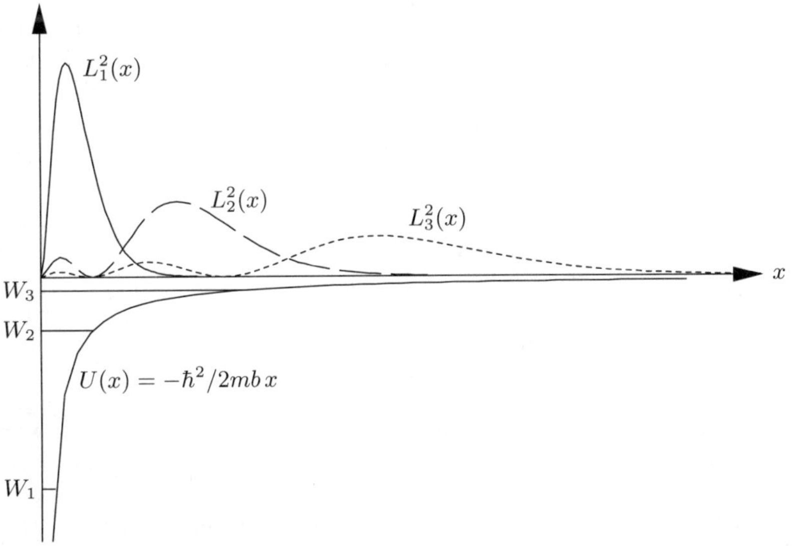

Fig. 5.2. Subband wave functions (squares) and energies for the $-1/x$ potential well for $b = 1$ nm

proposed by Fang and Howard to estimate the energy of the lowest subband for a more realistic potential shape characterizing the electrostatic behavior of a MOS capacitor [26].

Furthermore, the $-1/x$ potential is a nice example of a potential that has so-called *bound* states. A particle is said to be 'trapped' in a bound state if it occupies an energy eigenstate the wave function of which tends to zero far away from the well center. Physically, this means that the probability of finding the particle outside the well is rapidly decreasing to zero and the particle is bound to move in a relatively narrow region around the center of the potential well.
Bound states are often confused with quasi-bound states (or so-called virtual bound states) where the spatial localization is not prohibiting the leakage of particles out of the potential well. Contrary to bound states that have real (subband) wave functions, quasi-bound states may or may not carry current *in* the confinement direction.

Although it is impossible to solve Schrödinger's equation in closed form for an arbitrary potential shape, there is a simple way to identify bound states in one or more realistic potential wells within a semiconductor structure. For the sake of simplicity we have restricted the discussion to the case of one-dimensional confinement, say in the x-direction. The starting point is once again the one-dimensional Schrödinger equation

$$\frac{\mathrm{d}^2\chi_l(x)}{\mathrm{d}x^2} + \frac{2m}{\hbar^2}\left[W_l - U(x)\right]\chi_l(x) = 0\,. \tag{5.29}$$

In principle $U(x)$ may be a rather complicated profile consisting of several potential wells and barriers, but for our purposes they usually have one common feature: the potential shape becomes flat in at least one direction far away from the 'active regions'. For instance, the inversion layer of a MOSFET is localized in a potential well near the interface, but the potential may safely be assumed to be constant in the bulk part. Similarly, the potential is considerably varying along the channel between source and drain, but its values are fixed at the source and drain contacts. So, assuming in general that $U(x)$ is approaching an asymptotic value $U(\infty)$ for $x \to \infty$, we may extract the asymptotic behavior of $\chi_l(x)$ after substitution of

$$U(x) \to U(\infty)$$

into (5.29). This leaves us with the asymptotic equation

$$\frac{\mathrm{d}^2\chi_l(x)}{\mathrm{d}x^2} + \frac{2m}{\hbar^2}\left[W_l - U(\infty)\right]\chi_l(x) = 0 \tag{5.30}$$

the solution of which is entirely dependent on the sign of $W_l - U(\infty)$.

If $W_l - U(\infty) > 0$ there exists a real wavenumber $k_l > 0$ such that

$$k_l = \sqrt{\frac{2m}{\hbar^2}\left[W_l - U(\infty)\right]}\,, \tag{5.31}$$

whereas the asymptotic limit of $\chi_l(x)$ takes the general form

$$\chi_l(x) \overset{x\to\infty}{\longrightarrow} A_l \exp(\mathrm{i}k_l x) + B_l \exp(-\mathrm{i}k_l x) . \tag{5.32}$$

The details of the pre-factors A_l and B_l of course depend on the boundary conditions and the explicit solutions, but due to the plane-wave form of the asymptotic limit, the wave functions $\chi_l(x)$ will not decay for $x \to \pm\infty$ and in principle two independent solutions may be obtained for each W_l. Stated otherwise, the eigenstates $\{|\chi_l\rangle\}$ are not bound states but may be be considered quasi-bound states and they are making part of a doubly degenerate energy spectrum.

On the other hand, if $W_l - U(\infty) < 0$ there exists a real number $\alpha_l > 0$ such that

$$\alpha_l = \sqrt{\frac{2m}{\hbar^2}\left[U(\infty) - W_l\right]} \tag{5.33}$$

and

$$\chi_l(x) \overset{x\to\infty}{\longrightarrow} C_l \exp(\alpha_l x) + D_l \exp(-\alpha_l x) . \tag{5.34}$$

Mathematically, we have still two linearly independent solutions but now $\exp(\alpha_l x)$ will blow up at infinity so that we need to set $C_l = 0$ for any meaningful wave function. As a consequence, this part of the spectrum becomes non-degenerate meaning that to each energy W_l corresponds exactly one wave function $\chi_l(x)$. The latter is of course not strictly proven by the considerations about the asymptotic behavior, but the proof does hardly require more effort and goes as follows. Suppose that both $\chi_{1l}(x)$ and $\chi_{2l}(x)$ are solutions of (5.29) belonging to the eigenvalue W_l. Then

$$\begin{aligned} \frac{\mathrm{d}^2\chi_{1l}(x)}{\mathrm{d}x^2} + \frac{2m}{\hbar^2}\left[W_l - U(x)\right]\chi_{1l}(x) &= 0 \\ \frac{\mathrm{d}^2\chi_{2l}(x)}{\mathrm{d}x^2} + \frac{2m}{\hbar^2}\left[W_l - U(x)\right]\chi_{2l}(x) &= 0 . \end{aligned} \tag{5.35}$$

Multiplying the first equation by $\chi_{2l}(x)$ and the second one by $\chi_{2l}(x)$ and subtracting we obtain

$$\frac{\mathrm{d}^2\chi_{1l}(x)}{\mathrm{d}x^2}\chi_{2l}(x) - \frac{\mathrm{d}^2\chi_{2l}(x)}{\mathrm{d}x^2}\chi_{1l}(x) = 0 \tag{5.36}$$

or

$$\frac{\mathrm{d}}{\mathrm{d}x}\left[\frac{\mathrm{d}\chi_{1l}(x)}{\mathrm{d}x}\chi_{2l}(x) - \frac{\mathrm{d}\chi_{2l}(x)}{\mathrm{d}x}\chi_{1l}(x)\right] = 0 , \tag{5.37}$$

which means that the Wronskian of $\chi_{1l}(x)$ and $\chi_{2l}(x)$ must be a constant and since both wave functions should vanish at infinity this constant must be zero. Hence $\chi_{1l}(x)$ and $\chi_{2l}(x)$ are found to be linearly dependent.
Finally, all wave functions corresponding to $W_l - U(\infty) < 0$ are now decaying (at least) exponentially at infinity and therefore the particle is bound to a finite or semi-infinite region in x-space.

5.4 Potential Barriers

Not only potential wells but also potential barriers deserve a lot of attention in a study of transport in advanced micro-devices. Potential barriers are often huge on purpose so as to prevent carriers from migration into specific areas of the device. However, in quantum devices the transport is based on purely quantum mechanical phenomena such as tunneling, and in the latter case the potential barriers are thin and relatively low in order to realize a substantial flow of tunneling carriers.

Here, we will not systematically discuss various kinds of potential barriers that may occur in real devices. Interested readers who wish to get into sophisticated calculations may consult the wide variety of good textbooks on quantum mechanics.
The working example we will go through below, is mainly meant to illustrate some of the practical problems one has to face when single particle wave functions have to be calculated. It may be viewed as a simplified model of a one-dimensional barrier one may typically encounter in ballistic devices, resonant tunneling transistors etc. The following potential profile is proposed:

$$U(x) = \begin{cases} -\Delta_1 & \text{for } x < 0 \\ 0 & \text{for } 0 \leq x \leq a \\ -\Delta_2 & \text{for } x > a \end{cases} , \tag{5.38}$$

where $\Delta_2 > \Delta_1 > 0$ and the x-direction is once again assumed to be the transport direction. This profile is illustrated in Fig. 5.3 and could be seen as a crude model for the conduction band profile along the channel of a ballistic MOSFET with length a and applied drain voltage $eV_D = \Delta_2 - \Delta_1$. The source and drain contacts would then roughly correspond to the regions $x < 0$ and $x > a$ respectively. Since the dimensions of the source and drain contacts are taken to be very long compared to a, we expect to discover a continuous spectrum. Therefore we will drop the discrete subband index l and determine later how to label the energy eigenstates. The one-dimensional

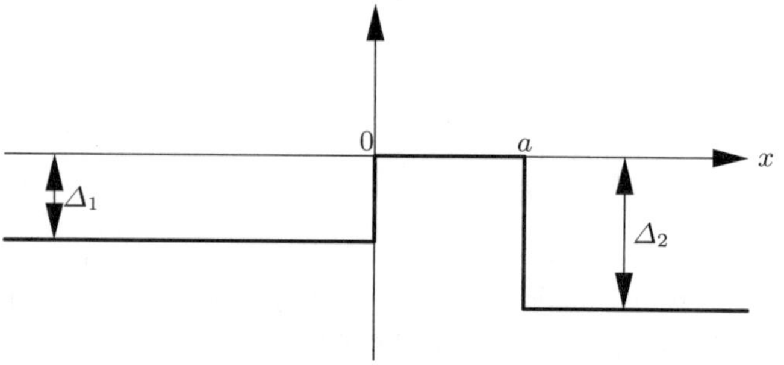

Fig. 5.3. Idealized potential energy profile of a ballistic transistor channel

Schrödinger equation is still given by (5.29) where now the piecewise constant profile (5.38) has to be substituted.

Let us first look for solutions in the energy interval $[0, +\infty]$. As suggested on the figure, we are taking the entire x-axis as the spatial domain. This implies that we do not have energy quantization imposed by boundary conditions. We could of course assign a large but finite length L to the contacting regions and try to solve for the energies and wave functions on the interval $[-L, L+a]$. Then, if we were to filter traveling solutions by adopting periodic boundary conditions

$$\chi(-L) = \chi(L+a) \tag{5.39}$$

we would end up with a transcendental eigenvalue equation which is just the equivalent of (5.16) that determined the quantized levels for the infinite square well in a quantum wire. However, if L becomes extremely large – representing the size of the macroscopic remainder of the circuit – the distance between subsequent levels vanishes anyway and so we can as well start our investigations for an infinite interval from the very beginning.
As a consequence, the energy eigenvalues are now making part of a continuum and the full expression of the wave function that corresponds to a particular energy eigenvalue W reads:

$$\chi(x) = \begin{cases} A_1\, \mathrm{e}^{\mathrm{i}k_1 x} + B_1\, \mathrm{e}^{-\mathrm{i}k_1 x} & \text{for } x < 0 \\ A\, \mathrm{e}^{\mathrm{i}kx} + B\, \mathrm{e}^{-\mathrm{i}kx} & \text{for } 0 \leq x \leq a \\ A_2\, \mathrm{e}^{\mathrm{i}k_2 x} + B_2\, \mathrm{e}^{-\mathrm{i}k_2 x} & \text{for } x > a \end{cases} , \tag{5.40}$$

where k_1, k and k_2 are three positive wave numbers respectively assigned to each of the subintervals, such that

$$W = -\Delta_1 + \frac{\hbar^2 k_1^2}{2m} = \frac{\hbar^2 k^2}{2m} = -\Delta_2 + \frac{\hbar^2 k_2^2}{2m} . \tag{5.41}$$

The above relations may be used to express k_1 and k_2 in terms of k which may be chosen to be the independent wave number:

$$k_j = \sqrt{k^2 + \frac{2m\Delta_j}{\hbar^2}} \quad , j = 1, 2 . \tag{5.42}$$

Within the continuous spectrum $W > 0$ we may conveniently distinguish between left and right traveling particles. Since the potential profile is piecewise constant, we may identify right traveling particles as being described by a plane wave propagating to the right when $x > a$. Let us start considering right traveling particles and set $B_2 = 0$. As well-known for any other cases of piecewise constant potentials, we need to 'match' the wave functions at all points where the potential profile changes abruptly, i.e. for $x = 0$ and $x = a$. 'Matching' means that we make sure that charge and current densities carried by a wave function are always continuous quantities, which requires that both the wave function and its derivative be continuous in the points $x = 0$

and $x = a$. This leaves us with four equations relating all five coefficients A_1, B_1, A, B and A_2:

$$\begin{aligned} A_1 + B_1 &= A + B \\ k_1(A_1 - B_1) &= k(A - B) \end{aligned} \tag{5.43}$$

and

$$\begin{aligned} A\,\mathrm{e}^{\mathrm{i}ka} + B\,\mathrm{e}^{-\mathrm{i}ka} &= A_2\,\mathrm{e}^{\mathrm{i}k_2a} \\ k\left(A\,\mathrm{e}^{\mathrm{i}ka} - B\,\mathrm{e}^{-\mathrm{i}ka}\right) &= k_2A_2\,\mathrm{e}^{\mathrm{i}k_2a} \,. \end{aligned} \tag{5.44}$$

Anticipating transport formulas to be discussed in later chapters, it is convenient and also appealing to express A_1, B_1, A and B in terms of the 'transmission coefficient' A_2:

$$\begin{aligned} A_1 &= \tfrac{1}{2}\,\mathrm{e}^{\mathrm{i}k_2a}\left[\left(1 + \tfrac{k_2}{k_1}\right)\cos ka - \mathrm{i}\left(\tfrac{k}{k_1} + \tfrac{k_2}{k}\right)\sin ka\right] A_2 \\ B_1 &= \tfrac{1}{2}\,\mathrm{e}^{\mathrm{i}k_2a}\left[\left(1 - \tfrac{k_2}{k_1}\right)\cos ka + \mathrm{i}\left(\tfrac{k}{k_1} + \tfrac{k_2}{k}\right)\sin ka\right] A_2 \\ A &= \tfrac{1}{2}\,\mathrm{e}^{\mathrm{i}(k_2-k)a}\left(1 + \tfrac{k_2}{k}\right) A_2 \\ B &= \tfrac{1}{2}\,\mathrm{e}^{\mathrm{i}(k_2+k)a}\left(1 - \tfrac{k_2}{k}\right) A_2 \,. \end{aligned} \tag{5.45}$$

Up to an overall constant absorbed into A_2 the wave functions are now completely determined by (5.40) and (5.45). If we wish to calculate interesting quantities like current and charge density we have to determine at least the modulus of this overall constant. (The phase is irrelevant and can simply be set equal to zero.) For that purpose we need to impose a proper normalization for the wave functions.

Delta Function Normalization. Having dealt already with the continuous eigenvalue spectrum of the position operator in the previous chapter we should be surprised to face a similar problem here. The normal way to proceed would be to take the 'norm' of any eigenstate $\langle\chi_q\rangle$ to be one and to account for orthogonality of eigenstates belonging to different energy eigenvalues: in other words, we would prefer to have an orthonormal basis of eigenstates:

$$\langle\chi_{k'}|\chi_k\rangle = \delta_{k'k} \,. \tag{5.46}$$

In practice however, the scalar product $\langle\chi_{k'}|\chi_k\rangle$ is evaluated in a convenient basis, say the coordinate representation which yields in our case (infinite interval on the x-axis):

$$\langle\chi_{k'}|\chi_k\rangle = \int_{-\infty}^{+\infty} \mathrm{d}x\; \chi^*_{k'}(x)\chi_k(x) \,, \tag{5.47}$$

which should equal the Kronecker delta $\delta_{k'k}$.

For $k' \neq k$ everything goes smoothly since also in a continuous spectrum the orthogonality of eigenstates belonging to different energy eigenvalues is preserved. However, if $k' = k$ we have a problem, since now the squared norm reads

$$\langle \chi_{k'} | \chi_k \rangle = \int_{-\infty}^{+\infty} \mathrm{d}x \, |\chi_k(x)|^2 \tag{5.48}$$

and *for a continuous spectrum* this integral generally does not exist. The latter becomes immediately clear in our case when we take a closer look at the actual contribution from the interval $[a, +\infty]$ to the integral of (5.48)

$$\begin{aligned} \int_a^{+\infty} \mathrm{d}x \, |\chi_k(x)|^2 &= \lim_{L\to\infty} \int_a^{L+a} \mathrm{d}x \, |\chi_k(x)|^2 \\ &= |A_2|^2 \lim_{L\to\infty} \int_a^{L+a} \mathrm{d}x \, |\mathrm{e}^{\mathrm{i}k_2 x}|^2 \\ &= |A_2|^2 \lim_{L\to\infty} L = +\infty \, . \end{aligned} \tag{5.49}$$

Without going into details we wish to point out that the above 'catastrophe', generally occuring for continuous spectra, is due to the asymptotic behaviour of the wave functions: contrary to the case of bound states, at large distances these wave functions are approaching plane waves or at least a class of functions that cannot be *quadratically integrated* over an infinite interval. As a general recipe, they should rather be normalized according to what is sometimes called the 'delta-function normalization'

$$\int_{-\infty}^{+\infty} \mathrm{d}x \, \chi_{k'}^*(x) \chi_k(x) = \delta(k' - k) \, . \tag{5.50}$$

Proceeding along these lines, we only need to evaluate the integral

$$\int_{-L}^{L+a} \mathrm{d}x \, \chi_{k'}^*(x) \chi_k(x)$$

and carefully take the limit $L \to \infty$. This is all straightforward but a little cumbersome and therefore we will first introduce some convenient shorthand notation.

First it should be noted that the coefficients $A_1, A \ldots$ are all dependent on the wave number q which determines the continuous energy eigenvalue $W(k) = \hbar^2 k^2 / 2m$ and the same applies of course to k_1 and k_2. This also means that when two different wave numbers k and k' come into the game, the distinction must be reflected in all above mentioned parameters as well. In this light we will adopt the following abbreviations:

$$A = A(k), \; A' = A(k'), \; k_1' = k_1(k'), \; \ldots \tag{5.51}$$

and further introduce the function

$$\delta_L(k) = -\frac{\mathrm{i}}{\pi} \, \frac{\mathrm{e}^{\mathrm{i}kL} - 1}{k} \, . \tag{5.52}$$

The latter converges to the delta functions when $L \to \infty$ as can be seen from the mathematical interlude on generalized functions.

Starting with the partial interval $[a, a+L]$ we easily obtain:

$$\int_a^{L+a} \mathrm{d}x\ \chi_{k'}^*(x)\chi_k(x) = A_2'^* A_2 \int_a^{L+a} \mathrm{d}x\ \exp(\mathrm{i}(k_2 - k_2')x)$$
$$= -\frac{\mathrm{i}A_2'^* A_2}{k_2 - k_2'} \exp\left(\mathrm{i}(k_2 - k_2')a\right) \left[\exp\left(\mathrm{i}(k_2 - k_2')L\right) - 1\right]$$
$$= \pi\ A_2'^* A_2\ \exp\left(\mathrm{i}(k_2 - k_2')a\right)\ \delta_L(k_2 - k_2')\ . \tag{5.53}$$

Taking the limit $L \to \infty$ we obtain

$$\int_a^{+\infty} \mathrm{d}x\ \chi_{k'}^*(x)\chi_k(x)\ =\ \pi\ |A_2|^2 \delta(k_2 - k_2')\ . \tag{5.54}$$

Remembering that k is the independent wave number, we need to express $\delta(k_2' - k_2)$ in terms of $\delta(k' - k)$. Defining

$$\begin{aligned} F(k') &= k_2' - k_2 \\ &= k_2(k') - k_2(k) \end{aligned} \tag{5.55}$$

it is clear that $k' = k$ is the unique solution of $F(k') = 0$ so that

$$\begin{aligned} \delta(k_2' - k_2) &= \delta(F(k')) \\ &= \frac{\delta(k' - k)}{\left|\dfrac{\partial F(k')}{\partial k'}\right|_{k'=k}} \\ &= \frac{k_2}{k}\ \delta(k' - k)\ , \end{aligned} \tag{5.56}$$

which leaves us with

$$\int_a^{+\infty} \mathrm{d}x\ \chi_{k'}^*(x)\chi_k(x)\ =\ \pi\ \frac{k_2}{k}\ |A_2|^2 \delta(k' - k)\ . \tag{5.57}$$

Similarly, the contribution from the interval $[-\infty, 0]$ can be evaluated as follows:

$$\int_{-L}^{0} \mathrm{d}x\ \chi_{k'}^*(x)\chi_k(x) = \int_{-L}^{0} \mathrm{d}x \left[A_1'^* \mathrm{e}^{-\mathrm{i}k_1'x} + B_1'^* \mathrm{e}^{\mathrm{i}k_1'x}\right]\left[A_1 \mathrm{e}^{\mathrm{i}k_1 x} + B_1 \mathrm{e}^{-\mathrm{i}k_1 x}\right]$$
$$= \pi \left[A_1'^* A_1 \delta_L(k_1 - k_1')\ +\ B_1'^* B_1 \delta_L(k_1' - k_1)\ +\ A_1'^* B_1 \delta_L(-k_1 - k_1')\right.$$
$$\left. +\ A_1 B_1'^* \delta_L(k_1 + k_1')\right]\ . \tag{5.58}$$

The cross terms do not contribute since they are converging to $\delta(-k_1 - k_1')$ and $\delta(k_1 + k_1')$ while both k_1 and k_1' are positive. Taking the limit $L \to \infty$ for the diagonal terms we find

$$\int_{-\infty}^{0} \mathrm{d}x\ \chi_{k'}^*(x)\chi_k(x)\ =\ \pi\left[|A_1|^2 + |B_1|^2\right] \delta(k_1' - k_1)\ . \tag{5.59}$$

Expressing $\delta(k_1' - k_1)$ similarly in terms of $\delta(k' - k)$ according to

$$\delta(k_1' - k_1) = \frac{k_1}{k}\, \delta(k' - k) \tag{5.60}$$

we end up with

$$\int_{-\infty}^{0} \mathrm{d}x\, \chi_{k'}^*(x)\chi_k(x) \;=\; \pi\, \frac{k_1}{k}\, \left[|A_1|^2 + |B_1|^2\right] \delta(k' - k)\,. \tag{5.61}$$

As long as the interval $[-L, L+a]$ is considered finite, we should also include the contribution from $[0, a]$. However, since the latter is of the order of a whereas the contributions from the 'large' intervals are proportional to L, we may skip $\int_0^a \mathrm{d}x\, \chi_{k'}^*(x)\chi_k(x)$ and write

$$\int_{-\infty}^{+\infty} \mathrm{d}x\, \chi_{k'}^*(x)\chi_k(x) \;=\; \frac{\pi}{k}\left\{ k_1\, \left[|A_1|^2 \,+\, |B_1|^2\right] \,+\, k_2\, |A_2|^2\right\} \delta(k' - k) \tag{5.62}$$

Following the delta function normalization, we must identify the left-hand side of (5.62) as $\delta(k' - k)$ and therefore we have

$$\frac{\pi}{k}\left\{ k_1\, \left[|A_1|^2 \,+\, |B_1|^2\right] \,+\, k_2\, |A_2|^2\right\} \;=\; 1\,. \tag{5.63}$$

From (5.45) we derive

$$|A_1|^2 \,+\, |B_1|^2 \;=\; \frac{1}{2}|A_2|^2 \left[\left(1 + \frac{k_2^2}{k_1^2}\right)\cos^2 ka \,+\, \left(\frac{k^2}{k_1^2} + \frac{k^2}{k^2}\right)\sin^2 ka\right]. \tag{5.64}$$

Substituting the latter result into (5.63), we finally obtain a closed-form expression for the transmission probability $T_2(k) = |A_2(k)|^2$:

$$T_2(k) = \frac{k}{\pi k_2}\left[1 + \frac{1}{2}\left(\frac{k_1}{k_2} + \frac{k_2}{k_1}\right)\cos^2 ka + \frac{1}{2}\left(\frac{k^2}{k_1 k_2} + \frac{k_1 k_2}{k^2}\right)\sin^2 ka\right]^{-1} \tag{5.65}$$

In a completely analogous manner the transmission probability can be calculated for left travelling particles by simply taking $B_1 = 0$ in (5.40). The result is

$$\begin{aligned} T_1(k) &= |A_1(k)|^2 \\ &= \frac{k}{\pi k_1}\left[1 + \frac{1}{2}\left(\frac{k_1}{k_2} + \frac{k_2}{k_1}\right)\cos^2 ka + \frac{1}{2}\left(\frac{k^2}{k_1 k_2} + \frac{k_1 k_2}{k^2}\right)\sin^2 ka\right]^{-1} \end{aligned} \tag{5.66}$$

The last example on potential barriers is rather simple from the mathematical point of view, but it reveals how carefully one should treat the energy spectrum for physical systems which are subdivided into several spatial sub-domains or intervals. Such a subdivision naturally arises in almost every

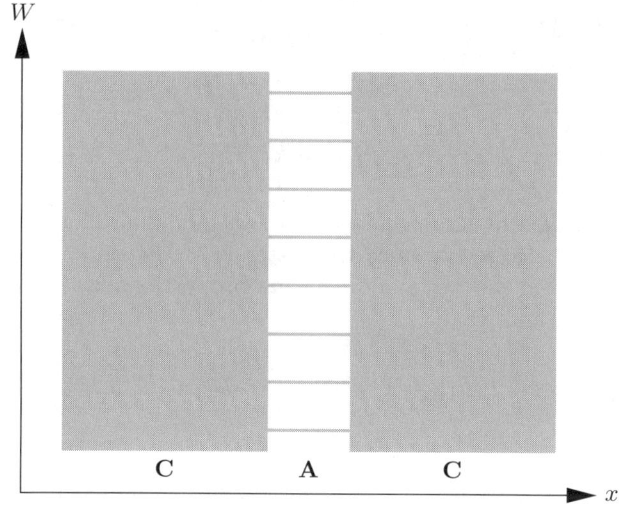

Fig. 5.4. Erroneous energy spectrum of an active area adjacent to huge contact regions

semiconductor structure which consists of at least 3 basic parts: 2 contacts (C) (source, drain, contacting 2DEG's in a quantum point contact, emitter and collector of a resonant tunnel diode, etc.) and a relatively narrow active area (A) (MOSFET channel, potential well between two barriers, etc.). Many qualitative as well as quantitative studies are suggesting (forcing?) the description of the energy spectrum as depicted in Fig. 5.4. In words: one assigns a continuous spectrum to each contact and a discrete spectrum to the active area *as if the geometrical subdivision implied that all three regions were also physically isolated!*. It does not take much effort to understand that this picture is wrong and quite misleading: 'energy' and 'position' are non-commuting operators and can thus never be determined simultaneously to the same accuracy so that one cannot label energy eigenstates with position (eigen)values. Stated otherwise, there are no separate sets of energies and states for different regions in space: there is one Fock space – and consequently only one energy spectrum (possibly containing continuous and discrete energy eigenvalues as well as resonances) and only one complete set of associated eigenstates – for the entire physical system. The only exception one could imagine is a complete physical isolation of the sub-regions being separated by infinitely high potential barriers. In such an extreme case no sub-system could feel the influence of one of its neighbors and a complete decoupling would occur, thereby ruling out any possibility of generating transport. In any other case, we have therefore to study the spectrum of the entire physical system. The following model gives a simplified description of two contacts and a potential well which are connected through two infinitely narrow potential barriers as is illustrated in Fig. 5.5. The delta-function nature of the barriers is not really

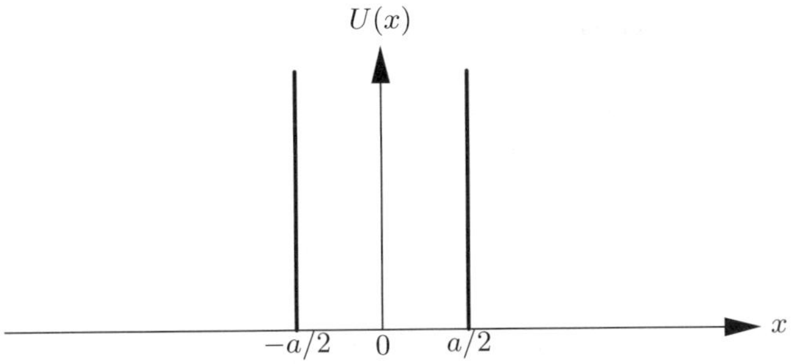

Fig. 5.5. Two identical delta-function barriers

required but it simplifies the calculations and does not alter the qualitative considerations. The potential profile is given by [27]

$$U(x) \;=\; \Lambda\,\left[\delta\left(x+\frac{a}{2}\right)\;+\;\delta\left(x-\frac{a}{2}\right)\right] \quad ; \; \Lambda > 0\,. \tag{5.67}$$

It is left as an exercise to show that the continuous energy spectrum of the right travelling particles is completely determined by

$$W(k) \;=\; \frac{\hbar^2 k^2}{2m} \qquad \text{for all } k > 0 \tag{5.68}$$

and

$$\chi(x) \;=\; \begin{cases} A_1\,\mathrm{e}^{\mathrm{i}kx} \;+\; B_1\,\mathrm{e}^{-\mathrm{i}kx} & \text{for } x < -\frac{a}{2} \\ A\,\mathrm{e}^{\mathrm{i}kx} \;+\; B\,\mathrm{e}^{-\mathrm{i}kx} & \text{for } -\frac{a}{2} < x < \frac{a}{2} \\ A_2\,\mathrm{e}^{\mathrm{i}kx} & \text{for } x > \frac{a}{2} \end{cases} \tag{5.69}$$

with

$$\begin{aligned} A_1(k) &= \left[1+\frac{\mathrm{i}k_0}{k}+\frac{k_0^2}{4k^2}\left(\mathrm{e}^{2\mathrm{i}ka}-1\right)\right] A_2(k) \\ B_1(k) &= -\mathrm{i}\frac{k_0}{k}\left(\cos ka \;+\; \frac{k_0}{2k}\sin ka\right) A_2(k) \\ A(k) &= \left(1+\mathrm{i}\,\frac{k_0}{2k}\right)\; A_2(k) \\ B(k) &= -\mathrm{i}\frac{k_0}{2k}\;\mathrm{e}^{\mathrm{i}ka}\; A_2(k) \\ k_0 &\equiv \frac{2m\Lambda}{\hbar^2}\,. \end{aligned} \tag{5.70}$$

It is immediately clear that there is only a continuum of states – one for each positive k – and all states are extending over the entire interval. There are no mysterious 'well states'. On the other hand, the situation of course differs considerably from the free particle case since the potential barriers are strikingly affecting the spectrum:

- considered as a function of energy, the states have a non-uniform tunneling probability enabling particles that are occupying these states to propagate from the right to the left (and vice versa);
- a countable subset of the continuum states consists of *resonant states* for which the transmission from region 1 to region 2 has reached a maximum and while reflections are suppressed in these regions.

In the case of right traveling states ($B_2 = 0$), the resonance condition would be

$$B_1 = 0 , \tag{5.71}$$

which, according to (5.70), amounts to

$$\cos ka + \frac{k_0}{2k} \sin ka = 0 . \tag{5.72}$$

In other words, the resonances correspond to a discrete subset of positive wavenumbers $k = k_n$ solving the transcendental equation

$$\tan k_n a = -\frac{2k_n}{k_0} \quad ; \quad n = 1, 2, 3, \ldots . \tag{5.73}$$

The resonances can be studied conveniently by analyzing the relative transmission probability $T(k) = |A_2(k)/A_1(k)|^2$ for a particle to tunnel through the barriers when it is injected into either of the two contact regions:

$$T(k) = \frac{4k^4}{4k^4 + k_0^2 \left(2k \cos ka + k_0 \sin ka\right)^2} . \tag{5.74}$$

Now suppose for one moment that the barrier height Λ (or k_0) tends to infinity. Then the three regions would fall apart and the wave functions within the isolated quantum well would be simple sines vanishing at $x = \pm a/2$:

$$\chi_n(x) = \sqrt{\frac{2}{a}} \sin k_n(x) . \tag{5.75}$$

We may therefore conclude that the eigenvalues of the *isolated potential well* approximately coincide with the energies of the resonant states of the *permeable well* when the barrier height tends to infinity.

However, several remarks and warnings are in order:

1. The numerical coincidence of the energy eigenvalues for the bound states of the impenetrable well and a subset of the travelling states of the permeable well must not make us believe that after all we can associate a separate set of states with the well: even the resonant states do extend over the entire interval and moreover, they are genuine extended, travelling states for which the reflected and transmitted waves are *not* combining into standing waves (see the coefficients $A(k_n)$ and $B(k_n)$).

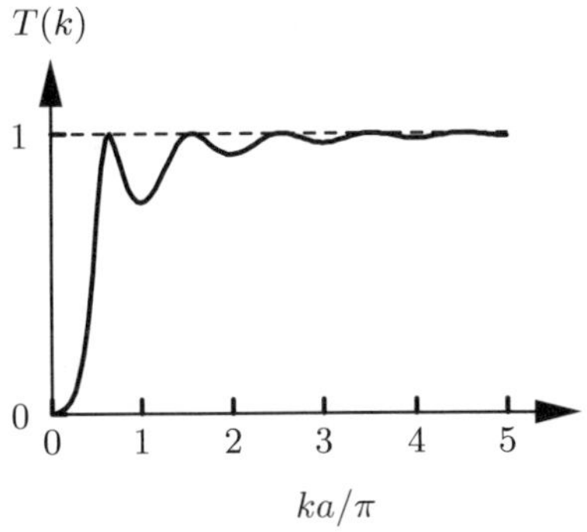

Fig. 5.6. Transmission coefficient versus wavenumber for $a = 5$ nm, $k_0 = 36$ nm^{-1}

2. The resonant states are not the only 'transport states': although the resonance peaks may be quite sharp (depending on the precise value of k_0) also the states in the neighborhood of the resonance peaks may contribute to the tunneling currents (see Fig. 5.6). Moreover, as the energy increases, also the transmission minima start increasing from zero so that the non-resonant contribution may be substantial.

In this light, we wish to conclude the present section with a final remark on the tunneling phenomenon. 'Tunneling' is a typical quantum mechanical phenomenon: it could be defined as propagation of a particle through a space region which is not accessible in classical mechanics. Mostly, this space region consists of one or more thin barriers within which the particle's wave functions do not drop to zero. So, tunneling is a spatial process describing the migration of a particle that resides in a tunneling state (such as the aforementioned resonance states). Tunneling generally does *not* require a change of state because there exists no natural 'tunneling' operator that invokes transitions between states – except for some more advanced processes such as phonon-assisted tunneling. Particles are thus not tunneling 'from reservoir states' say, to 'well states'!

5.5 Electromagnetic Fields

So far, the interaction of charged particles with electromagnetic fields has not been accounted for. In general, any electromagnetic field may be characterized by an electric field vector $\boldsymbol{E}(\boldsymbol{r},t)$ and a magnetic field induction $\boldsymbol{B}(\boldsymbol{r},t)$. Although both vector fields are unique and measurable observables, it proves quite convenient to deal directly with the electric scalar potential $V(\boldsymbol{r},t)$ and the magnetic vector potential $\boldsymbol{A}(\boldsymbol{r},t)$ from which $\boldsymbol{E}(\boldsymbol{r},t)$ and $\boldsymbol{B}(\boldsymbol{r},t)$ may be extracted according to

$$\boldsymbol{E}(\boldsymbol{r},t) = -\nabla V(\boldsymbol{r},t) - \frac{\partial \boldsymbol{A}(\boldsymbol{r},t)}{\partial t} \tag{5.76}$$

$$\boldsymbol{B}(\boldsymbol{r},t) = \nabla \times \boldsymbol{A}(\boldsymbol{r},t) \,. \tag{5.77}$$

The fields V and $\boldsymbol{A}$ are not uniquely defined since a so-called gauge transformation

$$\boldsymbol{A}'(\boldsymbol{r},t) = \boldsymbol{A}(\boldsymbol{r},t) + \nabla\chi(\boldsymbol{r},t) \tag{5.78}$$

$$V'(\boldsymbol{r},t) = V(\boldsymbol{r},t) - \frac{\partial\chi(\boldsymbol{r},t)}{\partial t}, \tag{5.79}$$

which is characterized by the gauge field $\chi(\boldsymbol{r},t)$ leaves the field strengths $\boldsymbol{E}$ and $\boldsymbol{B}$ invariant.

The question now arises how to incorporate the scalar and vector potentials into the wave function of a particle carrying an electric charge q. In classical mechanics the interaction between the electromagnetic field and the charged particle emerges in the electric force $q\boldsymbol{E}$ and the velocity dependent Lorentz force $q\boldsymbol{v} \times \boldsymbol{B}$ which may be derived from the (non-relativistic) Lagrangian

$$L = \frac{1}{2}mv^2 + q(\boldsymbol{v}\cdot\boldsymbol{A} - V), \tag{5.80}$$

which provides the so-called 'minimal coupling' between the particle and the electromagnetic field that gives rise to the standard force term $q(\boldsymbol{E}+\boldsymbol{v}\times\boldsymbol{B})$. Hence, the canonical momentum $\boldsymbol{p} = \partial L/\partial\boldsymbol{v}$ is given by:

$$\boldsymbol{p} = m\boldsymbol{v} + q\boldsymbol{A}. \tag{5.81}$$

Clearly, the 'kinetic momentum' $m\boldsymbol{v}$ differs from the the canonical momentum by the electromagnetic field contribution $q\boldsymbol{A}$:

$$m\boldsymbol{v} = \boldsymbol{p} - q\boldsymbol{A}. \tag{5.82}$$

Accordingly, the classical Hamiltonian for a free point charge in the presence of an electromagnetic field $(V, \boldsymbol{A})$ is calculated from the usual prescription

$$H_{\mathrm{CL}} = \boldsymbol{v}\cdot\frac{\partial L}{\partial\boldsymbol{v}} - L = \boldsymbol{v}\cdot\boldsymbol{p} - L. \tag{5.83}$$

Therefore,

$$H_{\mathrm{CL}} = \frac{1}{2m}(\boldsymbol{p} - q\boldsymbol{A})^2 + qV, \tag{5.84}$$

which may formally be obtained from the field free term $p^2/2m$ by carrying out the 'minimal' substitution

$$\boldsymbol{p} \to \boldsymbol{p} - q\boldsymbol{A} \tag{5.85}$$

and adding the potential energy $qV(\boldsymbol{r})$.

The classical Hamiltonian (5.84) can now easily be transformed into the coordinate representation of its quantum mechanical counter part. Invoking the correspondence principle for the canonical momentum $\boldsymbol{p}$, we finally obtain:

$$H_{\mathrm{QM}} = \frac{1}{2m}(-\mathrm{i}\hbar\nabla - q\boldsymbol{A})^2 + qV \,. \tag{5.86}$$

At first sight, a paradox appears because the above Hamiltonian and the corresponding Schrödinger equation are affected by the electromagnetic field only through the gauge dependent potentials V and $\boldsymbol{A}$, while it is expected that the energy eigenvalues and all other physically observable quantities should be 'gauge-invariant'. i.e. not depending on a particular choice of the gauge field $\chi(\boldsymbol{r}, t)$.

The paradox is easily resolved by the observation that any gauge transformation of the form (5.78–5.79) should be accompanied by a similar transformation of the wave function:

$$\Phi'(\boldsymbol{r}, t) = \Phi(\boldsymbol{r}, t) \exp\left(\frac{\mathrm{i}q}{\hbar}\chi(\boldsymbol{r}, t)\right) \,. \tag{5.87}$$

It is left as an exercise for the reader to prove that the full gauge transformation (5.78, 5.79, 5.87) leaves the energy eigenvalues of the stationary Schrödinger equation invariant indeed.

Similarly, the expression for the current (5.3) carried by a state $|\Phi\rangle$ should appropriately be extended by adding the vector potential contribution:

$$\begin{aligned} \boldsymbol{J}(\boldsymbol{r}, t) = & -\frac{\mathrm{i}q\hbar}{2m}\left[\Phi^*(\boldsymbol{r}, t)\nabla\Phi(\boldsymbol{r}, t) - (\nabla\Phi^*(\boldsymbol{r}, t))\,\Phi(\boldsymbol{r}, t)\right] \\ & -\frac{q^2}{m}\boldsymbol{A}(\boldsymbol{r}, t)|\Phi(\boldsymbol{r}, t)|^2 \,. \end{aligned} \tag{5.88}$$

5.6 Spin

In many cases, the dynamics of single particle is not only determined by the evolution of its momentum and energy, but should also be characterized in terms of other quantum mechanical degrees of freedom for which no classical analog exists. 'Spin' is such an example, representing a quantum mechanical type of angular momentum that gives rise to a direct interaction of the particle with a magnetic field, other than the Lorentz force that couples through the magnetic field through the electric charge. For instance, neutrons are so-called spin 1/2 particles and can still feel the influence of a magnetic field through their spin coordinates.

Although the origin of the term 'spin' reminds us of a lot of exciting toys like spinning wheels and gyroscopes, spin is not straightforwardly identical to physical rotations. The latter are typically described by the *o*rbital angular momentum $\boldsymbol{L} = \boldsymbol{r} \times \boldsymbol{p}$ which of course does have a classical analog. If we are in the coordinate representation, we can easily work out the quantization of $\boldsymbol{L}$, i.e. making it an operator by the usual replacement $\boldsymbol{p} = -\mathrm{i}\hbar\nabla$. It is left as an exercise to show that the components of $\boldsymbol{L}$ are then satisfying the following commutation rules:

$$\begin{aligned}
[L_x, L_y] &= \mathrm{i}\hbar L_z \\
[L_y, L_z] &= \mathrm{i}\hbar L_y \\
[L_z, L_x] &= \mathrm{i}\hbar L_z
\end{aligned} \tag{5.89}$$

or in a more compact notation:

$$\boldsymbol{L} \times \boldsymbol{L} = \mathrm{i}\hbar \boldsymbol{L} . \tag{5.90}$$

Anyway, as was already stated, $\boldsymbol{L}$ differs quite a lot from the spin angular momentum. Nevertheless, if we get inspired by the formal commutation rules (5.89) in which nothing further refers to the particularity of the x-representation, we might go back to the 'third recipe', use our imagination and boldly state that the spin operator should obey the same relations:

$$\boldsymbol{S} \times \boldsymbol{S} = \mathrm{i}\hbar \boldsymbol{S} . \tag{5.91}$$

Now, it turns out that this works fine since, as a general result, in quantum mechanics all kinds of angular momenta no matter how they may differ internally are found to share the very same algebraic structure defined by the relations (5.89) and (5.91).

Due to these commutation rules, the three components of $\boldsymbol{S}$ cannot be simultaneously diagonalized. On the other hand, the square of the magnitude $\boldsymbol{S}^2 = S_x^2 + S_y^2 + S_z^2$ does commute with all components of $\boldsymbol{S}$:

$$\left[\boldsymbol{S}^2, \boldsymbol{S}\right] = \boldsymbol{0} . \tag{5.92}$$

Consequently we may always construct a set of eigenstates simultaneously diagonalizing $\boldsymbol{S}^2$ and S_z where the z-axis is in principle any axis in space that conveniently serves as a reference axis.

Physically, this means that once we have chosen the z-axis, we can only obtain well-defined measured values for the magnitude of the spin and its projection onto the z-axis. The corresponding eigenstates are then conveniently written as

$$\boldsymbol{S}^2 \; |SM\rangle = S(S+1)\hbar^2 \; |SM\rangle \tag{5.93}$$

$$S_z \; |SM\rangle = M\hbar \; |SM\rangle \; , \tag{5.94}$$

where S can generally be any positive integer or any positive, integer multiple of $1/2$ and M increases from $-S$ to $+S$ with steps equal to 1:

$$\begin{aligned}
S &= 0,\; 1/2,\; 1,\; 3/2,\; 2,\; 5/2, \ldots \\
M &= -S,\; -S+1,\; \ldots,\; S-1,\; S .
\end{aligned} \tag{5.95}$$

Finally, it is now clear that each wave function is not only depending on space coordinates $\boldsymbol{r}$ but should also contain the coordinates S and M thereby referring to the spin state. This is generally not too difficult since in principle one knows the type of particles one is dealing with. Anticipating the discussion of next chapter, we assert that in the case of electrons S must equal $1/2$ so that there are only two (orthogonal) eigenstates of S_z available:

the 'spin-up' state $|1/2, +1/2\rangle$ represented by the spinor $\begin{pmatrix}1\\0\end{pmatrix}$

the 'spin-down' state $|1/2, -1/2\rangle$ represented by the spinor $\begin{pmatrix}0\\1\end{pmatrix}$

Taking the two spinors as a natural basis for the spin dependent part of Fock space, we may conveniently represent the spin vector operator by the relation

$$\boldsymbol{S} = \frac{\hbar}{2}\boldsymbol{\sigma} \,. \tag{5.96}$$

where $\boldsymbol{\sigma} = (\sigma_x, \sigma_y, \sigma_z)$ are the well-known Pauli spin matrices:

$$\sigma_x = \begin{pmatrix}0 & 1\\1 & 0\end{pmatrix}, \quad \sigma_y = \begin{pmatrix}0 & -\mathrm{i}\\ \mathrm{i} & 0\end{pmatrix}, \quad \sigma_z = \begin{pmatrix}1 & 0\\0 & -1\end{pmatrix} \,. \tag{5.97}$$

It is left as an exercise to the reader to prove the following identities:

$$\sigma_x^2 = \sigma_y^2 = \sigma_z^2 = \mathbf{I} \tag{5.98}$$

$$(\boldsymbol{\sigma}\cdot\boldsymbol{u})(\boldsymbol{\sigma}\cdot\boldsymbol{v}) = \boldsymbol{u}\cdot\boldsymbol{v} + \mathrm{i}\boldsymbol{\sigma}\cdot(\boldsymbol{u}\times\boldsymbol{v}) \,, \tag{5.99}$$

where $\boldsymbol{u}$ and $\boldsymbol{v}$ are ordinary vectors in $\mathbf{R}^3$ or any vector operators that are not related to spin (such as position and momentum) and $\mathbf{I}$ is the two-dimensional unit matrix.

To conclude this chapter we show how the Pauli matrices can be used to incorporate the direct interaction between the spin of a particle and a magnetic field into the Schrödinger equation.

As was anticipated before, Fock space has to be extended upon the introduction of spin in order to account for the additional spin degrees of freedom and the language of spinors offers a suitable tool to do so. In practice, the wave function of a spin-1/2 particle may conveniently be represented by a two-component spinor by projecting the state vector on the full set of eigenstates of the position operator and the eigenstates of one of the components of the spin operator (usually taken to be s_z)

$$\langle \boldsymbol{r}, s|\Phi\rangle = \Phi(\boldsymbol{r}, s) = \begin{pmatrix}\Phi_1(\boldsymbol{r})\\ \Phi_2(\boldsymbol{r})\end{pmatrix} \,. \tag{5.100}$$

Considered an operator in the coordinate representation, the one-particle Hamiltonian is not only to act on the x-, y- and z-dependent parts of the wave function, but should also affect the spin. Since the latter is introduced through the spinor formalism, it should be clear that the one-particle Hamiltonian has to take the form of a 2×2 matrix. Fortunately, it is quite easy to construct the four entries of this matrix on exploiting the fact that the Pauli matrices together with the unit matrix $\mathbf{I}$ are the so-called generators of the SU_2 group, which implies that all 2×2 matrices in $\mathbf{C}^2$ can be expressed as a *unique* linear combination of σ_x, σ_y, σ_z and $\mathbf{I}$.

If no magnetic field is turned on (and if no other spin dependent interactions are present), the spinor part of the wave function should remain unaltered under the action of the Hamiltonian and the corresponding 2×2 matrix would be proportional to the the unit matrix, thereby reflecting two-fold spin degeneracy:

$$H = \begin{pmatrix} \frac{p^2}{2m} + U(\boldsymbol{r}) & 0 \\ 0 & \frac{p^2}{2m} + U(\boldsymbol{r}) \end{pmatrix} = \left(\frac{p^2}{2m} + U(\boldsymbol{r})\right)\mathsf{I} , \qquad (5.101)$$

with

$$\boldsymbol{p} = -\mathrm{i}\hbar\nabla \quad . \qquad (5.102)$$

As expected, the energy spectrum derived from the Schrödinger equation $H\Phi_n = E_n\Phi_n$ is the same regardless whether the $\{\Phi_n\}$ constitute a set of one- or two-component eigenfunctions.

Furthermore, using the identity (5.99) for $\boldsymbol{u} = \boldsymbol{v} = \boldsymbol{p}$, we may rewrite (5.101) as

$$H = \frac{1}{2m}(\boldsymbol{\sigma}\cdot\boldsymbol{p})(\boldsymbol{\sigma}\cdot\boldsymbol{p}) + U(\boldsymbol{r})\mathsf{I} \quad . \qquad (5.103)$$

However, if an electromagnetic field is turned on, the interaction of the spin and the magnetic field will become explicit upon the minimal substitution of

$$\boldsymbol{p} \to \boldsymbol{p} - q\boldsymbol{A} \qquad (5.104)$$

in (5.103).

Indeed, deriving the identity

$$[\boldsymbol{\sigma}\cdot(\boldsymbol{p} - q\boldsymbol{A})][\boldsymbol{\sigma}\cdot(\boldsymbol{p} - q\boldsymbol{A})] = (\boldsymbol{p} - q\boldsymbol{A})^2 - \hbar q\boldsymbol{\sigma}\cdot\nabla\times\boldsymbol{A} \qquad (5.105)$$

from (5.99) with $\boldsymbol{u} = \boldsymbol{v} = -\mathrm{i}\hbar\nabla$, we obtain

$$\begin{aligned} H &= \frac{1}{2m}[\boldsymbol{\sigma}\cdot(\boldsymbol{p} - q\boldsymbol{A})][\boldsymbol{\sigma}\cdot(\boldsymbol{p} - q\boldsymbol{A})] + (U + qV)\mathsf{I} \\ &= \left(\frac{1}{2m}(\boldsymbol{p} - q\boldsymbol{A})^2 + U + qV\right)\mathsf{I} - \frac{\hbar q}{2m}\boldsymbol{\sigma}\cdot\boldsymbol{B} \quad . \end{aligned} \qquad (5.106)$$

This is the well-known Pauli–Schrödinger Hamiltonian governing both the orbital motion of a charged particle in a magnetic field and the direct interaction between the spin and the magnetic field.

For an electron ($q = -e$) the corresponding Pauli–Schrödinger equation can be brought to its standard form

$$\begin{aligned} \mathrm{i}\hbar\frac{\partial\Phi(\boldsymbol{r},t)}{\partial t} &= \frac{1}{2m}(-\mathrm{i}\hbar\nabla + e\boldsymbol{A}(\boldsymbol{r},t))^2\Phi(\boldsymbol{r},t) \\ &\quad + [U(\boldsymbol{r}) - eV(\boldsymbol{r},t) + g\mu_B\boldsymbol{B}(\boldsymbol{r},t)\cdot\boldsymbol{S}]\,\Phi(\boldsymbol{r},t) \end{aligned} \qquad (5.107)$$

by omitting the unit matrix symbol I and introducing the Landé factor g and the Bohr magneton

$$\mu_B = \frac{e\hbar}{2m} \; . \tag{5.108}$$

It should be noted that within the non-relativistic treatment presented in this section, the Landé factor exactly reduces to 2, whereas relativistic quantum electrodynamics accounts for the hyperfine level splitting leading to the corrected value $g = 2.0023$, for an electron.

Exercises

5.1. Prove equation (5.3).

5.2. Show that (5.15) carries a total current I given by

$$I = \frac{q}{L_x} \frac{\hbar k_x}{m} \; .$$

5.3. Why does (5.18) not depend on x?

5.4. Prove equation (5.68).

5.5. Prove that the full gauge transformation (5.78, 5.79, 5.87) leaves the energy eigenvalues of the stationary Schrödinger equation invariant.

5.6. Prove equation (5.89).

5.6. Prove the spinor identities (5.99).

6. Second Quantization

In the preceding chapters we have mostly paid attention to the quantum mechanics of single particles. For that purpose we have worked the so-called first quantization scheme replacing classical quantities by suitable quantum mechanical operators acting on Fock space vectors. Apart from our limited capability of solving Schrödinger's equation in closed from, all this worked rather well not only from the conceptual point of view, but also when practical calculations were to be addressed. However, the major goal of transport theory is to characterize macroscopic transport behaviour (transport coefficients, drift velocities, signal and noise currents etc.) from the microscopic migration of a huge number of particles. Systematic studies of quantum mechanical many-particle systems however are most conveniently carried out within the framework of *second quantization.* Before getting into the formalism of second quantization we may wonder whether it is really necessary to introduce another formalism in order to determine the dynamical evolution of two or more particles.

It should be noticed first that 'second quantization' is not an another quantization of physical quantities that were already quantized within the 'first quantization' scheme. There is only one quantum mechanics and both first and second quantization are just providing the tools and the framework to develop quantum mechanics for one-particle and many-particle systems respectively. Now the previous question has a simple answer: no. It is not strictly necessary to invoke 'second quantization' if many-particle problems are to be tackled, but in general the practical calculations required to arrive at the solutions turn out to be rather cumbersome if not unrealistic. The real gain in adopting the formalism of second quantization is that it provides us with an elegant framework to carry out specific calculations while the quantum-statistical properties of the many-particle ensembles are automatically taken care of, as will become clear in the next sections.

6.1 Identical Particles

The need of applying second quantization techniques to solve problems in transport theory is clearly illustrated by the extremely huge number of particles under investigation: we are not dealing with just two or three particles travelling in free configuration space, but rather with particle concentrations varying between 10^{10} cm^{-3} (semiconductors) and more than 10^{20} cm^{-3} (metals)[1]. Now, if the dimension of Fock space is already infinite for a one-particle system, it becomes something like ∞^N for an N-particle system since the global Fock space emerges as a direct product of N one-particle Fock spaces. This brings us to the assertion that each wave function describing a N-particle states should be regarded as a product of N one-particle wave functions, at least if they do not interact. Indeed, suppose that $H_1(\boldsymbol{r}, \boldsymbol{p})$ is the Hamiltonian 'felt' by one particle, i.e.

$$H_1(\boldsymbol{r}, \boldsymbol{p}) = \frac{p^2}{2m} + U(\boldsymbol{r}) \,, \tag{6.1}$$

where the momentum operator takes the form $-i\hbar\boldsymbol{\nabla}$ of the coordinate representation.

The Hamiltonian describing a system of N identical particles generally reads

$$\begin{aligned} H(\boldsymbol{r}_1, \ldots, \boldsymbol{r}_N, \boldsymbol{p}_1, \ldots, \boldsymbol{p}_N) &= \textstyle\sum_{j=1}^{N} \left[\frac{p_j^2}{2m} + U(\boldsymbol{r}_j)\right] + H_{\text{int}} \\ &= \textstyle\sum_{j=1}^{N} H_1(\boldsymbol{r}_j, \boldsymbol{p}_j) \,+\, H_{\text{int}} \,, \end{aligned} \tag{6.2}$$

where the interaction term H_{int} consists of all two-particle interactions of the form $v(|\boldsymbol{r}_j - \boldsymbol{r}_l|)$. Now, suppose that we managed to solve for the energy spectrum and the corresponding eigenfunctions of the one-particle Hamiltonian, i.e.

$$H_1 \phi_k(\boldsymbol{r}) \;=\; \varepsilon_k \phi_k(\boldsymbol{r}) \tag{6.3}$$

then we may construct the following product wave functions:

$$\Phi_{k_1, \ldots k_N}(\boldsymbol{r}_1, \ldots, \boldsymbol{r}_N) \;=\; \phi_{k_1}(\boldsymbol{r}_1) \ldots \phi_{k_N}(\boldsymbol{r}_N) \,. \tag{6.4}$$

The quantum numbers $\{k_1, k_2, \ldots k_N\}$ refer to the first, second ... Nth electron respectively and can only take the allowed k-values defined in (6.3).

Considering first the case of a non-interacting system we set $H_{\text{int}} = 0$. Bearing in mind that $H_1(\boldsymbol{r}_j, \boldsymbol{p}_j)$ is only acting on the wave functions of the jth particle, we may easily evaluate the action of the total Hamiltonian on the product wave function (6.4):

[1] Quantum dots may contain less than 10 electrons, but considered a single entity together with the connecting leads, it is still a many-particle system.

$$
\begin{aligned}
H\Phi_{k_1,\dots k_N}(\boldsymbol{r}_1,\dots,\boldsymbol{r}_N) \\
&= \sum_{j=1}^{N} \phi_{k_1}(\boldsymbol{r}_1)\,\dots\,\left[H_1(\boldsymbol{r}_j,\boldsymbol{p}_j)\phi_{k_j}(\boldsymbol{r}_j)\right]\dots\,\phi_{k_N}(\boldsymbol{r}_N) \\
&= \sum_{j=1}^{N} \phi_{k_1}(\boldsymbol{r}_1)\dots\,\varepsilon_{k_j}\phi_{k_j}(\boldsymbol{r}_j)\,\dots\,\phi_{k_N}(\boldsymbol{r}_N) \\
&= \left(\sum_{j=1}^{N}\varepsilon_{k_j}\right)\phi_{k_1}(\boldsymbol{r}_1)\,\dots\,\phi_{k_N}(\boldsymbol{r}_N) \\
&= \left(\sum_{j=1}^{N}\varepsilon_{k_j}\right)\phi_{k_1\dots k_N}(\boldsymbol{r}_1\dots\boldsymbol{r}_N)\,. \qquad (6.5)
\end{aligned}
$$

This proves at least that the states $\{\Phi_{k_1,\dots k_N}\}$ are energy eigenstates of the total Hamiltonian H anyway and that the corresponding, *global* energy eigenvalues are given by

$$
E_{k_1,\dots k_N} = \varepsilon_{k_1} + \varepsilon_{k_2} + \dots + \varepsilon_{k_N}\,. \qquad (6.6)
$$

The ability of expressing the global energy eigenvalues $E_{k_1,\dots k_N}$ as simple sums of one-particle energies is of course due to the assumed absence of any inter-particle interaction and the answer would be entirely different (and in most cases unavailable in closed form) if H_{int} weren't zero. But even for such an oversimplified non-interacting system, there arises a problem with the 'global' wave functions $\{\Phi_{k_1,\dots,k_N}(\boldsymbol{r}_1,\dots,\boldsymbol{r}_N)\}$.

Contrary to classical mechanics, if two or more identical particles are sharing the same physical configuration space, their wave functions will overlap and there is no way to make any distinction between these moving particles. So, quantum mechanics is clearly prohibiting to distinguish between identical particles and it is good to realize that this compelling feature imposes severe restrictions on the quantum mechanical description for all particle numbers exceeding 1. An immediate consequence is the apparent symmetry of the many-particle system under arbitrary exchanges of particles or particle coordinates and this kind of symmetry should be reflected explicitly in the many-particle wave functions.

However, going back to the global wave functions proposed in (6.4) we see that we can distinguish between, say particle 1 and particle 2 by simply *exchanging* these two particles; representing the exchange operation by P_{12} we would obtain another wave function

$$
P_{12}\Phi_{k_1,\dots k_N}(\boldsymbol{r}_1,\dots,\boldsymbol{r}_N) = \phi_{\mathbf{k_2}}(\boldsymbol{r}_1)\phi_{\mathbf{k_1}}(\boldsymbol{r}_2)\dots\phi_{k_N}(\boldsymbol{r}_N)\,, \qquad (6.7)
$$

which is clearly different from (6.4), although it has the same energy $E_{k_1,\dots k_N}$. This can however be remedied in a simple way as follows.

Suppose for simplicity that $N = 2$. Then, instead of considering the two wave functions $\Phi_{k_1,k_2}(\boldsymbol{r}_1, \boldsymbol{r}_2)$ and $\Phi_{k_2,k_1}(\boldsymbol{r}_1, \boldsymbol{r}_2)$ which still allow us to distinguish between the two particles, we could as well replace the two wave functions by their sum and difference so as to obtain the so-called symmetric and anti-symmetric wave functions

$$\Phi^{(S)}_{k_1,k_2}(\boldsymbol{r}_1, \boldsymbol{r}_2) = \frac{1}{\sqrt{2}} \left[\Phi_{k_1,k_2}(\boldsymbol{r}_1, \boldsymbol{r}_2) + \Phi_{k_2,k_1}(\boldsymbol{r}_1, \boldsymbol{r}_2)\right] \tag{6.8}$$

$$\Phi^{(A)}_{k_1,k_2}(\boldsymbol{r}_1, \boldsymbol{r}_2) = \frac{1}{\sqrt{2}} \left[\Phi_{k_1,k_2}(\boldsymbol{r}_1, \boldsymbol{r}_2) - \Phi_{k_2,k_1}(\boldsymbol{r}_1, \boldsymbol{r}_2)\right] , \tag{6.9}$$

where the factors $1/\sqrt{2}$ are taking care of the normalization.

Clearly, these two types of wave functions are now invariant under the exchange of the two particles (up to a minus sign). The latter may be expressed mathematically by simply noting that $\Phi^{(S)}$ and $\Phi^{(A)}$ are the two eigenvectors of P_{12} with eigenvalues $+1$ and -1 respectively. Moreover, remembering that $\Phi^{(S)}$ and $\Phi^{(A)}$ have also the same energy, we might conclude that the energy spectrum is now doubly degenerate. This conclusion however is wrong: the apparent degeneracy is not physical because identical particles should rather be regarded as belonging to either of two classes: bosons or fermions. Bosons (photons, phonons, Cooper pairs in superconductors, He^4 atoms, etc.) belong to the symmetric class whereas fermions (electrons, protons, positrons, neutrons, muons, etc.) should always be described by anti-symmetric wave functions. Moreover, it follows from relativistic quantum mechanics that all bosons and fermions have integer spin and half-integer spin quantum numbers respectively.

Let us now find out how to proceed for one particular class, say fermions. It follows for $N = 2$ that we should retain only the anti-symmetric wave functions $\{\Phi^{(A)}_{k_1,k_2}(\boldsymbol{r}_1, \boldsymbol{r}_2)\}$ and it is straightforward to show that for arbitrary N the appropriate anti-symmetric wave functions are given by

$$\begin{aligned}\Phi_{k_1,\ldots k_N}(\boldsymbol{r}_1, \ldots, \boldsymbol{r}_N) \\ = \frac{1}{\sqrt{N!}} \sum_{\mathbf{P}} (-1)^{\mathbf{P}} \phi_{\mathbf{P}k_1}(\boldsymbol{r}_1)\phi_{\mathbf{P}k_2}(\boldsymbol{r}_2)\ldots\phi_{\mathbf{P}k_N}(\boldsymbol{r}_N) ,\end{aligned} \tag{6.10}$$

where $\{\mathbf{P}k_1,\ \mathbf{P}k_2,\ \ldots\mathbf{P}k_N\}$ represents one of the $N!$ permutations of $\{k_1,\ k_2,\ \ldots k_N\}$. Reflecting the anti-symmetrical wave functions, the summation in (6.10) reduces to the expansion of a determinant which is called after Slater, yielding

$$\Phi_{k_1,\ldots k_N}(\boldsymbol{r}_1, \ldots, \boldsymbol{r}_N) = \frac{1}{\sqrt{N!}} \begin{vmatrix} \phi_{k_1}(\boldsymbol{r}_1) & \phi_{k_1}(\boldsymbol{r}_2) & \ldots & \ldots & \phi_{k_1}(\boldsymbol{r}_N) \\ \phi_{k_2}(\boldsymbol{r}_1) & \phi_{k_2}(\boldsymbol{r}_2) & \ldots & \ldots & \phi_{k_2}(\boldsymbol{r}_N) \\ \cdot & \cdot & & & \cdot \\ \cdot & \cdot & & & \cdot \\ \cdot & \cdot & & & \cdot \\ \phi_{k_N}(\boldsymbol{r}_1) & \phi_{k_N}(\boldsymbol{r}_2) & \ldots & \ldots & \phi_{k_N}(\boldsymbol{r}_N) \end{vmatrix} \tag{6.11}$$

It's worth noting that a Slater determinants may 'visualize' the Pauli principle stating that two fermions can never occupy the same single-particle state. Indeed, the latter would give rise to the occurrence of two identical rows or columns in the determinant, which then of course would vanish. (Strictly speaking however, this conclusion is valid only if we incorporate also the spin quantum numbers $\{M_1,\ M_2,\ \ldots,\ M_N\}$ into the single-particle quantum numbers $\{k_1,\ k_2,\ \ldots,\ k_N\}$).

However, from the computational point of view, remembering that N is generally extremely large, we are led to the conclusion that Slater determinants are hardly workable tools when it comes to practical calculations. In addition, the previous discussion has dealt only with non-interacting particles and so it becomes clear that a more sophisticated tool is required to study interacting systems while keeping track of the boson/fermion statistics. This tool is provided by the formalism of second quantization in the form of field operators.

6.2 Field Operators

Being defined in the many-particle Fock space, field operators provide a workable procedure to construct in a systematic way quantum mechanical operators assigned to all kinds of many-particle quantities such as energy, momentum, angular momentum, stress, current and charge density, energy flux etc. The field operator concept can be introduced in different ways, but here we have adopted a simple algebraic approach, introduced by B. Robertson in 1973 [28].

Since both fermion and boson systems are addressed, we recall that rectangular brackets are customary to denote a commutator whereas anti-commutators are represented by $\{,\}$, i.e. $\forall A, B$:

$$\begin{aligned}[A,B] &\equiv AB\ -\ BA \\ \{A,B\} &\equiv AB\ +\ BA\ .\end{aligned} \tag{6.12}$$

6.2.1 Definition

We postulate the following set of commutation rules which may be regarded as an algebraic definition of the field operators $\psi(\boldsymbol{r})$ and $\psi^\dagger(\boldsymbol{r})$:

$$\left.\begin{aligned}[\psi(\boldsymbol{r}),\psi(\boldsymbol{r}')] &= 0 \\ \left[\psi^\dagger(\boldsymbol{r}),\psi^\dagger(\boldsymbol{r}')\right] &= 0 \\ \left[\psi(\boldsymbol{r}),\psi^\dagger(\boldsymbol{r}')\right] &= \delta(\boldsymbol{r}-\boldsymbol{r}')\end{aligned}\right\}\quad \text{for bosons} \tag{6.13}$$

and

$$\begin{aligned} \{\psi(\boldsymbol{r}), \psi(\boldsymbol{r}')\} &= 0 \\ \{\psi^\dagger(\boldsymbol{r}), \psi^\dagger(\boldsymbol{r}')\} &= 0 \qquad \text{for fermions} \\ \{\psi(\boldsymbol{r}), \psi^\dagger(\boldsymbol{r}')\} &= \delta(\boldsymbol{r}-\boldsymbol{r}') . \end{aligned} \tag{6.14}$$

6.2.2 Field Operators, Wave Functions and Topology

Before investigating some essential properties of field operators and the physical quantities that may relate to them, we should make sure not to mix up field operators with wave functions. Several reasons can be imagined to explain the confusion that sometimes arises in this context.

1. Both field operators and wave functions – at least in the position representation – are explicitly depending on a position vector $\boldsymbol{r}$.
2. Often the same Greek letter ψ is used to denote both of them (in this course we have mostly used ϕ and Φ to denote wave functions).
3. The equations of motion satisfied by field operators is not a Schrödinger equation but in many cases it looks very similar.
4. Field operators may be expanded in wave functions and therefore one may see both of them appear in the same equation.

On the other hand, we should realize that wave functions are complex functions describing single-particle features while field operators are operators acting in the Fock space spanned by many-particle states.

Since we will have to deal with the explicit position dependence of the field operators and all extracted operators, it proves convenient to formulate some basic assumptions concerning the topology of configuration space and to introduce some appropriate and unambiguous notation.

1. Unless stated otherwise, we assume that all particles under study are moving in a three-dimensional space volume Ω. Depending on the topology considered, Ω may be either a finite, closed or an infinite, open sub-set of $\mathbf{R}^3$.
2. Topologically, Ω is a so-called connected region, which means that for any two point P_1 and P_2 *within* Ω you can always travel from P_1 to P_2 without leaving Ω. The region Ω will taken to be simply connected if we are dealing with systems that may exchange particles with the environment, for instance when addressing open-ended conductors where charge carriers are entering and leaving the conductor arbitrarily far away. Otherwise, Ω will assumed to be multiply connected, which means that it contains one or more 'holes' and that no closed contour encircling a hole can be shrunk to a single point without crossing the boundary of Ω.
   The simplest electric circuit – whether it is a mesoscopic ring or a macroscopic circuit consisting of battery connected to a resistor or any two-terminal device – contains exactly one hole. A resistor network on the

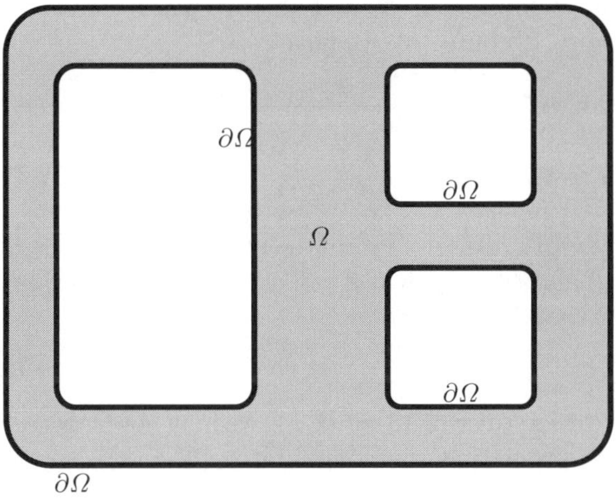

Fig. 6.1. Multiply connected region Ω containing two holes. $\partial\Omega$ denotes the boundary surface

other hand would be topologically equivalent to the multiple hole region sketched in Fig. 6.1.

3. Unless Ω coincides with $\mathbf{R}^3$, it is assumed to be bounded by a one or more simple, closed and orientable surfaces constituting the boundary $\partial\Omega$ in each point of which an outward pointing normal can be defined.
4. The volume element is generally denoted by $\mathrm{d}\tau$ and the surface element on $\partial\Omega$ is the vector $\mathrm{d}\boldsymbol{S}$ directed along the outward pointing normal.
5. If Ω is closed, all particles are confined to Ω and no charges are leaking away from Ω to the outside world.

6.2.3 Field Operators in Fock Space

From a straightforward application of the operator algebra defined by the commutation rules (6.14) and (6.13) we will learn that $\psi(\boldsymbol{r})$ and $\psi^\dagger(\boldsymbol{r})$ are acting as lowering and raising operators in Fock space. First, we construct a linear operator $\hat{N}$ that will be identified as the particle number operator:

$$\hat{N} = \int_\Omega \mathrm{d}\tau \, \psi^\dagger(\boldsymbol{r})\psi(\boldsymbol{r}) \, . \tag{6.15}$$

Obviously, $\hat{N}$ is a Hermitian operator and all its eigenvalues are therefore real. Clearly, as is the case for all vector spaces on which linear operators are defined, the null vector of Fock space is a trivial eigenstate with eigenvalue zero, but we will assume that there exists at least one non-trivial eigenstate $|n\rangle$ with an eigenvalue n such that

$$\hat{N} \, |n\rangle = n \, |n\rangle \, . \tag{6.16}$$

It follows directly from the definition of $\hat{N}$ and (6.16) that n cannot be negative. Indeed, projecting (6.16) onto $\langle n|$, we have

$$\left\langle n|\hat{N}|n\right\rangle = n\,\langle n|n\rangle$$

$$\int_{\Omega} \mathrm{d}\tau\ \langle n|\psi^{\dagger}(\boldsymbol{r})\psi(\boldsymbol{r})|n\rangle = n\,\langle n|n\rangle\ . \tag{6.17}$$

Here, $\langle n|n\rangle$ and $\langle n|\psi^{\dagger}(\boldsymbol{r})\psi(\boldsymbol{r})|n\rangle$ respectively represent the squared magnitudes of the state vectors $|n\rangle$ and $\psi(\boldsymbol{r})\ |n\rangle$ which can only take positive values or zero. Therefore, we infer that

$$n \geq 0\ . \tag{6.18}$$

With the help of the commutation rules (6.13) or (6.14) we may now construct another eigenstate of $\hat{N}$ by evaluating the commutator $[\hat{N}, \psi(\boldsymbol{r})]$, say for fermions

$$\begin{aligned}
\left[\hat{N},\, \psi(\boldsymbol{r})\right] &= \int_{\Omega} \mathrm{d}\tau'\ \left[\psi^{\dagger}(\boldsymbol{r}')\psi(\boldsymbol{r}'),\, \psi(\boldsymbol{r})\right] \\
&= \int_{\Omega} \mathrm{d}\tau'\ \left(\psi^{\dagger}(\boldsymbol{r}')\left\{\psi(\boldsymbol{r}'),\, \psi(\boldsymbol{r})\right\} - \left\{\psi^{\dagger}(\boldsymbol{r}'),\, \psi(\boldsymbol{r})\right\}\psi(\boldsymbol{r}')\right) \\
&= -\int_{\Omega} \mathrm{d}\tau'\ \delta(\boldsymbol{r}' - \boldsymbol{r})\,\psi(\boldsymbol{r}') \\
&= -\,\psi(\boldsymbol{r})\ ,
\end{aligned} \tag{6.19}$$

where we have used the identity

$$[AB, C] = A\{B, C\} - \{A, C\}\,B \tag{6.20}$$

to disentangle products of three field operators. Letting act both sides of (6.19) upon $|n\rangle$, we obtain

$$\left[\hat{N},\, \psi(\boldsymbol{r})\right]\ |n\rangle = -\,\psi(\boldsymbol{r})\ |n\rangle \tag{6.21}$$

or

$$\begin{aligned}
\hat{N}\left(\psi(\boldsymbol{r})\ |n\rangle\right) &= \psi(\boldsymbol{r})\,\hat{N}\ |n\rangle - \psi(\boldsymbol{r})\ |n\rangle \\
&= n\,\psi(\boldsymbol{r})\,|n\rangle - \psi(\boldsymbol{r})\ |n\rangle \\
&= (n-1)\,\psi(\boldsymbol{r})\ |n\rangle\ .
\end{aligned} \tag{6.22}$$

Consequently, $\psi(\boldsymbol{r})\ |n\rangle$ is either zero or another non-trivial eigenstate this time belonging to the eigenvalue $n-1$. In other words, if we let act $\psi(\boldsymbol{r})$ on an arbitrary eigenstate of $\hat{N}$, it produces another eigenstate with an eigenvalue exactly 1 less than the original one, which shows that $\psi(\boldsymbol{r})$ is a lowering operator. Applying $\psi(\boldsymbol{r})$ k several times with different position arguments, we arrive at

$$\hat{N}\left(\psi(\boldsymbol{r}_k)\psi(\boldsymbol{r}_{k-1})\ldots\psi(\boldsymbol{r}_1)\ |n\rangle\right) = (n-k)\,\psi(\boldsymbol{r}_k)\psi(\boldsymbol{r}_{k-1})\ldots\psi(\boldsymbol{r}_1)\ |n\rangle\ . \tag{6.23}$$

For a given eigenvalue n, we may identify $\{n-1, n-2, \ldots, n-k\}$ as the set of k subsequent, lower eigenvalues and obtain the corresponding eigenstates by applying ψ k times on the state vector $|n\rangle$. It should be noted that the position vectors $\boldsymbol{r}_1, \boldsymbol{r}_2, \ldots, \boldsymbol{r}_k$ may generally run continuously over all points of the region Ω, which amounts to a substantial degeneracy of the spectrum of $\hat{N}$.

The sequence of eigenvalues will inevitably be terminated when $k = n$ and the state $|0\rangle$ is reached, because otherwise we would be able to yield eigenstates with negative eigenvalues $n-k$ and contradict (6.18).

Putting everything together, we conclude that the eigenvalues of $\hat{N}$ constitute a ladder of positive integers $\{n\}$ while the lowest eigenstate satisfying

$$\psi(\boldsymbol{r})\,|0\rangle = 0 \quad \forall \boldsymbol{r} \in \Omega \tag{6.24}$$

corresponds to eigenvalue 0. Clearly, $|0\rangle$ describes a state with no particles at all, and is therefore called the vacuum state.

While $\psi(\boldsymbol{r})$ is generally removing a particle at position $\boldsymbol{r}$ when it was originally in the state acted upon by $\psi(\boldsymbol{r})$, it can be understood in a similar way that the action of $\psi^\dagger(\boldsymbol{r})$ will always result in the creation of a particle at some position $\boldsymbol{r}$. Moreover, it is now possible to construct explicitly quantum states carrying a well-defined number of particles just by repeated application of $\psi^\dagger(\boldsymbol{r})$ and/or $\psi(\boldsymbol{r})$ (either with the same or different position arguments) onto the vacuum state. Some examples are discussed below.

One-Particle States. Clearly, $\psi^\dagger(\boldsymbol{r})\,|0\rangle$ is a general one-particle state, representing the creation of a particle at a point $\boldsymbol{r}$, since

$$\hat{N}\,\psi^\dagger(\boldsymbol{r})\,|0\rangle = \psi^\dagger(\boldsymbol{r})\,|0\rangle \tag{6.25}$$

while the corresponding eigenvalue equals 1, as expected. Moreover, although we don't have yet a general recipe to construct all relevant operators in second quantized form, we may be tempted to consider the localized state $\psi^\dagger(\boldsymbol{r})\,|0\rangle$ as an eigenstate of a position vector operator $\hat{\boldsymbol{r}}$ acting in Fock space. Concretely, the analogy with first quantization would suggest the assertion

$$\hat{\boldsymbol{r}}\left(\psi^\dagger(\boldsymbol{r})\,|0\rangle\right) = \boldsymbol{r}\left(\psi^\dagger(\boldsymbol{r})\,|0\rangle\right) \tag{6.26}$$

which can be proven to be correct indeed.

Two-Particle States. Similarly, if $\psi^\dagger$ is acting twice with different position arguments $\boldsymbol{r}$ and $\boldsymbol{r}'$ on $|0\rangle$, we obtain a two-particle state:

$$\hat{N}\left(\psi^\dagger(\boldsymbol{r})\psi^\dagger(\boldsymbol{r}')\,|0\rangle\right) = 2\left(\psi^\dagger(\boldsymbol{r})\psi^\dagger(\boldsymbol{r}')\,|0\rangle\right) . \tag{6.27}$$

This example further shows that the basic commutation rules are explicitly reflecting either boson or fermion statistics. For instance, if the two particles are bosons, interchanging the positions $\boldsymbol{r}$ and $\boldsymbol{r}'$ does not change the state:

$$\left[\psi^\dagger(\boldsymbol{r}), \psi^\dagger(\boldsymbol{r}')\right] = 0 \quad \Rightarrow \quad \psi^\dagger(\boldsymbol{r})\psi^\dagger(\boldsymbol{r}')\,|0\rangle = \psi^\dagger(\boldsymbol{r}')\psi^\dagger(\boldsymbol{r})\,|0\rangle \; . \tag{6.28}$$

In the case of fermions however, we would end up with a change of sign,

$$\psi^\dagger(\boldsymbol{r})\psi^\dagger(\boldsymbol{r}')\,|0\rangle = -\,\psi^\dagger(\boldsymbol{r}')\psi^\dagger(\boldsymbol{r})\,|0\rangle \tag{6.29}$$

and, in particular, the case $\boldsymbol{r} = \boldsymbol{r}'$ would predict

$$\left(\psi^\dagger(\boldsymbol{r})\right)^2 |0\rangle = 0 \tag{6.30}$$

which again reflects Pauli's principle preventing two fermions from occupying the same localized state. Large numbers of bosons, on the other hand, may generally occupy the same quantum state. If this happens on a macroscopic scale for the ground state of the system, we are dealing with 'Bose condensation', a phenomenon which is responsible for superfluidity and superconductivity.

We might also wish to treat mixtures of bosons and fermions, for example when we want to study electron–phonon or electron–photon interactions. Obviously, this can only be accomplished if we appropriately define separate sets of field operators for both kinds of particles.

Normalization. Another consequence of the basic commutation rules is the natural occurrence of delta normalization for the case of the above discussed states.

For fermions, we have

$$\begin{aligned}\left\langle 0|\psi(\boldsymbol{r}')\psi^\dagger(\boldsymbol{r})|0\right\rangle &= \left\langle 0|\{\psi(\boldsymbol{r}'),\psi^\dagger(\boldsymbol{r})\}|0\right\rangle - \left\langle 0|\psi^\dagger(\boldsymbol{r})\psi(\boldsymbol{r}')|0\right\rangle \\ &= \left\langle 0|\{\psi(\boldsymbol{r}'),\psi^\dagger(\boldsymbol{r})\}|0\right\rangle \\ &= \delta(\boldsymbol{r}-\boldsymbol{r}')\,\langle 0|0\rangle \\ &= \delta(\boldsymbol{r}-\boldsymbol{r}')\end{aligned} \tag{6.31}$$

under the assumption that $\langle 0|0\rangle = 1$. Similarly, we may derive

$$\begin{aligned}\left\langle 0|\psi(\boldsymbol{r}_1)\psi(\boldsymbol{r}_2)\psi^\dagger(\boldsymbol{r}_2')\psi^\dagger(\boldsymbol{r}_1')|0\right\rangle &= \delta(\boldsymbol{r}_1-\boldsymbol{r}_1')\delta(\boldsymbol{r}_2-\boldsymbol{r}_2') \\ &\quad -\delta(\boldsymbol{r}_1-\boldsymbol{r}_2')\delta(\boldsymbol{r}_2-\boldsymbol{r}_1') \; .\end{aligned} \tag{6.32}$$

6.2.4 The Connection to First Quantization

In spite of their intrinsic elegance, the localized one- and two-particle states discussed in the previous section and, more generally, all many-particle states should exhibit a one-to-one correspondence to the many-particle wave functions arising in a first quantization approach, if the equivalence of the two pictures is to be demonstrated. Also the connection of the field operator concept to the many-particle wave functions and the many-particle state vectors in Fock space deserves further investigation.

In Sect. 6.1 we have introduced many-particle wave functions for a system of N particles and we will now illustrate how to combine them with the system's field operators to yield an explicit expression for the many-particle state vectors. A rigorous and more elaborate treatment may be found in standard publications on many-particle quantum mechanics [17, 20, 28, 29, 30]. The states we considered in the previous section are actually genuine state vectors of many-particle Fock space, but they only represent a restricted, though gigantic set of localized states in which all particles are located at sharply defined positions $\boldsymbol{r}_1, \ldots, \boldsymbol{r}_N$ in configuration space. In order to formally construct a general N-particle Fock state $|\Phi_N\rangle$, for instance, an energy eigenstate of the many-particle Hamiltonian, we follow the following straightforward recipe:

- construct a basis of localized many-particle states through multiple application of the field operator;
- multiply the localized many-particle state vectors with the first-quantized wave functions and integrate over all position coordinates.

i.e.

$$
\begin{aligned}
|\Phi_N\rangle = & \frac{1}{\sqrt{N!}} \int_\Omega \mathrm{d}\tau_1 \int_\Omega \mathrm{d}\tau_2 \ldots \int_\Omega \mathrm{d}\tau_N \\
& \times \Phi_N(\boldsymbol{r}_1, \boldsymbol{r}_2, \ldots, \boldsymbol{r}_N)\, \psi^\dagger(\boldsymbol{r}_N) \ldots \psi^\dagger(\boldsymbol{r}_2)\, \psi^\dagger(\boldsymbol{r}_1)\; |0\rangle \;, \qquad (6.33)
\end{aligned}
$$

where the pre-factor is a normalization constant, ensuring that $\langle\Phi_N|\Phi_N\rangle = 1$. At first glimpse, this result may look disappointing, since we intended to get rid of the cumbersome many-particle wave functions $\Phi_N(\boldsymbol{r}_1, \boldsymbol{r}_2, \ldots, \boldsymbol{r}_N)$ that are now reappearing in the explicit construction of Fock states. However, as stated previously, (6.33) is merely shown to illustrate the link between first and second quantization – and, if wanted to check formally the equivalence of both representations of many-particle quantum mechanics. After all, if we propose to formulate quantum mechanics in the framework of second quantization, we need to make sure that the latter incorporates the same physics as the first quantization scheme. For instance, providing nothing but an alternative description, the second-quantized Hamiltonian $\hat{H}$ that will be introduced in the next section should give rise to the same energy spectrum as its first-quantized twin brother. In other words, the Schrödinger equation in Fock space

$$
\hat{H}\,|\Phi_{N\lambda}\rangle \;=\; E_{N\lambda}\,|\Phi_{N\lambda}\rangle \qquad (6.34)
$$

is required to be generate the same energy spectrum $\{E_{N\lambda}\}$ as its first-quantized counterpart and should therefore be equivalent to

$$
H\Phi_{N\lambda}(\boldsymbol{r}_1, \boldsymbol{r}_2, \ldots, \boldsymbol{r}_N) \;=\; E_{N\lambda}\Phi_{N\lambda}(\boldsymbol{r}_1, \boldsymbol{r}_2, \ldots, \boldsymbol{r}_N) \qquad (6.35)
$$

provided that the Fock states $\{|\Phi_{N\lambda}\rangle\}$ and the N-particle wave functions $\{\Phi_{N\lambda}(\boldsymbol{r}_1, \boldsymbol{r}_2, \ldots, \boldsymbol{r}_N)\}$ are related through (6.33).

Moreover, as both the wave functions and the state vector are labeled by the particle number N apart from other quantum numbers λ, it is suggested that

$$\hat{N}\ |\Phi_N\rangle\ =\ N\ |\Phi_N\rangle \tag{6.36}$$

holds, which would also require explicit verification.

The above-mentioned equivalence as well as the validity of (6.36) and many other assertions can all be justified in a formal way with the help of the field operator algebra. For more elaborate calculations, we again refer to the paper by Robertson [28].

6.2.5 How to Construct the Operators

Starting from the number operator $\hat{N}$ we can easily infer that $\psi^\dagger(\boldsymbol{r})\psi(\boldsymbol{r})$ should be regarded as an operator for the particle concentration. As such, it immediately provides an explicit form for the electron charge density operator $\varrho(\boldsymbol{r})$:

$$\varrho(\boldsymbol{r})\ =\ -e\,\psi^\dagger(\boldsymbol{r})\psi(\boldsymbol{r})\ . \tag{6.37}$$

However, this identification is rather *ad hoc* and we really need a more systematic approach to construct second-quantized operators. We will propose the appropriate procedure from an intuitive point of view, but, as was announced in the previous section, the formal proofs and calculations will be omitted. In transport theory, we will have to deal mainly with

- one-particle operators: momentum, density, current, free particle Hamiltonian
- two-particle operators: interactions between two identical particles or two particles of different kinds.

It should be noticed that the term 'one-particle operator' does not necessarily refer to single particle quantum mechanics: in a many-particle framework it simply addresses microscopic processes in which the creation and annihilation of only one particle are involved. A similar remark can be made for two-particle operators.

In order to avoid confusion due to the use of inappropriate symbols, we will adopt the following notation for the remainder of this section. For an arbitrary physical quantity $\mathcal{A}$ we denote the corresponding operator as

A_1: in a single-particle system,
A: in a many-particle system, described in first quantization,
$\hat{A}$: in a many-particle system, described in second quantization.

One-Particle Operators. In the case of *global* one-particle operators, the procedure to construct a second-quantized operator for some observable $\mathcal{A}$ is straightforward:

- squeeze the single-particle operator A_1 between two field operators – a creator and a destructor – at some space point
- integrate over all space points within Ω

or,

$$\hat{A} = \int_{\Omega} \mathrm{d}\tau \, \psi^\dagger(\boldsymbol{r}) \, A_1(\boldsymbol{r}, \boldsymbol{p}) \, \psi(\boldsymbol{r}) \, . \tag{6.38}$$

Mathematically, $\boldsymbol{p} = -\mathrm{i}\hbar\nabla$ represents the momentum operator that we encountered in single-particle quantum mechanics and $A_1(\boldsymbol{r}, \boldsymbol{p}) \; \psi(\boldsymbol{r})$ should thus be interpreted as the action of $A_1(\boldsymbol{r}, \boldsymbol{p})$ upon the dependence of $\psi(\boldsymbol{r})$ on $\boldsymbol{r}$. Such an operation can of course only be carried out explicitly when the abstract field operator is expanded in concrete basis functions as will be explained later. Physically, one may regard the action of the operator $\psi^\dagger(\boldsymbol{r}) \; A_1(\boldsymbol{r}, \boldsymbol{p}) \; \psi(\boldsymbol{r})$ on an N-particle Fock state $|\Phi\rangle$ as a three-step process:

- $\psi(\boldsymbol{r})$ 'probes' the presence of a particle at $\boldsymbol{r}$ in the state $|\Phi\rangle$ and removes one particle whenever present;
- A_1 acts on the dependence of the resulting $(N-1)$-particle state on $\boldsymbol{r}$;
- $\psi^\dagger(\boldsymbol{r})$ puts a particle back at $\boldsymbol{r}$, thereby restoring the original number of particles, N.

Examples of Global One-Particle Operators

1. Hamiltonian (without particle-particle interactions).
   The second quantized Hamiltonian for a many-particle system moving in a potential energy $U(\boldsymbol{r})$ is given by

$$\hat{H} = \int_{\Omega} \mathrm{d}\tau \, \psi^\dagger(\boldsymbol{r}) \left(\frac{p^2}{2m} + U(\boldsymbol{r}) \right) \psi(\boldsymbol{r}) \, . \tag{6.39}$$

   For the sake of comparison, we recall that the first-quantized version of the non-interacting Hamiltonian was found to be

$$H = \sum_{j=1}^{N} H_1(\boldsymbol{r}_j, \boldsymbol{p}_j) = \sum_{j=1}^{N} \left(\frac{p_j^2}{2m} + U(\boldsymbol{r}_j) \right) \, . \tag{6.40}$$

2. Total momentum operator

$$\hat{\boldsymbol{P}} = \int_{\Omega} \mathrm{d}\tau \, \psi^\dagger(\boldsymbol{r}) \, \boldsymbol{p} \, \psi(\boldsymbol{r}) \, . \tag{6.41}$$

3. Number operator

$$\hat{N} = \int_{\Omega} \mathrm{d}\tau\, \psi^{\dagger}(\boldsymbol{r})\, \psi(\boldsymbol{r})\,. \tag{6.42}$$

Apart from the global one-particle operators, one often has to deal with *local* one-particle operators which in most cases appear as the densities of the global operators. Explicit construction may or may not be straightforward, and one is often thrown on physical intuition, first principles, classical analogies, constitutive equations etc. for justification of the proposed forms.

Examples of Local One-Particle Operators

1. Electron charge density
   The charge density appears simply as the density of the number operator and, according to an earlier remark, it yields for the electron charge density:

$$\hat{\varrho}(\boldsymbol{r}) = -e\, \psi^{\dagger}(\boldsymbol{r})\, \psi(\boldsymbol{r})\,. \tag{6.43}$$

2. Electron current density
   Similarly, up to some pre-factors taking care of units and dimensionality, the (electron) current density may be regarded as the density associated with the total momentum $\hat{\boldsymbol{P}}$. So we are encouraged to try

$$\begin{aligned} \hat{\boldsymbol{J}}(\boldsymbol{r}) &= -e\, \psi^{\dagger}(\boldsymbol{r})\, \boldsymbol{v}\, \psi(\boldsymbol{r}) \\ &= -e\, \psi^{\dagger}(\boldsymbol{r})\, \frac{-\mathrm{i}\hbar\nabla + e\boldsymbol{A}}{m}\, \psi(\boldsymbol{r}) \end{aligned} \tag{6.44}$$

   which more or less looks like a charge density times a velocity as we might expect. However, taking a closer look at (6.44), we see that the proposed current density operator is not Hermitian! So we will use the brute force approach and make $\hat{\boldsymbol{J}}$ Hermitian, by adding the Hermitian conjugated operator and halving the sum:

$$\hat{\boldsymbol{J}}(\boldsymbol{r}) = -\frac{e}{2} \left[\psi^{\dagger}(\boldsymbol{r})\, \boldsymbol{v}\, \psi(\boldsymbol{r}) + \left(\boldsymbol{v}^{*} \psi^{\dagger}(\boldsymbol{r}) \right) \psi(\boldsymbol{r}) \right]\,. \tag{6.45}$$

   Strictly speaking, we should also check whether this expression satisfies the continuity equation $\nabla \cdot \hat{\boldsymbol{J}} = -\partial \hat{\varrho} / \partial t$. This however will be shifted to chapter 10 on the use of quantum mechanical balance equations.

Question. *If hermiticity raises a problem when* $\hat{\boldsymbol{J}}$ *is constructed, why didn't it already show up when the momentum* $\hat{\boldsymbol{P}}$ *was constructed since, after all,* $\hat{\boldsymbol{P}} = -m/e \int_{\Omega} \mathrm{d}\tau \hat{\boldsymbol{J}}$ *?*

Two-Particle Operators. As can be expected, the two-particle processes involve the creation and annihilation of two particles, which is formally reflected in the presence of four field operators. Starting from a classical potential energy $v(\boldsymbol{r}, \boldsymbol{r}')$ associated with two interacting particles at positions $\boldsymbol{r}$ and $\boldsymbol{r}'$, the interaction Hamiltonian for inter-particle interaction takes the following general form[2]:

$$\hat{H}' = \frac{1}{2} \int_{\Omega} \mathrm{d}\tau \int_{\Omega} \mathrm{d}\tau' \, \psi^\dagger(\boldsymbol{r}')\psi^\dagger(\boldsymbol{r}) \, v(\boldsymbol{r}, \boldsymbol{r}') \, \psi(\boldsymbol{r})\psi(\boldsymbol{r}') \,, \tag{6.46}$$

which compares to its first-quantized analog

$$H' = \frac{1}{2} \sum_{j,k=1, j\neq k}^{n} v(\boldsymbol{r}_j, \boldsymbol{r}_k) \,. \tag{6.47}$$

Note the order of the arguments $\boldsymbol{r}$ and $\boldsymbol{r}'$ in (6.46). In many cases we will have to deal with bare or screened Coulomb interactions

$$v(\boldsymbol{r}, \boldsymbol{r}') = v(|\boldsymbol{r} - \boldsymbol{r}'|) = \pm \frac{e^2}{4\pi\varepsilon|\boldsymbol{r} - \boldsymbol{r}'|} \,, \tag{6.48}$$

where the plus sign covers all repulsive interactions such as electron–electron and hole–hole scattering while the minus sign similarly addresses attractive electrostatic interactions, including electron–hole attraction and excitons.

Some remarks are in order:
Various other Coulomb interactions (like between charge carriers and dopants) are not mentioned here, not only because they are not representing identical particles but also because the corresponding scattering agents are mainly treated as classical (i.e. non-quantized) particles.[3]

The operators discussed in the previous examples did conserve the number of particles. This feature is clearly reflected in formal structure of the two-particle operators that are containing an equal number of both creation and annihilation operators. It is therefore not a surprise that these operators commute with the total number operator $\hat{N}$ which is the mathematical expression of particle conservation. The latter happens to be a natural physical constraint for fermions that are confined to a closed volume Ω, but it generally does not apply to all boson systems. For instance, electron–phonon and electron–photon scattering are described by interaction Hamiltonians that do conserve the number of electrons but not the number of phonons or photons. The latter is due to the specific form of the two generic interaction processes

- annihilation and creation of an electron (local counting), absorption of a phot(n)on: $\psi^\dagger(\boldsymbol{r})\psi(\boldsymbol{r})\chi(\boldsymbol{r})$
- creation and annihilation of an electron, (local counting), emission of a phot(n)on: $\psi^\dagger(\boldsymbol{r})\psi(\boldsymbol{r})\chi^\dagger(\boldsymbol{r})$

---

[2] The pre-factor 1/2 prevents double counting of the interactions.
[3] One important exception is the quantized radiation field.

where $\chi(\boldsymbol{r})$ and $\chi^\dagger(\boldsymbol{r})$ are the phot(n)on field operators. Denoting the coupling constant by η we may concisely write the global interaction in the form of a Hermitian second-quantized operator

$$\hat{H}_{EP} = \int_\Omega \mathrm{d}\tau\, \psi^\dagger(\boldsymbol{r})\psi(\boldsymbol{r}) \left[\eta\, \chi(\boldsymbol{r}) + \eta^* \chi^\dagger(\boldsymbol{r})\right] . \tag{6.49}$$

6.3 More Creation and Annihilation Operators

So far, the dependence of the field operators on the position vector $\boldsymbol{r}$ has been rather abstract. It gave us a pictorial interpretation of one- and two-particle processes and an appropriate environment to carry out formal statements and relations, but it is not really helpful when the real computational work has to start. For instance, how should we calculate terms like $\nabla\psi(\boldsymbol{r})$ or $\nabla^2\psi(\boldsymbol{r})$ which are prominently present in the expressions for second-quantized operators entering the basic Hamiltonian?

The answer to this question is surprisingly simple: by expanding the abstract field operators in a Fourier series of single-particle wave functions. From the physical point of view, it may indeed look amazing that the information gathered about single-particle energy spectra and the related eigenfunctions are so directly relevant to the study of many-particle systems, even interacting ones! However, from the mathematical point of view, the two environments have clearly one common feature: both field operators and single-particle wave functions are depending on exactly one position vector $\boldsymbol{r}$ (whereas the cumbersome many-particle wave functions have as many position vector arguments as there are particles). Moreover, the stationary solutions of a single-particle Schrödinger equation are known to constitute a complete set of basis functions on the domain Ω. Therefore, we can in principle use this set to construct a Fourier series for any function defined on Ω, including the field operators, provided we can give a meaningful interpretation of the corresponding Fourier coefficients.

Concretely, let us assume that we have solved the single-particle Schrödinger equation for some relevant potential profile – that does of course not address any many-particle interaction,

$$\left(-\frac{\hbar^2}{2m}\nabla^2 + U(\boldsymbol{r})\right)\phi_k(\boldsymbol{r}) = \varepsilon_k\phi_k(\boldsymbol{r}) . \tag{6.50}$$

We further assume that the orthogonal, complete set $\{\phi_k(\boldsymbol{r})\}$ is has been made orthonormal after proper normalization:

$$\langle\phi_{k'}|\phi_k\rangle = \int_\Omega \mathrm{d}\tau\, \phi^*_{k'}(\boldsymbol{r})\phi_k(\boldsymbol{r}) = \delta_{kk'} . \tag{6.51}$$

If the spectrum has a continuous part, we must accordingly invoke delta normalization and replace the sums by integrals.

Now, we may expand the field operator $\psi(\boldsymbol{r})$ in a Fourier series based on the complete, orthonormal basis $\{\phi_k(\boldsymbol{r})\}$:

$$\psi(\boldsymbol{r}) = \sum_k c_k \, \phi_k(\boldsymbol{r}) \, , \tag{6.52}$$

where the Fourier coefficients $\{c_k\}$ are formally obtained from

$$c_k = \int_\Omega \mathrm{d}\tau \, \phi_k^*(\boldsymbol{r}) \, \psi(\boldsymbol{r}) \, . \tag{6.53}$$

Obviously, these are not the usual Fourier coefficients which are known to be complex numbers or so-called c-numbers. Because of $\psi(\boldsymbol{r})$ being an operator, the set $\{c_k\}$ constitutes another set of operators. While the field operator $\psi(\boldsymbol{r})$ annihilates a particle at some position $\boldsymbol{r}$, c_k should rather be regarded as annihilating a particle in a single-particle energy eigenstate $|\phi_k\rangle$. Analogously, the Hermitian conjugate, appearing in

$$\psi^\dagger(\boldsymbol{r}) = \sum_k c_k^\dagger \, \phi_k^*(\boldsymbol{r}) \tag{6.54}$$

is now seen to create a particle in the *single-particle* state $|\phi_k\rangle$. Moreover, acting on an arbitrary N*-particle* state $|\Phi_N\rangle$, the operator $c_k^\dagger c_k$ is counting how many particles of the kth single-particle state indexsingle-particle state are found to reside in the state $|\Phi_N\rangle$. Therefore the representation provided by the $\{c_k\}$ and $\{c_k^\dagger\}$ is also called *the occupation number representation.*

This interpretation is further corroborated by the observation that the operators $\{c_k\}$ and $\{c_k^\dagger\}$ are obeying the same type commutation relations as the field operators, as can be extracted from (6.53) and its Hermitian conjugate

$$\begin{aligned} \left\{c_k, \, c_{k'}^\dagger\right\} &= \int_\Omega \mathrm{d}\tau \int_\Omega \mathrm{d}\tau' \, \{\psi(\boldsymbol{r}), \, \psi^\dagger(\boldsymbol{r}\,')\} \\ &= \int_\Omega \mathrm{d}\tau \, \phi_{k'}^*(\boldsymbol{r}) \, \phi_k(\boldsymbol{r}) \\ &= \delta_{kk'} \, . \end{aligned} \tag{6.55}$$

So far, the electron spin has been incorporated in the set of single-particle quantum numbers k. However, in view to future calculations, it is convenient (and less confusing) to include the spin quantum numbers explicitly. Since the electron spin is $1/2$, its z-component has only 2 eigenvalues: $\hbar/2$ and $-\hbar/2$ which may be denoted shortly by an integer index σ taking only the values $\pm\, 1$. From now on, we will explicitly insert σ into the set of electron single-particle quantum numbers.

For the sake of convenience, we will express in the following sections the many-particle Hamiltonian and other operators in the occupation number representation.

6.3.1 The Electron Hamiltonian

The bare electron Hamiltonian (without inter-particle interactions) was given in (6.39) and may be rewritten

$$\begin{aligned}\hat{H}_E &= \int_\Omega \mathrm{d}\tau\, \psi^\dagger(\boldsymbol{r}) \left(-\frac{\hbar^2}{2m}\nabla^2 + U(\boldsymbol{r})\right) \psi(\boldsymbol{r}) \\ &= \sum_{kk'\sigma} c^\dagger_{k'\sigma} c_{k\sigma} \int_\Omega \mathrm{d}\tau\, \phi^*_{k'}(\boldsymbol{r}) \underbrace{\left(-\frac{\hbar^2}{2m}\nabla^2 + U(\boldsymbol{r})\right) \phi_k(\boldsymbol{r})}_{\varepsilon_k \phi_k(\boldsymbol{r})} \\ &= \sum_{kk'\sigma} c^\dagger_{k'\sigma} c_{k\sigma}\, \varepsilon_k \int_\Omega \mathrm{d}\tau\, \phi^*_{k'}(\boldsymbol{r})\phi_k(\boldsymbol{r}) = \sum_{kk'\sigma} c^\dagger_{k'\sigma} c_{k\sigma}\, \varepsilon_k\, \delta_{kk'} \\ &= \sum_{k\sigma} \varepsilon_k\, c^\dagger_{k\sigma} c_{k\sigma} \,. \end{aligned} \tag{6.56}$$

6.3.2 The Number Operator

Similarly, we can easily derive an expression for the total number operator:

$$\hat{N} = \sum_{k\sigma} c^\dagger_{k\sigma} c_{k\sigma} \,. \tag{6.57}$$

6.3.3 Charge and Current Density

The expressions for charge and current density are obtained along the same lines:

$$\begin{aligned}\hat{\varrho}(\boldsymbol{r}) &= -e \sum_{kk'\sigma} \phi^*_{k'}(\boldsymbol{r})\phi_k(\boldsymbol{r})\, c^\dagger_{k'\sigma} c_{k\sigma} \\ \hat{\boldsymbol{J}}(\boldsymbol{r}) &= \sum_{kk'\sigma} \left[\frac{\mathrm{i}e\hbar}{2m} \left(\phi^*_{k'}(\boldsymbol{r})\nabla\phi_k(\boldsymbol{r}) - \left(\nabla\phi^*_{k'}(\boldsymbol{r})\right)\phi_k(\boldsymbol{r})\right) \right. \\ &\qquad \left. - \frac{e^2}{m}\phi^*_{k'}(\boldsymbol{r})\phi_k(\boldsymbol{r})\boldsymbol{A}(\boldsymbol{r}) \right] c^\dagger_{k'\sigma} c_{k\sigma} \,. \end{aligned} \tag{6.58}$$

6.3.4 Many-Particle Ground State of a Non-Interacting System

As a final illustration, we calculate the ground-state energy of a non-interacting N-electron system by using only commutator algebra and the knowledge provided by the single-particle wave functions. Since we are now dealing with a system of non-interacting electron – which is translated in the absence of off-diagonal terms in the Hamiltonian – we obtain by direct evaluation that not only the total number operator but also each individual counting operator $c^\dagger_{k\sigma}c_{k\sigma}$ does commute with $\hat{H}_E$. Consequently, each N-particle eigenstate may characterized by the sequence of occupation numbers $\{n_{k\sigma}\}$. Each occupation number $n_{k\sigma}$ is an eigenvalue of the counting operator $c^\dagger_{k\sigma}c_{k\sigma}$ and can therefore take only the values 0 or 1 since we are dealing with fermions. Writing the N-electron ground-state equation in Fock space as

$$\hat{H}_E\,|\Phi_{N0}\rangle \;=\; E_{N0}\,|\Phi_{N0}\rangle \tag{6.59}$$

or

$$\sum_{k\sigma}\varepsilon_k n_{k\sigma}\,|\Phi_{N0}\rangle \;=\; E_{N0}\,|\Phi_{N0}\rangle\;, \tag{6.60}$$

we now specify the single-electron wave functions for the simple case of N free electrons moving in a three-dimensional, rectangular box with volume $L_xL_yL_z$, subjected to periodic boundary conditions:

$$\varepsilon_k = \frac{\hbar^2k^2}{2m}$$
$$\phi_{\mathbf{k}}(\boldsymbol{r}) = \frac{1}{\sqrt{L_xL_yL_z}}\mathrm{e}^{\mathrm{i}\mathbf{k}\cdot\mathbf{r}}\;,$$

Defining the Fermi wavenumber k_F to be the highest value of $k = |\boldsymbol{k}|$ which still contributes to $|\Phi_{N0}\rangle$, we can write

$$\sum_{k\le k_\mathrm{F}}\varepsilon_k\,|\Phi_{N0}\rangle \;=\; E_{N0}\,|\Phi_{N0}\rangle \tag{6.61}$$

since all contributing occupation numbers are equal to one.

Clearly this yields

$$E_{N0} \;=\; \sum_{\sigma,k\le k_\mathrm{F}}\varepsilon_k = 2\sum_{k\le k_\mathrm{F}}\varepsilon_k \tag{6.62}$$

and similarly

$$N \;=\; 2\sum_{k\le k_\mathrm{F}}1\;. \tag{6.63}$$

Restricting our discussion to the case of a large box ($L_x, L_y, L_z \to \infty$), we may convert the sums into integrals according to the usual prescription in three-dimensional configuration space

$$\sum_{\mathbf{k}}(\ldots) \;=\; \frac{L_xL_yL_z}{(2\pi)^3}\int_{\text{all space}}\mathrm{d}^3k\,(\ldots) \tag{6.64}$$

we obtain all familiar results, such as

$$E_{N0} \;=\; \frac{3}{5}\,N\,\varepsilon_\mathrm{F}\;,\quad \varepsilon_\mathrm{F} \;=\; \frac{\hbar^2k_\mathrm{F}^2}{2m}\;,\quad k_\mathrm{F} \;=\; \left(3\pi^2 n_0\right)^{1/3}\;, \tag{6.65}$$

where $n_0 = N/L_xL_yL_z$ is the uniform electron concentration.

Summary

In this chapter, we have used the second quantization formalism to express various transport quantities in terms of creation and annihilation operators. Doing so, we have encountered the number of particles as being an integer-valued eigenvalue of the number operator. This should not create the false impression that the notion of 'number of particles' is only meaningful for many-particle states that are simultaneously diagonalizing the total number operator. As we have introduced the concept of expectation value in single-particle quantum mechanics, we may equally estimate the 'expected value' of a physical quantity in any state $|\Phi\rangle$ of many-particle Fock space by calculating the appropriate many-particle expectation value. For instance, when a particular phonon mode q is involved in the electron–phonon scattering processes, the expectation value $\langle\Phi|a_q^\dagger a_q|\Phi\rangle$ provides an estimate of the number of phonons residing in the state $|\Phi\rangle$ although the latter is not an eigenstate of the phonon number operator $a_q^\dagger a_q$. More general, the toolbox of statistical mechanics will provide adequate procedures required to calculate all statistical-equilibrium and non-equilibrium averages of all operator combinations, such as $c_{k'\sigma}^\dagger c_{k\sigma}$, underlying the basic physical quantities that are relevant to quantum transport.

Exercise

6.1. Find an analogous expression for the Slater determinant (6.11) for the case of bosons.

7. Equilibrium Statistical Mechanics

7.1 The Entropy Principle

In this section we review the entropy principle, which may serve as a basis for deriving various assemble density functions. The principle has a long history and has been the topic of controversy over many years. Although the starting point is rather simple, it lasted a few decades before the correct mathematical formulation was obtained and that the disputes were settled. The source of the confusion has been that in statistics there is a subtle difference between frequency (the occurrence rate of events) and the probability of events. The debate concerned the question, whether on can assign probabilities without measuring the frequencies. According to one school, the probability should represent the frequency in a random experiment, whereas the other school assigns to the concept of probability a state of knowledge about a system. One may say that the first school adapt an objective point of view, whereas the the latter one is an 'subjective' school. Mathematically, both approaches are equivalent, but the subjective interpretation is wider, since it allows to assign frequency ratios as representing probabilities.

The entropy principle is the precise mathematical formulation of the interpretation following the 'subjective' approach. According to the pioneers in statistics (Bernoulli, Laplace, Bayes), one may assign equal a priori probability to events A and B if one has no reason to assume that one outcome will be more likely than the other. Our state of knowledge is expressed by the rule: $p(A) = p(B)$. This principle has many names, e.g. to mention a few:

Principle of Insufficient Reason,
Principle of Indifference,
Desideratum of Consistency,
Principle of Equal a priori Probability.

The use of this principle in physics started with Maxwell, Boltzmann and Gibbs who attempted to derive the thermodynamic equations from a microscopic modeling of matter. For that purpose they assumed, that matter constituted a large number of particles, such that a huge phase space was spanned. Since it is not possible by any means (nor would be of interest) to determine the exact position of an actual system in this phase space, a so called micro-state, one must assume that all micro-states, which are compatible with the macroscopic restrictions of the system, such as the volume of the container, or the total energy of the system, should be considered with equal a priori probability. The micro-canonical ensemble density function is an explicit result justified by this way of reasoning.

The entropy principle has been strongly advocated by E.T. Jaynes [31] in a series of papers and conference presentations. The following example for constructing a 'gambling theory', nicely illustrates his view. Unless, we have evidence that a dice has been biased (for instance by loading one face with led) we must assume that the faces are upwards with equal probability, i.e. $p(1) = p(2) = p(3) = p(4) = p(5) = p(6) = \frac{1}{6}$. This fair dice has the property that the 'amount of uncertainty' is the highest for predicting the up-face in a throw.

Information theory provides a unique and unambiguous means to express the amount of uncertainty in a quantitative way. The appropriate expression was discovered by Shannon, an engineer working in communication technology. Suppose a receiver collects messages and he knows that a fixed number of messages are possible. The probability that the ith message comes in, is p_i. The quantity that measures in unique way the amount of uncertainty which will be the next message, is the information-entropy function, which is defined as

$$S = -k \sum_i^N p_i \log p_i \; . \tag{7.1}$$

This function is maximal if all p_i are equal. Of course one can construct many functions, S, which have this property, so the question arises, why to select this one. However, only this choice of S provides a quantity which is extensive, in other words, of we combine two sets A and B, then $S = S(A) + S(B)$, since the combined system has $m = m_1 m_2$ states, and therefore

$$S(AB) = \sum_{ij}^{m_1 m_2} p_i p_j f(p_i p_j) = \sum_i^{m_1} p_i f(p_i) + \sum_j^{m_2} p_j f(p_j) \; . \tag{7.2}$$

The function S measures the information that we have of a system, since if we have more knowledge of the system, e.g. a number of states are excluded $(p_k = 0, k = k_1, ..k_q)$ then S decreases. Therefore, *information* I can be quantitatively expressed by

$$I = S_{\max} - S \; . \tag{7.3}$$

It is not a coincidence that the expression that was obtained by Shannon, is identical in form to the formula that Gibbs wrote down for the entropy of the probability distribution of micro-states in phase space. Both results starts from the desire to avoid unjustified bias, i.e. avoidance of including 'information' which is not available and to arrive at a quantity which is extensive.

From now on we define the entropy of a system as the normalized sum or integral

$$S = -k_{\mathrm{B}} \int \frac{\prod \mathrm{d}p\mathrm{d}q}{h^{3N}} \, \varrho(p,q,t) \, \log \varrho(p,q,t) \; , \tag{7.4}$$

where h is a unit cell size in the (pq)-phase space. The dimension of h is the same as Planck's constant. There is even more to say. By comparing the entropy value obtained in this way, a remarkable good agreement is found with phenomenological values from thermodynamic experiments, provided one takes for h also the numerical value of the Planck's constant. Of course, it is not correct to conclude, that from phase space considerations alone, one could deduce quantum mechanics. On the contrary, quantum mechanics provides an explanation for the size of the unit cell in phase space.

7.2 The Canonical and Grand-Canonical Ensembles

In equilibrium the density function is not explicitly time-dependent. Moreover, the entropy principle requires that ϱ should be such that S is maximal, i.e.

$$\delta S = -k_{\mathrm{B}} \delta \left(\int \frac{\prod^{N} \mathrm{d}p\mathrm{d}q}{h^{3N}} \, \varrho(p,q,t) \, \log \varrho(p,q,t) \right) = 0 \; . \tag{7.5}$$

Furthermore, we want to keep the expectation values for the energy to be fixed, since the energy is a conserved quantity.

$$\delta \langle H_N \rangle = \delta \left(\frac{\int \prod^{N} \mathrm{d}p\mathrm{d}q \; H_N(p,q) \, \varrho(p,q,t)}{\int \prod^{N} \mathrm{d}p\mathrm{d}q \; \varrho(p,q,t)} \right) = 0 \; . \tag{7.6}$$

We consider maximization of S with respect to varying the probability distribution ϱ in phase space. Since the integrated distribution is constant, i.e. the number of ensemble points, we also have

$$\delta \left(\int \prod^{N} \mathrm{d}p\mathrm{d}q \; \varrho(p,q,t) \right) = 0 \; . \tag{7.7}$$

We can easily find ϱ by introducing Lagrange multipliers for the two constraints, and obtain

$$\varrho(p,q) = C \exp\left(-\beta \, H_N\right) \; . \tag{7.8}$$

This result is known as the density function for the *canonical* ensemble and the partition function is

$$Z_N = \int \prod^{N} \mathrm{d}p\mathrm{d}q \exp\left(-\beta H_N\right) . \tag{7.9}$$

The partition function Z_N is actually the integral of the density function over the phase space. The normalization can be chosen such that the *outcome* of the integral is equal to one ($Z_N = 1$). However, the result of the integration is of minor importance. The description of physics is contained in the functional dependence of the integral on the phase-space coordinates. In other words: it matters *how* the ensemble points are distributed in the phase space. This is a similar manner of using mathematics as is done with path integrals. The outcome of a path integral is also an irrelevant issue. (Occasionally the path integral is only heuristically defined and a final number can not even be provided!) However, it does matter how the path integral depends in its integration variables.

As a further extension of these ideas one may relax the number of particles contained in the statistical system. Just as with the total energy, one could argue that the exact number of particles has not be determined, but merely the expectation value of the number of particles. Since the number of particles may vary between zero and infinity, we need to consider a Gibbs space which is the Cartesian product space of the canonical Gibbs spaces for the one, two, up to infinite number of particles. The density function is now defined on this product space. In order to obtain a clear view of the properties of this space, we consider some mathematical stuff that is used for spanning such spaces with coordinates.

The vector space $\mathbf{R}^{(N)}$ can be considered as the set of all real N-tuples $\boldsymbol{x} = (x_1, x_2, ..., x_N)$. Furthermore, any vector in $\mathbf{R}^{(N)}$ can be written as $\boldsymbol{x} = \sum_i x_i \boldsymbol{e}_i$, where $\boldsymbol{e}_i = (0, .., 0, 1, 0, .., 0)$ is the unit vector along the i-th direction. Therefore, one may regard $\mathbf{R}^{(N)}$ as the *direct sum* of N one-dimensional subspaces.

$$\mathbf{R}^{(N)} = \sum_{i=1}^{N} \oplus \mathbf{R}^{(1)} . \tag{7.10}$$

Of course, there are many ways to decompose a vector space into a set of mutually orthogonal subspaces.
Alternatively, $\mathbf{R}^{(N)}$ is also the Cartesian product of N one-dimensional spaces $\mathbf{R}^{(1)}$,

$$\mathbf{R}^{(N)} = \mathbf{R}^{(1)} \times \mathbf{R}^{(1)} \times .... \times \mathbf{R}^{(1)} \tag{7.11}$$

since a Cartesian product of two sets A and B is defined as all ordered pairs (a, b) with $a \in A$ and $b \in B$. Finally, it is also possible to *define* a vector space V as a direct sum of vector spaces $V_1, V_2, ..., V_N$, i.e.

$$V = \sum_{i=1}^{N} \oplus V_i \tag{7.12}$$

as the set of all ordered tuples $\boldsymbol{x} = (\boldsymbol{x}_1, \boldsymbol{x}_2, ..., \boldsymbol{x}_N)$ with $\boldsymbol{x}_i \in V_i$.

Since all multiplications are done with respect to the same field, it follows that

$$V_1 \times V_2 \times ... \times V_N = \sum_{i=1}^{N} \oplus V_i \ . \tag{7.13}$$

We are now ready to define the 'working space' of the grand-canonical ensemble.

$$\Gamma_{GC} = \Gamma_{(0)} \times \Gamma_{(1)} \times \Gamma_{(2)} .. \times \Gamma_{(\infty)} = \sum_{N=0}^{\infty} \oplus \Gamma_{(N)} \ . \tag{7.14}$$

The density function describes the density of ensemble points on this space. The distribution of ensemble points mimics the distribution of stars and galaxies in the universe. Most of the space is empty and the ensemble points cluster on sheets in this space. This is because a classical physical system has always an integer number of particles, although we do not know how many particles there are in the system under consideration.

Let us consider the artificial case of a one-particle system with a one-dimensional phase space, i.e. $\dim \Gamma_{(1)} = 1$. The two-particle phase space is $\Gamma_{(2)} = \Gamma_{(1)} \times \Gamma_{(1)}$. The phase space of zero particles is zero dimensional. The 'grand-canonical' phase space up to two particles is $\Gamma_{GC} = \Gamma_{(0)} \times \Gamma_{(1)} \times \Gamma_{(2)}$, and $\dim \Gamma_{GC} = 3$. However, the physical system has either zero, one or two particles and therefore the density function in this three-dimensional space is zero, if $\boldsymbol{x} = (\boldsymbol{x}_0, \boldsymbol{x}_1, \boldsymbol{x}_2)$ is not at the $\Gamma_{(0)}$-point or on the $\Gamma_{(1)}$-line or in the $\Gamma_{(2)}$-plane. As a consequence, the integration of the density function reduces to a sum over all these subspaces

$$\int_{\Gamma_{GC}} \mathrm{d}x \, \varrho = \sum_{N=0}^{\infty} \int_{\Gamma_N} \prod^{N} \mathrm{d}p\mathrm{d}q \, \varrho(p, q, N) \ . \tag{7.15}$$

The entropy definition is generalized as follows

$$S = -k_{\mathrm{B}} \sum_{N=0}^{\infty} \int \frac{\prod^N \mathrm{d}p\mathrm{d}q}{h^{3N}} \varrho_N(p, q, t) \log \varrho_N(p, q, t) \ . \tag{7.16}$$

One additional constraint, namely the constraint that determines the expected number of particles, needs to be included, i.e.

$$\delta \left(\frac{\sum_N \int \prod \mathrm{d}p\mathrm{d}q \, N \, \varrho_N(p, q, t)}{\sum_N \int \prod \mathrm{d}p\mathrm{d}q \varrho_N(p, q, t)} \right) = 0 \ , \tag{7.17}$$

as well as an additional Lagrange multiplier. The resulting density function becomes

$$\varrho_N(p, q) = C \exp\left(-\beta \left(H_N - \mu N\right)\right) \ . \tag{7.18}$$

This density function is known as the the density function for the *grand canonical* ensemble, and μ is the chemical potential. Since the counter for the number of particles is a discrete variable, the grand partition function is

$$\mathcal{Q} = \sum_{N=0}^{\infty} z^N Z_N \ , \tag{7.19}$$

where $z = \mathrm{e}^{\beta\mu}$ is the *fugacity*. The usefulness of the grand canonical ensemble becomes clear if we construct density functions for non-equilibrium systems. For, instance a straightforward generalization is the density function for which the chemical potential and temperature are space-time dependent, i.e.

$$\mathcal{Q} = \sum_{N=0}^{\infty} \int \prod^{N} \mathrm{d}p\mathrm{d}q \exp\left(- \int \mathrm{d}\tau \beta(\boldsymbol{x}, t) \left(h(\boldsymbol{x}, t) - \mu(\boldsymbol{x}, t)\, n(\boldsymbol{x}, t) \right) \right) \ . \tag{7.20}$$

It is clear that one could guess a formula which is constructed from equilibrium statistical mechanics by replacing global by local quantities, but great care is needed to give an appropriate interpretation to the local variables $\beta(\boldsymbol{x}, t)$ and $\mu(\boldsymbol{x}, t)$. Furthermore it should be emphasized that $\mathcal{Q} = \mathcal{Q}(t)$. An intuitive picture is provided by subdividing space into cells and by fixing the expectation value for the energy and number of particles in each cell. In this way, $\boldsymbol{x}$ is a pointer to discriminate the various cells.

7.3 Quantum Statistical Physics

We constructed the density functions for *classical* systems by using the entropy principle for the distribution of the ensemble members in *phase-* or *Gibbs space*. Since quantum-mechanical descriptions are all based on the association of physical states to vectors in a *Hilbert-* or *Fock space*, it is to be expected that quantum-statistical physics starts with a reconsidering the association of Fock-space vectors to physical states. The microscopic evolution of a classical system is described by a single point in phase space. This situation corresponds to a density function

$$\varrho(p, q, t) = \prod_{i=1}^{3N} \left(\delta(p_i - p_i(t))\delta((q_i - q_i(t))\right) \ . \tag{7.21}$$

Statistical considerations are included by 'smearing out' this spiky distribution function where each point represents a copy of the physical system. The microscopic evolution of a quantum system is described by a unique vector in Fock space, i.e. $|\Phi\rangle$. In order to include statistical considerations, we may proceed by smearing the description over a number of states, i.e.

$$|\Psi\rangle = \sum_n c_n |\Phi_n\rangle \ . \tag{7.22}$$

What discriminates this state from just another microscopic deterministic quantum state which can be always constructed by the superposition principle? In order to answer this question we have a look at the expectation value of some operator A in this state. The outcome is

$$\langle A\rangle = \frac{\langle\Psi|A|\Psi\rangle}{\langle\Psi|\Psi\rangle} = \frac{\sum_n\sum_m c_n^* c_m \langle\Phi_n|A|\Phi_n\rangle}{\sum_n\sum_m c_n^* c_m \langle\Phi_n|\Phi_n\rangle} \ . \tag{7.23}$$

The coefficients c_n and c_m contain the correlation between all the states $|\Phi_n\rangle$. In order to include a statistical description of a physical system using vectors in Fock space we need to postulate that a priori knowledge of the relative phases of the coefficients is not accessible. For the coefficients c_n this means that

$$c_n^* c_m = \delta_{n,m} f(n) \ . \tag{7.24}$$

This equation is known as the random phase postulate. Without further knowledge of the function $f(n)$ it is not possible to calculate expectation values. The micro-canonical way to determine $f(n)$ uses the eigenstates of the Hamiltonian, i.e.

$$H|\Phi_n\rangle = E_n |\Phi_n\rangle \tag{7.25}$$

and *postulates* that

$$c_n^* c_n = \begin{cases} 1 & E < E_n < E + \Delta \ , \\ 0 & \text{otherwise.} \end{cases} \tag{7.26}$$

An ensemble interpretation can still be given. One may imagine that each eigenstate which is compatible with the energy constraint corresponds to a copy of physical system. Each copy is described in the same Fock space and all copies are incoherent[1]. Since we only need the norm of c_n we can introduce the probabilities $p_n = c_n^* c_n$, which are real and positive. Furthermore, the expectation value of the operator A is

$$\langle A\rangle = \frac{\sum_n p_n \langle\Phi|A|\Phi\rangle}{\sum_n p_n} = \frac{\mathrm{Tr}(\varrho A)}{\mathrm{Tr}\varrho} \ , \tag{7.27}$$

where we define the *density operator*[2]

$$\varrho = \sum_n p_n |\Phi_n\rangle\langle\Phi_n| \ . \tag{7.28}$$

[1] Incoherence refers to the view that the phases of the ket vectors that represent the different copies, are not correlated.

[2] The density operator must not be confused with the electron charge density operator!

By introducing the density operator we have obtained a description of the physical system without reference to a particular basis in Fock space. Of course the appearance in one basis may be much more simple as another one, such as in the above situation where the density operator is diagonal in the basis spanned by the eigenstates of the Hamiltonian. In the Schrödinger picture where the basis is fixed in time and the states evolve according to the Schrödinger equation, the density operator is

$$\begin{aligned}\varrho_S(t) &= \textstyle\sum_n p_n \, |\Phi_n(t)\rangle \langle\Phi_n(t)| \\ &= \textstyle\sum_n p_n \, \mathrm{e}^{-\frac{\mathrm{i}}{\hbar}Ht} \, |\Phi_n(0)\rangle \langle\Phi_n(0)| \, \mathrm{e}^{\frac{\mathrm{i}}{\hbar}Ht} \\ &= U(t)\varrho_S(0)U^\dagger(t)\end{aligned} \tag{7.29}$$

and the equation of motion of the density operator is

$$\mathrm{i}\hbar\frac{\partial\varrho}{\partial t} = [H, \varrho] \; . \tag{7.30}$$

This result is the *quantum* Liouville equation. In the Heisenberg picture the basis rotates according to the evolution operator $U^\dagger$ and therefore the states of the system are fixed in time with respect to this basis. As a consequence, in the Heisenberg picture the density operator is

$$\begin{aligned}\varrho_H(t) &= \varrho_H(0) \\ \dot{\varrho}_H &= 0 \; .\end{aligned} \tag{7.31}$$

7.4 Quantum Ensembles

The canonical ensemble is described by the density operator whose matrix elements in the basis of eigenstates of the Hamiltonian, are

$$\varrho_{mn} = \delta_{mn}\mathrm{e}^{-\beta E_n} \; , \qquad \beta = \frac{1}{k_\mathrm{B}T} \; . \tag{7.32}$$

This formula can be viewed as a postulate which is constructed by using the classical canonical formula as a guide to find an analogous result for quantum statistical mechanics. It is also possible to derive this formula from the entropy principle, but then one must postulate that the entropy operator in quantum mechanics is

$$S = -k_\mathrm{B} \ln \varrho \; . \tag{7.33}$$

The density operator is found by maximizing $\langle S \rangle$ with the constraint that H is fixed. The canonical density operator is

$$\varrho = \sum_n \mathrm{e}^{-\beta E_n} \, |\Phi_n\rangle \langle\Phi_n| = \mathrm{e}^{-\beta H} \sum_n |\Phi_n\rangle \langle\Phi_n| = \mathrm{e}^{-\beta H} \; . \tag{7.34}$$

The partition function is

$$Z = \mathrm{Tr}\, \mathrm{e}^{-\beta H} \tag{7.35}$$

and the expectation value of the observable A is

$$\langle A \rangle = \frac{\mathrm{Tr} A\, \mathrm{e}^{-\beta H}}{Z} \,. \tag{7.36}$$

The grand-canonical ensemble is given by the density operator

$$\varrho = C z^N \mathrm{e}^{-\beta H_N} \,. \tag{7.37}$$

The operator acts on the space which is the direct sum of the zero-, one-, two-, ..., ∞-particle Fock spaces and N is the number operator and H_N the Hamilton operator for N particles. The grand partition function is

$$\mathcal{Q} = \sum_{N=0}^{\infty} z^N Z_N \tag{7.38}$$

and the expectation value of the observable A is

$$\langle A \rangle = \frac{1}{Q} \sum_{N=0}^{\infty} \langle A \rangle_N \,, \tag{7.39}$$

where $\langle A \rangle_N$ is the expectation value for A in the canonical ensemble for N particles.

We have provided a scheme for calculating expectation, provided the (grand) partition is known. However, it would be desirable if we could re-establish the relation of good-old phenomenological equilibrium thermodynamics. The rules of thermodynamics can indeed be reconstructed by making the appropriate identifications.

- Entropy – this is the expectation value of the entropy operator, i.e. $S = \langle S \rangle$.
- Internal energy – this is the expectation value of the Hamiltonian, i.e. $U = \langle H \rangle$.
- Free energy – this is given as $F = U - TS$.
- Particle number – this is the expectation value of the number operator $\langle N \rangle$.
- Grand potential – this is given by $\Omega = -kT \ln \mathcal{Q}$.

7.5 Photons and Phonons – Some Partition Functions

The general formalism, which was presented above, will here be applied to a few specific cases. First we will consider a gas of photons. A *photon* is a particle with spin $\hbar$ and no mass. There are only two spin orientations possible. One way to look at photons is by interpreting photons as the quantized objects which are obtained by applying the general quantization prescriptions to the Maxwell equations. So the wave equations of electrodynamics are submitted to the rules of quantum mechanics. The Maxwell equations in vacuum are

$$\begin{aligned}\nabla \cdot \boldsymbol{E} &= \frac{\varrho}{\varepsilon_0} \\ \nabla \times \boldsymbol{E} &= -\frac{\partial \boldsymbol{B}}{\partial t} \\ \nabla \cdot \boldsymbol{B} &= 0 \\ \nabla \times \boldsymbol{B} &= \mu_0 \boldsymbol{J} + \varepsilon_0 \mu_0 \frac{\partial \boldsymbol{E}}{\partial t} \,.\end{aligned} \tag{7.40}$$

The constant $\varepsilon_0 = 8.8544 \times 10^{-12}\ \mathrm{N^{-1}m^{-2}C^2}$, is the permittivity of the vacuum and the constant $\mu_0 = 1.2566 \times 10^{-6}\ \mathrm{m\,kgC^{-2}}$, is the permeability of the vacuum. The product $\varepsilon_0\mu_0 = c^{-2}$, where $c = 2.9979 \times 10^8$ m/s is the speed of light.

From the third Maxwell equation it follows that there exists a vector field $\boldsymbol{A}$, such that

$$\boldsymbol{B} = \nabla \times \boldsymbol{A} \,. \tag{7.41}$$

Substitution in the second Maxwell equation gives

$$\nabla \times \left[\boldsymbol{E} + \frac{\partial \boldsymbol{A}}{\partial t} \right] = 0 \tag{7.42}$$

from which follows that there exists a scalar field ϕ such that

$$\boldsymbol{E} + \frac{\partial \boldsymbol{A}}{\partial t} = -\nabla \phi \,. \tag{7.43}$$

After substitution of this result into the first and last Maxwell equations we obtain

$$\nabla^2 \phi + \frac{1}{c}\frac{\partial}{\partial t}(\nabla \cdot \boldsymbol{A}) = -\frac{\varrho}{\varepsilon_0} \tag{7.44}$$

$$\nabla^2 \boldsymbol{A} - \frac{1}{c^2}\frac{\partial^2}{\partial t^2}\boldsymbol{A} - \nabla \cdot (\nabla \cdot \boldsymbol{A} + \frac{1}{c}\frac{\partial}{\partial t}\phi) = -\mu_0 \boldsymbol{J} \,. \tag{7.45}$$

Since the fields $\boldsymbol{A}$ and ϕ are not unique, we must perform a *gauge fixing*. A possible choice is to select a set of potentials such that

$$\nabla \cdot \boldsymbol{A} + \frac{1}{c}\frac{\partial \phi}{\partial t} = 0 \,. \tag{7.46}$$

This choice is known as the Lorentz or Feynman gauge. The field equations in this gauge are

$$\nabla^2 \phi - \frac{1}{c^2}\frac{\partial^2}{\partial t^2}\phi = -\frac{\varrho}{\varepsilon_0} \tag{7.47}$$

$$\nabla^2 \boldsymbol{A} - \frac{1}{c^2}\frac{\partial^2}{\partial t^2}\boldsymbol{A} = -\mu_0 \boldsymbol{J} \,. \tag{7.48}$$

It should be emphasized that the particular choice of the gauge fixing should not effect the physical outcomes. Such gauge invariance requirements are important cross checks whether one has obtained sensible results. For describing photons, which correspond to solutions of the Maxwell equations in free space, i.e. in absence of charges or currents, the *Coulomb* gauge or *radiation* gauge is often exploited. The gauge fixing condition is

$$\nabla \cdot \boldsymbol{A} = 0 \,. \tag{7.49}$$

The free-field equation for the potential ϕ becomes in this gauge

$$\nabla^2 \phi = 0 \,, \tag{7.50}$$

with solution $\phi = 0$. The vector field $\boldsymbol{A}$ satisfies the wave equation

$$\nabla^2 \boldsymbol{A} - \frac{1}{c^2}\frac{\partial^2}{\partial t^2}\boldsymbol{A} = 0 \,. \tag{7.51}$$

The radiation condition implies the the direction of the vector $\boldsymbol{A}$ is always perpendicular to the propagation vector $\boldsymbol{q}$ of the wave. A solution of the wave equation is

$$\boldsymbol{A}(\boldsymbol{r}, t) = \boldsymbol{\varepsilon} \exp\left(\mathrm{i}(\boldsymbol{q}\cdot\boldsymbol{r} - \omega t)\right) \,. \tag{7.52}$$

The polarization $\boldsymbol{\varepsilon}$ is perpendicular to $\boldsymbol{q}$ and therefore only two independent directions can be chosen. In order to count the number of possible modes it is convenient to consider a large box in space with volume $V = L^3$. Then the boundary conditions imply that

$$\boldsymbol{q} = \frac{2\pi \boldsymbol{n}}{L} \,, \qquad n_i = \ldots, -2, -1, 0, 1, 2, \ldots \,. \tag{7.53}$$

Applying the quantization rules to the wave equation results into quanta with energy $\hbar\omega$ and momentum $\hbar\boldsymbol{q}$ and polarization $|\boldsymbol{\varepsilon}| = 1$. The determination of the partition function consists of summing over all possible number of photons states. However, since photons have no rest mass, it does not take energy to add an additional photon or to remove one with arbitrary long wave length. Therefore , the photon chemical potential is always zero, or equivalently, the fugacity is one. A state is determined by saying how many photons are present for a particular $\boldsymbol{q}$ and $\boldsymbol{\varepsilon}$. This number varies from 0,1,2 ... to ∞. Let us denote this number by $n[\boldsymbol{q}, \boldsymbol{\varepsilon}]$

$$\begin{aligned} Z &= \sum_{n[\boldsymbol{q},\boldsymbol{\varepsilon}]} \exp\left(-\beta \sum_{\boldsymbol{q},\boldsymbol{\varepsilon}} E(n[\boldsymbol{q},\boldsymbol{\varepsilon}])\right) = \prod_{\boldsymbol{q},\boldsymbol{\varepsilon}} \sum_{n=0}^{\infty} \exp\left(-\beta\hbar\omega n\right) \\ &= \prod_{\boldsymbol{q},\boldsymbol{\varepsilon}} \frac{1}{1 - \mathrm{e}^{-\beta\hbar\omega}} \,. \end{aligned} \tag{7.54}$$

The sum over the polarization directions can be just converted into a power:

$$Z = \prod_{\boldsymbol{q}} \left(\frac{1}{1 - \mathrm{e}^{-\beta\hbar\omega}} \right)^2 . \tag{7.55}$$

With this equation one can easily derive Planck's formula for black-body radiation and related results, such as Stefan's law for the radiated energy per area. In particular, the internal energy is

$$U = -\frac{\partial}{\partial\beta} \ln Z = -\frac{\partial}{\partial\beta} \sum_{\boldsymbol{q}} \ln \left(\frac{1}{1 - \mathrm{e}^{-\beta\hbar\omega}} \right)^2 \tag{7.56}$$

which gives after the usual substitution $\sum_{\boldsymbol{q}} \to \frac{V}{(2\pi)^3} \int \mathrm{d}^3 k$,

$$u = \frac{U}{V} = \int_0^\infty \mathrm{d}\omega \frac{\hbar}{\pi^2 c^3} \frac{\omega^3}{\mathrm{e}^{\beta\hbar\omega} - 1} . \tag{7.57}$$

So much for photons: we continue with considering *phonons*, which are also bosons. The phonons are the quanta of vibration. A crystal consisting of N atoms has $3N$ eigen-frequencies, $\omega_1, \omega_2, \ldots$. Their values depends on the structure of the lattice. A simple model, due to Einstein assumes that all frequencies are equal. Debye improved the model by presenting a model with a continuous spectrum for ω. Again a state is fully determined by specifying the number of phonons n_i for each frequency ω_i. The energy for such a state is

$$E\{n_i\} = \sum_{i=1}^{3N} n_i \hbar\omega_i \tag{7.58}$$

and the partition function is

$$Z = \sum_{\{n_i\}} \exp\left(-\beta E\{n_i\}\right) = \prod_{i=1}^{3N} \frac{1}{1 - \mathrm{e}^{-\beta\hbar\omega_i}} . \tag{7.59}$$

From this partition function the thermodynamic laws for the specific heat and internal energy can be derived.

In general one has that the partition functions for non-interacting boson and fermions are

$$Z_N = \sum_{\{n_p\}} \exp\left(-\beta E\{n_p\}\right) \tag{7.60}$$

$$E\{n_p\} = \sum_p \varepsilon_p n_p \tag{7.61}$$

$$\sum_p n_p = N \tag{7.62}$$

$$\begin{aligned} n_p &= 0, 1, 2 \ldots \qquad && \text{bosons,} \\ n_p &= 0, 1 && \text{fermions.} \end{aligned}$$

In these equations the index p labels the eigenstates of the one-particle Hamiltonian. The constraints hamper the evaluation of the partition function. However, it is possible to evaluate the grand-partition functions since the constraints are relaxed. The results are:

$$\mathcal{Q} = \sum_{N=0}^{\infty} z^N Z_N = \prod_p \frac{1}{1 - z \exp(-\beta \varepsilon_p)} \qquad \text{bosons} \tag{7.63}$$

and

$$\mathcal{Q} = \sum_{N=0}^{\infty} z^N Z_N = \prod_p (1 + z \exp(-\beta \varepsilon_p)) \qquad \text{fermions.} \tag{7.64}$$

7.6 Preview of Non-equilibrium Theory

In this section we will present Kubo's derivation of electrical conductivity. The derivation will be done in a classical system, but the example is so illustrative for many refinements which followed, that Kubo's approach has dominated for a considerable period the research in non-equilibrium statistical physics[11, 32]. The derivation in classical mechanics points towards all the important ingredients, that will return while discussing the Mori's approach to non-equilibrium behavior of fluids and gases as well as the heat flow in solids [33]. The current density is

$$\boldsymbol{J} = -e\, n\, \boldsymbol{v} \, . \tag{7.65}$$

However, the interpretation of this equation requires some explanation. The velocity, as well as the density, should be regarded as averaged quantities, i.e.

$$\boldsymbol{v} = \frac{1}{N} \sum_{n=1}^{N} \frac{\boldsymbol{p}_n}{m} \, . \tag{7.66}$$

Let us assume that the current- and charge densities are constant, so $n = N/\Omega$. We define $\boldsymbol{I} \equiv \Omega\, \boldsymbol{J}$, where the total volume is Ω. Then the current density is the volume average of $\langle \boldsymbol{I} \rangle$, i.e.

$$\begin{aligned} \langle \boldsymbol{I} \rangle &= \Omega\, \boldsymbol{J} \\ &= \left\langle \Omega\, (-e) \left(\frac{N}{\Omega} \right) \left(\frac{1}{N} \sum_{n=1}^{N} \frac{\boldsymbol{p}_n}{m} \right) \right\rangle \\ &= \left\langle -e \sum_{n=1}^{N} \frac{\boldsymbol{p}_n}{m} \right\rangle . \end{aligned} \tag{7.67}$$

The expectation value also is

$$\langle \boldsymbol{I} \rangle = \left\langle -e \sum_{n=1}^{N} \frac{\boldsymbol{p}_n}{m} \right\rangle = -e \int \mathrm{d}q^{3N} \mathrm{d}p^{3N} \left(\sum_{n=1}^{N} \frac{\boldsymbol{p}_n}{m} \right) \varrho(p, q, t) \ . \tag{7.68}$$

The Hamiltonian is $H = H_0 + H_{\mathrm{int}}$ and the unperturbed Hamiltonian is

$$H_0 = \sum_{n=1}^{N} \frac{p_n^2}{2m} + V \ . \tag{7.69}$$

The perturbation is the potential of the force driving the current. In a dipole approximation we assume that the the force is generated by a constant external electric field, therefore

$$H_{\mathrm{int}} = +e \sum_{n=1}^{N} \boldsymbol{q}_n \cdot \boldsymbol{E} \ . \tag{7.70}$$

In order to do calculations, we need to have a clue of the temporal development of ϱ. The only known method to obtain some insight in this problem is to set up a perturbative calculation. For that purpose we refer to the first trick which was exploited by Kubo, but was used already many times before him in other domains, namely the adiabatic switching-on of the interaction part of the Hamiltonian. So instead of having an external field existing already at $t = -\infty$, Kubo assumed that the field at $t = -\infty$ is switched off and gradually reaches its full strength at $t = 0$. Mathematically, this situation can be achieved by replacing the interaction Hamiltonian by

$$H_{\mathrm{int}} = +e \sum_{n=1}^{N} \boldsymbol{q}_n \cdot \boldsymbol{E}\, \mathrm{e}^{\varepsilon t} \ . \tag{7.71}$$

Now all results will depend on ε. Unless there are good physical arguments to estimate this parameter, the ε dependence is unsatisfactory. Therefore, this parameter is merely a *regulator*, i.e. it is an intermediate quantity to help doing calculations and the value should be set at $\varepsilon = 0$ at the end of the calculation. For later comparison, it is preferred to omit the normalization of ϱ. Therefore we set

$$\langle \boldsymbol{I} \rangle = -e \frac{\int \mathrm{d}q^{3N} \mathrm{d}p^{3N} \left(\sum_{n=1}^{N} \frac{\boldsymbol{p}_n}{m} \right) \varrho(p, q, t)}{\int \mathrm{d}q^{3N} \mathrm{d}p^{3N}\ \varrho(p, q, t)} \ . \tag{7.72}$$

Since $H_{\mathrm{int}} \to 0$ for $t \to -\infty$, we find that

$$\varrho(p, q, t) = \exp(-\beta H_0) = \varrho_0 \ , \qquad t \to -\infty \ . \tag{7.73}$$

If time progresses, the interaction strength increases and we want to determine in a perturbative manner the impact of the interaction on ϱ. For that purpose we must solve the Liouville equation,

$$\frac{\partial \varrho}{\partial t} = [H(t), \varrho] = \mathcal{L}\,\varrho \tag{7.74}$$

perturbatively, with the boundary condition given above.
Now consider the discrepancy between the time-dependent solution $\varrho(t)$ and the equilibrium part, i.e. $\Delta\varrho(t) = \varrho(t) - \varrho_0$. Furthermore, we split the Liouville operator into two parts, i.e. $\mathcal{L} = \mathcal{L}_0 + \mathcal{L}_{\text{int}}$. Then the Liouville equation becomes

$$\begin{aligned}\frac{\partial}{\partial t}\Delta\varrho(t) &= (\mathcal{L}_0 + \mathcal{L}_{\text{int}})\,(\varrho_0 + \Delta\varrho(t)) \\ &= \mathcal{L}_0\Delta\varrho(t) + \mathcal{L}_{\text{int}}\varrho_0 + \text{higher order}\;.\end{aligned} \tag{7.75}$$

A perturbative solution can be obtained by considering

$$\varrho_D(t) = \exp\left(-\mathcal{L}_0 t\right)\Delta\varrho(t) \tag{7.76}$$

which satisfies the time-evolution equation

$$\frac{\partial}{\partial t}\Delta\varrho_D(t) = \exp\left(-\mathcal{L}_0 t\right)\mathcal{L}_{\text{int}}\,\varrho_0\;. \tag{7.77}$$

The evaluation of the part $\mathcal{L}_{\text{int}}\,\varrho_0$ is easily done by evaluation of the Poisson brackets. The result is

$$\mathcal{L}_{\text{int}}\,\varrho_0 = \beta\,\mathrm{e}^{\varepsilon t}\,\boldsymbol{E}\cdot\left(-e\sum_{n=1}^{N}\frac{\boldsymbol{p}_n}{m}\right)\;. \tag{7.78}$$

The next step is the integration of the time-evolution equation

$$\Delta\varrho_D(t) = \int_{-\infty}^{t}\mathrm{d}\tau\,\mathrm{e}^{\varepsilon\tau}\exp\left(-\mathcal{L}_0\tau\right)\beta\,\boldsymbol{E}\cdot\left(-e\sum_{n=1}^{N}\frac{\boldsymbol{p}_n}{m}\right)\varrho_0 \tag{7.79}$$

$$= \beta\,\exp\left(-\mathcal{L}_0 t\right)\int_0^{\infty}\mathrm{d}\tau\,\mathrm{e}^{-\varepsilon\tau}\left(\boldsymbol{E}\,\mathrm{e}^{\varepsilon t}\right)\cdot\boldsymbol{I}\left(-\tau\right)\varrho_0\;, \tag{7.80}$$

where we defined

$$\boldsymbol{I}\left(\tau\right) = \exp\left(-\mathcal{L}_0\tau\right)\boldsymbol{I}\;,\qquad \boldsymbol{I} = -e\sum_{n=1}^{N}\frac{\boldsymbol{p}_n}{m} \tag{7.81}$$

from which we obtain that

$$\Delta\varrho(t) = \beta\int_0^{\infty}\mathrm{d}\tau\,\mathrm{e}^{-\varepsilon\tau}\,\boldsymbol{I}\left(-\tau\right)\cdot\left(\boldsymbol{E}\,\mathrm{e}^{-\varepsilon t}\right)\varrho_0\;. \tag{7.82}$$

We have now an expression for $\varrho(t)$ and therefore we can evaluate any variable at any particular time. The variable of interest is the current, i.e.

$$\langle\boldsymbol{I}(t)\rangle = \frac{\int \mathrm{d}q^{3N}\mathrm{d}p^{3N}\,\boldsymbol{I}\,\left(\varrho_0 + \Delta\varrho(t)\right)}{\int \mathrm{d}q^{3N}\mathrm{d}p^{3N}\,\left(\varrho_0 + \Delta\varrho(t)\right)}\;. \tag{7.83}$$

In equilibrium the current is zero, furthermore the effective field is $\boldsymbol{E}(t) = \boldsymbol{E}\,\mathrm{e}^{\varepsilon t}$ and Ohm's law becomes

$$\langle I \rangle_\mu (t) = \Omega\, \sigma_{\mu\nu}\, E_\nu(t)\ , \tag{7.84}$$

where

$$\sigma_{\mu\nu} = \frac{\beta}{\Omega} \int_0^\infty \mathrm{d}\tau\, \mathrm{e}^{-\varepsilon\tau}\ \langle I_\mu(0)\, I_\nu(-\tau)\rangle_0\ . \tag{7.85}$$

This is Kubo's famous result for the derivation of the electric conductivity. There are a number of important observations to be made.

- The non-equilibrium problem was deduced to an equilibrium problem.
- The conductivity is a current–current time-correlation function.
- A time-smearing produces the time-independent (static) result.

We will see in the next chapter that these three observations are preserved in the refined models for non-equilibrium transport theories.
For weakly interacting particles we expect the time-correlation function to decay exponentially, i.e.

$$\langle I_\mu I_\nu(-\tau)\rangle_0 \simeq \langle I_\mu I_\nu\rangle_0\, \mathrm{e}^{-\frac{\tau}{\tau_0}} = \frac{1}{3}\delta_{\mu\nu}\, \langle I^2\rangle_0\, \mathrm{e}^{-\frac{\tau}{\tau_0}}\ . \tag{7.86}$$

This approximation gives $\sigma_{\mu\nu} = \delta_{\mu\nu}\, \frac{ne^2\tau_0}{m}$, which agrees with Drude's classical model of conductance.

Exercises

7.1. Show that the expectation value of the Hamiltonian, which is the internal energy, is given by

$$U = \langle H\rangle = -\frac{\partial}{\partial\beta} \ln Z\ .$$

7.2. Show that $\Omega = U - T\,S - \mu\,N$ and therefore

$$N = -\frac{\partial\Omega}{\partial\mu} \quad \text{and} \quad \mu = \left(\frac{\partial F}{\partial N}\right)_V\ .$$

7.3. Prove equation (7.57).

8. Non-equilibrium Statistical Mechanics

8.1 Definition of the Problem

The problem of transport, which is inherently statistical, becomes very subtle, if one wants to set up a formalism in which the starting point is the phase or Gibbs space. The key ingredient is of course that the density operator is now explicitly time dependent, in order to cover the changes that occur in time. In equilibrium statistical physics, the Liouville equation is

$$[H, \varrho] = 0 \tag{8.1}$$

and it is possible to find a density operator, namely $\varrho = \exp(-\beta H)$, which is an exact solution of this equation. For non-equilibrium statistical physics, the Liouville equation is

$$\mathrm{i}\hbar \frac{\partial \varrho}{\partial t} = [H, \varrho] \tag{8.2}$$

which is formally solved by

$$\varrho(t) = \exp\left(-\frac{\mathrm{i}}{\hbar} Ht\right) \varrho(0) \exp\left(\frac{\mathrm{i}}{\hbar} Ht\right) . \tag{8.3}$$

However, this solution follows in a microscopic sense the time evolution, and as such it does not correspond to an entropy-increasing solution. Indeed

$$S(t) = -k_\mathrm{B} \operatorname{Tr}\left[\varrho(t) \ln \varrho(t)\right] = S(0) . \tag{8.4}$$

As was emphasized in the introduction, transport phenomena are usually accompanied by dissipative effects, i.e. an increment of entropy should be one of the essential outcomes of a sensible non-equilibrium statistical theory of transport. So we are confronted with the following paradox:

We must construct a solution of the Liouville equation in order to build a consistent time-dependent transport theory, but if we do so, we do not include dissipative transport.

How do we get out of this contradictory situation? An important step towards a better understanding of the origin of dissipation, is to construct a model in which is shown explicitly how to irreversibility comes into the solution. Consider an electron coupled to N phonons. At $t = 0$ the N phonons are distributed according to an equilibrium distribution. Moreover, at $t = 0$, a constant electric field is switched on. The equations of motion can be solved exactly. This will be discussed in more detail in Chap. 14. The important observations from this model are:

- For N phonons the solution is fully reversible.
- If N becomes ∞, the evolution is irreversible.
- The entropy production is related to the dissipated power according to the rules of thermodynamics.

So in order to escape from the paradox, *infinitely* many degrees of freedom must be considered. There are several approaches to reach this goal. We will have a closer look to two methods, which both play an important role for describing transport in semiconductors and mesoscopic systems.

In the first approach one considers a finite system coupled to an infinite system. The latter one is usually a heat bath. In our model, if $N \to \infty$, we have one or more electrons (the finite system) coupled to a phonon heat bath. In this approach, where reservoirs or heat baths are coupled to the actual system of interest, we are dealing with *open* systems.

An alternative way to arrive at infinite systems is by *coarse graining*. The underlying idea is that the system itself has infinitely many degrees of freedom and even small volume elements have so many degrees of freedom, that one may quietly assume that each volume element by itself contains an infinite number of degrees of freedom. In fact, a field theory is the ideal candidate that obeys this property because each space point contains a field degree of freedom and any finite volume element has infinitely many space points [34]. Since a many-body system has a field-theoretical formulation, we will exploit this formulation to present the approach based on coarse graining. All the degrees of freedom in a volume element are then collected and treated as a single new degree of freedom. So coarse graining is a way of 'thinning' degrees of freedom. The latter is a very general method to manipulate a problem such that it becomes solvable and several methods of thinning exist. A method is selected depending on the problem under consideration and the properties that one wants to emphasize. A serious drawback of thinning is that it is often impossible to justify from first principles the description using a few effective variables, although the arguments are plausible. Consequently, one does not really provide an explanation but an approximate description. The situation becomes even more precarious if one cannot make quantitative statements about the accuracy of the effective models which arise after thinning. Fortunately, the models can be quite useful for practical applications, which justify further evaluation. A well-known thinning method results into the hydrodynamical model which we will now analyze.

8.2 Hydrodynamics

Instead of deriving the hydrodynamic model in full detail for all hydrodynamical variables, we will consider a simplified hydrodynamical problem, such that we do not dwell in notational confusion, but nevertheless see very clearly, at which step the approximations are made, in order to derive a macroscopic result from microscopic dynamics. The example that we will work out is the derivation of Fourier's law for the heat flow in a rigid system, i.e.

$$\lambda_0 \frac{\partial T}{\partial t} = -\nabla \cdot \boldsymbol{J}_{\mathrm{Q}} \tag{8.5}$$

$$\boldsymbol{J}_{\mathrm{Q}} = -\kappa_0 \nabla T \,, \tag{8.6}$$

where κ_0 is the thermal conductivity and λ_0 the heat capacity. The generalization to a full hydrodynamical description uses very similar steps, i.e. approximations, but is unfortunately complicated considerably due to the large amount of terms which have to be kept track of.

Consider a many-particle system, described by the Hamiltonian, which is expressed in terms of the particle field operators ψ and $\psi^\dagger$

$$H = \int \mathrm{d}^3 r \, \psi^\dagger \left(-\frac{\hbar^2}{2m} \Delta + V(\boldsymbol{r}) \right) \psi = \int \mathrm{d}^3 r \, h(\boldsymbol{r}) \,. \tag{8.7}$$

In the Schrödinger representation, ψ and $\psi^\dagger$ are time-independent.

The rigidity of the system is justified by the fact that the field operators are all anchored to a position in space. In a full hydrodynamic derivation we must allow expectation values of the local momentum to be non vanishing, but the anchoring of the basic dynamical variables remains in tact. The field operators ψ and $\psi^\dagger$ obey the commutation relations

$$[\psi(\boldsymbol{r}), \psi(\boldsymbol{r}')] = \left[\psi^\dagger(\boldsymbol{r}), \psi^\dagger(\boldsymbol{r}')\right] = 0 \tag{8.8}$$

$$\left[\psi(\boldsymbol{r}), \psi^\dagger(\boldsymbol{r}')\right] = \delta(\boldsymbol{r} - \boldsymbol{r}') \,. \tag{8.9}$$

These commutation relations correspond to bosons obeying the Schrödinger equation. For electrons and holes we must replace these relations by anti-commutation relations for fermions. For our pedagogical example we work out the bosonic case, in order not to obscure the details of our reasoning with the complication related to anti-commuting property of the variables. The introduction of the effective variables takes place by introducing space-time cells of size $\omega = \Delta x \Delta y \Delta z \Delta t = \Delta V \Delta t$. At each time step we define an effective cell Hamiltonian using the energy-density operator

$$h_j = \int_{\Delta V} \mathrm{d}^3 r \, h(\boldsymbol{r}) \,. \tag{8.10}$$

In the Heisenberg representation we can write down the microscopic conservation law for the energy, since for a closed system the total energy is conserved:

$$\frac{\partial}{\partial t} h(\boldsymbol{r}, t) + \nabla \cdot \boldsymbol{q}(\boldsymbol{r}, t) = 0 , \tag{8.11}$$

where $\boldsymbol{q}(\boldsymbol{r}, t)$ is a vector operator representing the Poynting vector for the microscopic energy flow.

The cell size ω serves as a communication scale between the microscopic and macroscopic dynamics. We will now apply the entropy principle at each time-plane, t_i, and for each cell ΔV_j. The result is a statistical operator $\varrho_0(t_i)$ which is formally equivalent to an equilibrium density operator for the space-time cell ω

$$\varrho_0(t_i) = Z_0^{-1}(t_i) \exp\left(- \sum_j \beta_j(t_i) h_j \right) \tag{8.12}$$

$$Z_0(t_i) = \text{Tr} \exp\left(- \sum_j \beta_j(t_i) h_j \right) . \tag{8.13}$$

On the macroscopic scale, ΔV is considered to be infinitesimal small. Therefore this operator is often written as

$$\varrho_0(t) = Z_0^{-1}(t) \exp\left(- \int \mathrm{d}^3 r \; \beta(\boldsymbol{r}, t) \, h(\boldsymbol{r}) \right) . \tag{8.14}$$

However, we should remember its meaning as a cell summation. The operator $\varrho_0(t)$ is a localized version of density operator ϱ that is used to describe equilibrium statistical physics.

The cell sizes or scales ΔV and Δt are chosen such that measurement instruments are not capable of detecting variations at smaller sizes. This makes it plausible that we use the cell variables h_j, instead of $h(\boldsymbol{r})$.

Starting with some density operator $\varrho_0(t_0)$ at t_0, the density operators $\varrho_0(t_i)$ do not correspond to the solution of the Liouville equation at times t_i. Nevertheless, the Lagrange multipliers $\beta_j(t_i)$ are chosen such that the expectation values after a time step Δt for the operators h_j correspond to the *exact* expectation values:

$$\langle h_j(t_i) \rangle_0 \; = \; \langle h_j(t_i) \rangle \; = \; h_j(t_i) = h(\boldsymbol{r}, t) , \tag{8.15}$$

where the last equality must be interpreted on the macroscopic scale, i.e. $h(\boldsymbol{r}, t)$ is a hydrodynamical variable.

In order to construct the hydrodynamical equation of motion, the microscopic evolution is considered in going from from t_i to t_{i+1}. However, at each time step, the initial density operator is taken to be $\varrho_0(t_i)$. Proceeding with the density operator $\varrho_0(t_{i+1})$ after a time interval Δt, instead of continuing with the exact density operator $\varrho(t_{i+1})$, we should realize that only the expectation values of $h_j(t_i)$ are faithfully described. All other correlations are neglected, which implies the loss of information and therefore irreversibility. In Fig. 8.1, the time evolution of ϱ according to the hydrodynamical modeling is sketched [35].

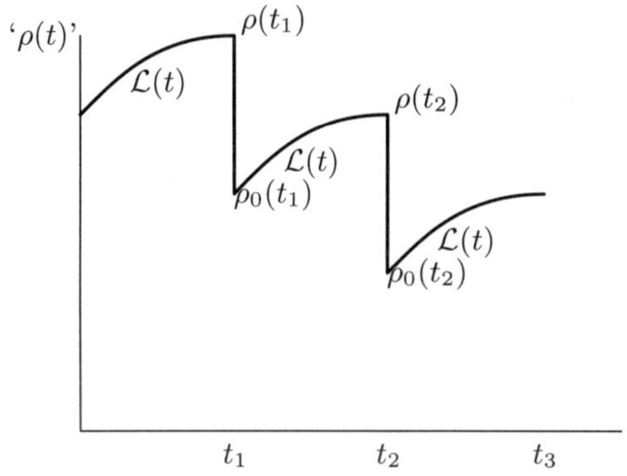

Fig. 8.1. Illustration of irreversible evolution of the density operator $\varrho(t)$

8.2.1 A First Glance at Entropy

The entropy in the local-equilibrium state at time t_i is

$$S(t_i) = -k_{\mathrm{B}} \langle \log \varrho_0(t_i) \rangle_0^{t_i} \ . \tag{8.16}$$

Using the microscopic time evolution from t_i to t_{i+1} we obtain

$$\begin{aligned} S(t_{i+1}) - S(t_i) &= -k_{\mathrm{B}} \operatorname{Tr} \varrho_0(t_{i+1}) \log \varrho_0(t_{i+1}) + k_{\mathrm{B}} \operatorname{Tr} \varrho_0(t_i) \log \varrho_0(t_i) \\ &= -k_{\mathrm{B}} \operatorname{Tr} \varrho(t_{i+1}) \log \varrho_0(t_{i+1}) + k_{\mathrm{B}} \operatorname{Tr} \varrho(t_i) \log \varrho(t_i) \\ &= k_{\mathrm{B}} \operatorname{Tr} \varrho(t_{i+1}) \log \varrho(t_{i+1}) \varrho_0^{-1}(t_{i+1}) \\ &\geq 0 \end{aligned} \tag{8.17}$$

since $\log x \geq 1 - \frac{1}{x}$. So the entropy always increases, with equality only when $\varrho_0 = \varrho$. Later we will give an explicit expression for the rate at which the entropy increases [33,35–39].

After each time step, the new β's depend on the old ones, i.e. we obtain a parameter-flow equation of the form:

$$\beta_l(t_{i+1}) = f_l(\beta_1(t_i), \beta_2(t_i)...) \ . \tag{8.18}$$

In steady state or equilibrium, the β's do not change in time anymore, so these values of β correspond to a fixed point of the parameter-flow equation.

$$\beta_l^*(t_{i+1}) = f_l(\beta_2^*(t_i), \beta_2^*(t_i)...) \ . \tag{8.19}$$

Linearization around the fixed point gives with $\Delta\beta_l = \beta_l - \beta^*$ the following equation [35]

$$\Delta\beta_l(t_{i+1}) = \sum_k \Lambda_{lk} \ \Delta\beta_k(t_i) \ , \qquad \Lambda_{lk} = \left(\frac{\partial f_l}{\partial \beta_k} \right)_{\beta=\beta^*} \tag{8.20}$$

or in matrix notation: $\Delta\beta(t_{i+1}) = \Lambda \Delta\beta(t_i)$. After n iterations, i.e. $t = n\,\Delta t$, we obtain

$$\Delta\beta_l(t) = \left(\exp \frac{t}{\Delta t} \log \Lambda \right)_{lk} \Delta\beta_k(0) \ . \tag{8.21}$$

For attractive fixed points, β, the eigenvalues of $\log \Lambda$ are negative and the system ends up in steady state or equilibrium. For marginal fixed points, we must consider second order terms in the parameter-flow equation. It may also happen that fixed points are repulsive, and one can imagine systems which are oscillating between two repulsive fixed points. These situations can occur for chemical systems (chemical clocks).

8.2.2 Deriving Fourier's Law

Let us now continue with deriving the heat equation and start with the derivation of the right-hand side of the equation. On the macroscopic time scale, changes are registered according to the Mori derivative, i.e.

$$\frac{\delta}{\delta t} f(t) = \frac{\delta}{\delta t_i} f(t_i) = \frac{f(t_{i+1}) - f(t_i)}{\Delta t} \ . \tag{8.22}$$

The expectation values for the energy operators h_j change in time as

$$\begin{aligned}
\frac{\delta}{\delta t} h_j(t_i) &= \frac{1}{\Delta t}\left(\langle h_j\rangle_0^{t_{i+1}} - \langle h_j\rangle_0^{t_i}\right) \\
&= \frac{1}{\Delta t}\left(\langle h_j\rangle^{t_{i+1}} - \langle h_j\rangle^{t_i}\right) \\
&= \frac{1}{\Delta t}\int_{t_i}^{t_{i+1}} \mathrm{d}t' \, \frac{\partial}{\partial t'} \langle h_j\rangle^{t'} \\
&= \frac{1}{\Delta t}\int_{t_i}^{t_{i+1}} \mathrm{d}t' \mathrm{i}\hbar \, \langle \, [H, h_j] \, \rangle^{t'} \\
&= -\frac{1}{\Delta t}\int_{t_i}^{t_{i+1}} \mathrm{d}t' \int_{\Delta V_j} \mathrm{d}^3x \, \nabla \cdot \langle \boldsymbol{q}(\boldsymbol{x})\rangle^{t'} \ .
\end{aligned} \tag{8.23}$$

With the use of the expansion

$$\boldsymbol{q}(\boldsymbol{r}+\boldsymbol{a}) = \boldsymbol{q}(\boldsymbol{r}) + \boldsymbol{a}\cdot\nabla\boldsymbol{q}(\boldsymbol{r}) + ... \tag{8.24}$$

we can interchange the divergence and the integration (see below) and obtain

$$\frac{\delta}{\delta t} h_j(t_i) = -\frac{1}{\Delta t}\int_{t_i}^{t_{i+1}} \mathrm{d}t' \, \nabla_0 \cdot \int_{\Delta V_j} \mathrm{d}^3r \, \langle \boldsymbol{q}(\boldsymbol{r})\rangle^{t'} \ , \tag{8.25}$$

where the divergence must be interpreted as follows:

$$\begin{aligned}
\nabla_0 \int_{\Delta V_j} \mathrm{d}^3r f(\boldsymbol{r}) &= \sum \hat{\boldsymbol{e}}_l \, \frac{1}{\Delta x_l}\left(\int_{\Delta V_j + \Delta x_l \hat{\boldsymbol{e}}_l} \mathrm{d}^3r \, f(\boldsymbol{r}) - \int_{\Delta V_j} \mathrm{d}^3r \, f(\boldsymbol{r})\right) \\
&= \sum \hat{\boldsymbol{e}}_l \, \frac{1}{\Delta x_l}\int_{\Delta V_j} \mathrm{d}^3r \, \left(f(\boldsymbol{r} + \Delta x_l \hat{\boldsymbol{e}}_l) - f(\boldsymbol{r})\right) \ .
\end{aligned} \tag{8.26}$$

The derivative ∇_0 is a spatial derivative on the macroscopic scale.

The approximation, $f(\boldsymbol{r}+\boldsymbol{a}) = f(\boldsymbol{r}) + \boldsymbol{a}.\nabla f(\boldsymbol{r})$, justifies the interchange, since to this order

$$\nabla_0 \int_{\Delta V_j} \mathrm{d}^3 r f(\boldsymbol{r}) = \int_{\Delta V_j} \mathrm{d}^3 r \nabla f(\boldsymbol{r}) \, . \tag{8.27}$$

This approximation implies that we restrict ourselves to long wave length variations in f.

We will now make explicit the dissipative contribution, which is generated by the difference between the exact and approximated density operator. For that purpose, we set

$$\phi(t_i) = \frac{1}{\Delta t} \int_{t_i}^{t_{i+1}} \mathrm{d}t' \varrho(t') \tag{8.28}$$

and

$$\Delta\varrho(t_i) = \phi(t_i) - \varrho_0(t_i) \, . \tag{8.29}$$

The hydrodynamical equation becomes

$$\frac{\delta}{\delta t} h(\boldsymbol{r},t) + \nabla_0 . \boldsymbol{q}(\boldsymbol{r},t) = -\nabla_0 . \int_{\Delta V_j} \mathrm{d}^3 r \, \langle \boldsymbol{q}(\boldsymbol{r}) \rangle_{\Delta}^{t} \, . \tag{8.30}$$

We have defined the expectation values with respect to $\Delta\varrho$ as

$$\langle O \rangle_{\Delta}^{t} = \mathrm{Tr}\, O \Delta\varrho(t) \, . \tag{8.31}$$

The expectation value of the energy-flux vector in the jth cell in the local-equilibrium state is

$$\boldsymbol{q}(\boldsymbol{r},t) = \mathrm{Tr}\,(\boldsymbol{q}_j \, \varrho_0(t_i)) = 0 \, , \tag{8.32}$$

since there is no preferred direction built into ϱ_0 . As a consequence we find that only the dissipative contribution survives.

Before continuing the evaluation, we will rewrite the operator $\hat{\phi}(t)$ such that the subtraction in (8.29) can be performed. For that purpose we define the time-smearing function

$$w_+(t) = \frac{1}{\Delta t} \theta(\Delta t - t) \tag{8.33}$$

such that

$$\phi(t) = \int_t^\infty \mathrm{d}t' w_+(t' - t) \varrho(t') \, . \tag{8.34}$$

Note that

$$\int_t^\infty \mathrm{d}t' w_+(t' - t) = 1 \, . \tag{8.35}$$

Let us also define the primitive function of w_+, i.e.

$$W_+(t) = \int_t^\infty \mathrm{d}t' w_+(t') \,. \tag{8.36}$$

These definitions allow us to perform partial integration and then

$$\begin{aligned}\phi(t) &= -W_+(t'-t)\varrho(t')\,|_t^\infty + \int_t^\infty \mathrm{d}t'\, W_+(t'-t)\left(\frac{\partial\varrho}{\partial t'}\right) \\ &= W_+(0)\varrho(t) + \int_0^\infty \mathrm{d}t'\, W_+(t')\left(\frac{\partial\varrho}{\partial t''}\right)_{t''=t'+t} \,.\end{aligned} \tag{8.37}$$

This provides us with the following result

$$\Delta\varrho(t_i) = -\frac{\mathrm{i}}{\hbar}\int_0^\infty \mathrm{d}t\, W_+(t)\exp\left(-\frac{\mathrm{i}}{\hbar}Ht\right)[H,\varrho_0(t_i)]\exp\left(\frac{\mathrm{i}}{\hbar}Ht\right) \,. \tag{8.38}$$

The next task that we are confronted with is to determine the commutator. For that purpose we use the *Kubo formula* [11, 32]

$$\left[A, \mathrm{e}^B\right] = \int_0^1 \mathrm{d}\lambda\, \mathrm{e}^{\lambda B}\,[A,B]\,\mathrm{e}^{(1-\lambda)B} \,. \tag{8.39}$$

Performing a number of steps similar to the ones that we have done before, namely using the microscopic conservation law and going to the long wave length approximation we find

$$\Delta\varrho(t_i) = -\sum_j \beta_j(t_i)\,\nabla_0 \cdot \int_{\Delta V_j} \mathrm{d}^3r' F\left(\boldsymbol{q}(\boldsymbol{r}'), t_i\right) \,, \tag{8.40}$$

where the function F is defined as

$$\begin{aligned}F\left(\boldsymbol{q}(\boldsymbol{r}), t_i\right) = &\int_0^\infty \mathrm{d}t W_+(t) \int_0^1 \mathrm{d}\lambda \\ &Z_0^{-1}(t_i)\,\mathrm{e}^{-\frac{\mathrm{i}}{\hbar}Ht}\mathrm{e}^{-\lambda B}\,\boldsymbol{q}(\boldsymbol{r})\,\mathrm{e}^{-(1-\lambda)B}\mathrm{e}^{\frac{\mathrm{i}}{\hbar}Ht}\end{aligned} \tag{8.41}$$

and the operator B is just the exponent of the local equilibrium density operator

$$B = \sum_j \beta_j(t_i) h_j \,. \tag{8.42}$$

Substitution of these results into the dissipative term of the hydrodynamical equation gives

$$\langle \boldsymbol{q}(\boldsymbol{r})\rangle_\Delta^{t_i} = -\sum_j \beta_j(t_i)\,\nabla_0 \cdot \int_{\Delta V_j} \mathrm{d}^3r'\,\mathrm{Tr}\,\boldsymbol{q}(\boldsymbol{r})\,F\left(\boldsymbol{q}(\boldsymbol{r}'), t_i\right) \,. \tag{8.43}$$

We will yet introduce another approximation. This time we assume the energy-flux operators are not appreciably correlated outside the cell. Thus on the macroscopic scale the correlations are local or like a delta function. So we set

$$\langle \boldsymbol{q}(\boldsymbol{r}) \rangle_{\Delta}^{t} = -\int \mathrm{d}^3 r' \, \beta(\boldsymbol{r}', t) \nabla_0 \, \kappa \, \delta_0(\boldsymbol{r}' - \boldsymbol{r})$$
$$= \kappa \, \nabla_0 \beta(\boldsymbol{r}, t) = -\frac{\kappa}{k_{\mathrm{B}} T^2} \nabla_0 T(\boldsymbol{r}, t) \tag{8.44}$$

and the thermal conductivity is given by

$$\kappa = \frac{1}{3} \mathrm{Tr} \int_0^\infty \mathrm{d}t W_+(t) \int \mathrm{d}^3 r \int_0^1 \mathrm{d}\lambda$$
$$\times \, \boldsymbol{q}(\boldsymbol{0}) \cdot \mathrm{e}^{-\frac{\mathrm{i}}{\hbar} H t} \mathrm{e}^{-\lambda B} \, \boldsymbol{q}(\boldsymbol{r}) \, \mathrm{e}^{-(1-\lambda) B} \mathrm{e}^{\frac{\mathrm{i}}{\hbar} H t} \, . \tag{8.45}$$

We assumed that the tensor κ is isotropic in space such that $\kappa_{ij} = \delta_{ij}\kappa$. So we are now half way in our derivation of Fourier's law, i.e. the right-hand side of this law is constructed.

$$\frac{\delta}{\delta t} h(\boldsymbol{r}, t) = \nabla_0 \cdot \left(-\frac{\kappa}{k_{\mathrm{B}} T^2} \nabla_0 T(\boldsymbol{r}, t) \right) \, . \tag{8.46}$$

We will next derive the left-hand side of this hydrodynamical equation of motion. This time we must focus on the changes in time of β:

$$\frac{\delta}{\delta t} h_j(t) = \frac{1}{\Delta t} \mathrm{Tr} h_j(\boldsymbol{x}) \, \Delta \varrho_0(t_i)$$
$$= \frac{1}{\Delta t} \left[\mathrm{Tr} \, h_j \, Z_0^{-1}(t_{i+1}) \exp\left(-\sum_l \beta_l(t_{i+1}) h_l \right) \right.$$
$$\left. - \mathrm{Tr} \, h_j \, Z_0^{-1}(t_i) \exp\left(-\sum_l \beta_l(t_i) h_l \right) \right] \, . \tag{8.47}$$

In order to evaluate $\Delta \varrho_0(t_i)$, we set

$$\beta_l(t_{i+1}) = \beta_l(t_i) + \Delta\beta_l(t_i) \tag{8.48}$$

and we use the following expansion

$$\exp(A + \Delta B) = \exp A$$
$$+ \int_0^1 \mathrm{d}\lambda \, \exp(\lambda A) \, \Delta B \, \exp(1-\lambda) A + O(\Delta^2) \, . \tag{8.49}$$

This approximation assumes that the temperature varies sufficiently slowly in time on the macroscopic scale such that $\Delta\beta^2 / \Delta t << 1$. The first term gives

$$Z_0^{-1}(t_{i+1}) \exp\left(-\sum_l \beta_l(t_i) h_l \right)$$
$$+ Z_0^{-1}(t_{i+1}) \int_0^1 \mathrm{d}\lambda \exp\left(-\lambda \sum_l \beta_l(t_i) h_l \right)$$
$$\times \left(-\sum_l \Delta\beta_l(t_i) h_l \right) \exp\left(-(1-\lambda) \sum_l \beta_l(t_i) h_l \right) . \tag{8.50}$$

In order to calculate the change in Z_0 we use

$$\text{Tr}\,\exp(A + \Delta B) = \text{Tr}\,\exp(A) + \text{Tr}\,\Delta B\,\exp(A) + O(\Delta^2) \tag{8.51}$$

from which we obtain

$$Z_0^{-1}(t_{i+1}) = Z_0^{-1}(t_{i+1}) - Z_0^{-2}(t_{i+1}) \\ \times \text{Tr}\left(-\sum_l \Delta\beta_l(t_i)h_l\right)\exp\left(-\sum_l \beta_l(t_i)h_l\right). \tag{8.52}$$

Once more we exploited the limited variation in $\Delta\beta$ by using

$$\frac{1}{x+\Delta x} = \frac{1}{x} - \frac{\Delta x}{x^2} + \ldots . \tag{8.53}$$

Putting everything together we find

$$\frac{\delta}{\delta t}h_j(t) = -\sum_l \frac{\Delta\beta_l(t_i)}{\Delta t}\,G(j,l)\,, \tag{8.54}$$

where

$$G(j,l) = \int_0^1 \mathrm{d}\lambda\,\langle \mathrm{e}^{\lambda B}h_j\mathrm{e}^{-\lambda B}h_l\,\rangle_0^{t_i} - \langle h_j\,\rangle_0^{t_i}\,\langle h_l\,\rangle_0^{t_i} \tag{8.55}$$

and B is defined in (8.42). Just as the vector operators for the energy flow , we assume now that the energy-density operators are not appreciably correlated outside the cell volumes. Then

$$G(j,l) = \lambda\,\delta_0^{(3)}(\boldsymbol{r} - \boldsymbol{r}') \tag{8.56}$$

and λ is is energy auto-correlation function defined as

$$\lambda = \int \mathrm{d}^3r \int_0^1 d\lambda\,\left(\langle \mathrm{e}^{\lambda B}\,h(\boldsymbol{r})\,\mathrm{e}^{-\lambda B}\,h_l(\mathbf{0})\rangle_0^{t_i} - \langle h(\boldsymbol{r})\rangle_0^{t_i}\,\langle h(\mathbf{0})\rangle_0^{t_i}\right). \tag{8.57}$$

On the macroscopic scale, we obtain

$$\begin{aligned}\frac{\delta}{\delta t}h_j(t) &= -\lambda\int \mathrm{d}^3r'\,\frac{\partial\beta(\boldsymbol{r}',t)}{\partial t}\delta_0^{(3)}(\boldsymbol{r} - \boldsymbol{r}') \\ &= -\lambda\frac{\partial\beta(\boldsymbol{r},t)}{\partial t} = \frac{\lambda}{k_\mathrm{B}T^2}\frac{\partial}{\partial t}T(\boldsymbol{r},t)\,.\end{aligned} \tag{8.58}$$

Thus summarizing we have arrived at the following result:

$$\frac{\lambda}{k_\mathrm{B}T^2}\frac{\partial}{\partial t}T(\boldsymbol{r},t) = \nabla_0\cdot\left(\frac{\kappa}{k_\mathrm{B}T^2}\,\nabla_0\,T(\boldsymbol{r},t)\right) \tag{8.59}$$

which is Fourier's law upon identifying the quantities $\lambda_0 = \lambda/k_\mathrm{B}T^2$, $\kappa_0 = \kappa/k_\mathrm{B}T^2$ and $\boldsymbol{J}_\mathrm{Q} = -\kappa_0\nabla_0 T(\boldsymbol{r},t)$.

We have derived the Fourier's law in great detail. All the approximative steps have been highlighted. The derivation provides us with microscopic expressions for the transport coefficients. It should be emphasized that the method is suitable for a generalization to derive a full hydrodynamic model that includes flows of particles and energy. The corresponding transport coefficients such as the viscosity will result again as auto-correlation functions. The equations analogous to Fourier's law are the transport equations, that have been presented in Chap. 2.

Fourier's law, as we have derived it, can be applied to describe the relaxation of a boson gas in a container, which is not at thermal equilibrium. It should be emphasized that the potential $V(\boldsymbol{r})$ contains a 'disturbing' part, such that the system is not straightforwardly integrable. In particular, for a constant potential $V(\boldsymbol{r}) = U_0$ in (8.7), the equations of motion can be solved exactly and irreversibility will never set in. This observation will manifests itself in the calculation of the transport coefficients. There is no clear criterion for choosing the time-smearing interval Δt. In fact, the thermal conductivity, κ is proportional to Δt. Since Δt is a quantity representing the time lap to irreversibility, the thermal conductivity κ is infinite for integrable systems. For non-integrable systems of systems with interactions, Δt is chosen such that $\Gamma \Delta t \gg 1$,and Γ is the decay width of the quasi-particles, due to the interactions.

There are several ways to include interactions. One may consider a self-interacting system of particles where the interactions are due to an interaction or scattering potential of the particles. Such potentials are usually used for describing irreversible behavior of gases and fluids. For closed systems this option is the only possibility for inducing irreversible behavior. In closed systems it is rather straightforward to identify the dynamical variables which will be used for setting up the local density operator. In general these observables are the local versions of the globally conserved observables. The latter ones are: the total energy, the total momentum, the total angular momentum and the total charge.

The dynamics of carriers in semiconductors can be described by viewing the system of carriers as an *open* system. The distortion, which is responsible for the irreversible behavior, is now provided by the interaction part of the Hamiltonian due to carrier–phonon scattering, carrier impurity scattering and possibly other interactions (defects, traps). Note that the self interactions are not excluded in open systems and in particular, for high-energetic electrons, the electron–electron scattering is an important mechanism in describing semiconductors. In this case it becomes more subtle to decide which observables will serve as input to the local density operator, because in open systems there will no longer be globally conserved observables. We will start from a non-perturbed lowest order approximation of non-interacting electrons and holes in the conduction and valence bands for defining $\hat{\varrho}_0$.

A complete theory of carrier transport in semiconductors, based on a hydrodynamical approach or long-wavelength approximation also requires a method how to evaluate the transport coefficients. For that purpose we will describe the Matsubara formalism [40]. However, it is instructive to have a closer look at the entropy first.

8.2.3 A Second Glance at Entropy

The entropy was redefined as the expectation value of $-\log \hat{\varrho}_0(t)$ on the hydrodynamical scale, since on the microscopic scale we had no change to obtain an increasing entropy. Since

$$\langle h_i \rangle^{t_n} = \langle h_i \rangle_0^{t_n} \ , \tag{8.60}$$

the second equality in the following equation follows:

$$S(t_n) = -\left\langle \log \varrho_0(t_n) \right\rangle_0^{t_n} = -\left\langle \log \varrho_0(t_n) \right\rangle^{t_n} \ . \tag{8.61}$$

The change of S in a time lap Δt is

$$\begin{aligned}\frac{\delta}{\delta t} S(t) &= -\frac{1}{\Delta t}\left(\left\langle \log \varrho_0(t_{n+1}) \right\rangle^{t_{n+1}} - \left\langle \log \varrho_0(t_n) \right\rangle^{t_n} \right) \\ &= -\frac{1}{\Delta t}\left(\left\langle \log \varrho_0(t_{n+1}) \right\rangle^{t_{n+1}} - \left\langle \log \varrho_0(t_n) \right\rangle^{t_{n+1}} \right. \\ &\qquad \left. + \left\langle \log \varrho_0(t_{n+1}) \right\rangle^{t_{n+1}} - \left\langle \log \varrho_0(t_n) \right\rangle^{t_n} \right) \ . \end{aligned} \tag{8.62}$$

The first difference is

$$-\frac{1}{\Delta t} \mathrm{Tr}\left(\varrho_0(t_{n+1}) \left[\log \varrho_0(t_{n+1}) - \log \varrho_0(t_n) \right] \right) \tag{8.63}$$

and is determined by the change in the Lagrange multiplier β. This part gives

$$\frac{\delta}{\delta t} \log Z_0(t) + \int \mathrm{d}^3 r \frac{\delta}{\delta t} \beta(\boldsymbol{r}, t) \left\langle h(\boldsymbol{r}) \right\rangle_0 = 0 \ . \tag{8.64}$$

The second difference is

$$\begin{aligned} -\frac{1}{\Delta t}\left(\int_{t_n}^{t_{n+1}} \mathrm{d}t \frac{\partial}{\partial t} \left\langle \log \varrho_0(t_n) \right\rangle^t \right) &= -\int \mathrm{d}^3 r \beta(\boldsymbol{r}, t) \nabla \cdot \left\langle \boldsymbol{q}(\boldsymbol{r}) \right\rangle_\Delta \\ &= -\int \mathrm{d}^3 r \nabla \cdot \left(\beta(\boldsymbol{r}, t) \left\langle \boldsymbol{q}(\boldsymbol{r}) \right\rangle_\Delta \right) \\ &\quad + \int \mathrm{d}^3 r \left\langle \boldsymbol{q}(\boldsymbol{r}) \right\rangle_\Delta \cdot \nabla \beta(\boldsymbol{r}, t) \\ &= \int \mathrm{d}^3 r \left(-\nabla_0 \cdot \boldsymbol{J}_\mathrm{S} + \sigma_\mathrm{S} \right) \ . \end{aligned} \tag{8.65}$$

The entropy flux $\boldsymbol{J}_\mathrm{S}$ in the long-wavelength approximation is given by

$$\boldsymbol{J}_\mathrm{S} = \beta(\boldsymbol{r}, t) \left\langle \boldsymbol{q}(\boldsymbol{r}) \right\rangle_\Delta^t \tag{8.66}$$

and the entropy source is

$$\sigma_{\mathrm{S}} = \langle \boldsymbol{q}(\boldsymbol{r}) \rangle_{\Delta}^{t} \cdot \nabla\beta(\boldsymbol{r}, t) \ . \tag{8.67}$$

The rate of entropy change is

$$\dot{S}(t) = \int \mathrm{d}^3 r \, \sigma_{\mathrm{S}}(\boldsymbol{r}, t) \ . \tag{8.68}$$

Using (8.44) we see that

$$\sigma_{\mathrm{S}} = \kappa \, (\nabla\beta)^2 \geq 0 \tag{8.69}$$

which means that the entropy will never decrease.

8.3 Matsubara Functions

The results for the thermal conductivity and for the heat capacitance are special cases of a very general pattern, which is found in a more complete analysis, which also includes, local carrier currents, momentum flow and charge flow. The transport coefficients are auto-correlation functions, which are evaluated using the local-equilibrium background density operator. We can introduce one additional approximation by assuming that the region of space over which the autocorrelation function is evaluated has no appreciable variation in the background temperature. The size of the region corresponds to a cell on the hydrodynamical scale. With this assumption, the evaluation of transport coefficients reduces to the determination of auto-correlation functions at equilibrium. The latter is indicated by the use of a $*$ in the formulas below. The thermal conductivity is

$$\kappa = \frac{1}{k_{\mathrm{B}} T^2} K(J_{\mathrm{Q}}, J_{\mathrm{Q}}) \ , \tag{8.70}$$

where

$$K(A, B) = \int_0^{\infty} \mathrm{d}t W_{+}(t) \int_0^1 \mathrm{d}\lambda \int \mathrm{d}^3 r \, \langle \, A(0) B(\boldsymbol{r}, t + \mathrm{i}\lambda\beta^*) \, \rangle_{\mathrm{con}}^{*} \ , \tag{8.71}$$

where the subscript 'con' reminds us that the contributions proportional to $\langle A \rangle \langle B \rangle$ should be subtracted.[1] The energy current is determined by the form of the Lagrangian density. For the Schrödinger equation we find

$$J_{\mathrm{Q}} = \boldsymbol{q}(\boldsymbol{r}, t) = \frac{\mathrm{i}\hbar}{2m} \left[(\nabla\psi^{\dagger}) H \psi - (H \psi^{\dagger}) \nabla\psi \right] \tag{8.72}$$

and H is the Hamilton operator $H = -\frac{\hbar^2}{2m}\nabla^2 + U$. We also have seen that the energy flux is proportional to the gradient of the thermodynamic variable $\beta = 1/T$, i.e.

[1] 'con' refers to 'connected'. A perturbation series expansion of $K(A, B)$ results into connected diagrams.

$$J_{\mathrm{Q}} = K(J_{\mathrm{Q}}, J_{\mathrm{Q}}) \nabla \beta . \tag{8.73}$$

If more observables are taken into account, then the final result will be of the form

$$J_{\mathrm{A}^i} = \sum_k K(A^i, A^j) \cdot \nabla \theta_{\mathrm{A}^j} , \tag{8.74}$$

where $\nabla \theta_{\mathrm{A}^j}$ are 'driving forces' for the thermodynamic fluxes. The third observation that we made was that the entropy increases according to

$$\frac{\delta S}{\delta t} = \int \mathrm{d}^3 r \, \boldsymbol{J}_{\mathrm{Q}} \cdot \nabla \beta . \tag{8.75}$$

This expression will generalize in the following sense:

$$\frac{\delta S}{\delta t} = \sum_{ij} \int \mathrm{d}^3 r \, \nabla \theta_{\mathrm{A}^i} \, K_{ij} \, \nabla \theta_{\mathrm{A}^j} , \tag{8.76}$$

where K is a positive-definite matrix. It is a subtle piece of work to transform these generic expressions into such a way that they can be identified with the known transport coefficients. In particular, one must separate the ballistic component from the thermodynamic component by boosting to co-moving coordinate frames. Here we just state that this can be done [36, 39].

There is also a subtle interplay going on between the construction of the current operators in the auto-correlation functions and the setup of the thermodynamic variable θ_j. For example, if the number operator participates in the construction of ϱ_0, then the associated Lagrange multiplier $\theta = \mu/T$. Such identifications make it difficult to write down straightforwardly the correct current operators as well as the final transport coefficients since they are linear combinations of the elementary auto-correlation functions.

Whatever, the final structure of the transport coefficients will be, we must be able to evaluate the auto-correlation functions given in (8.71). For that purpose we introduce the *Matsubara function.* Note that the matrix element $\langle \, , \rangle^*$ contains exponential factors with complex argument, i.e.

$$\langle A(0) B(\boldsymbol{r}, t + \mathrm{i}\lambda\beta^*) \rangle^* = \mathrm{Tr}\, A(0) \mathrm{e}^{(-\mathrm{i}t/\hbar - \lambda\beta^*)H} \, B(\boldsymbol{r}) \mathrm{e}^{(\mathrm{i}t/\hbar + \lambda\beta^*)H} . \tag{8.77}$$

It makes sense to combine the argument $(\mathrm{i}t/\hbar + \lambda\beta^*)$ into a single variable z, where z is a complex number. The evaluation of the matrix element can be done by finding an *analytic* expansion for

$$\langle A(0) B(\boldsymbol{r}, t + \mathrm{i}\lambda\beta^*) \rangle^* = \mathrm{Tr}\, A(0) \mathrm{e}^{-zH} \, B(\boldsymbol{r}) \mathrm{e}^{zH} \tag{8.78}$$

and substitute at the end of the calculation the value $z = -\mathrm{i}t/\hbar - \lambda\beta^*$. Perturbation theory allows us to determine the analytic structure of the matrix element, under the restriction that the result is approximately correct, due to the perturbative nature of the method. The t- and λ-integration in equation (8.71) require that the ordering of the operators is respected. The correct ordering should not be spoiled by the analytic continuation. The ordering becomes simple if we choose some line in the complex plane on which we define the ordering and evaluate the result along this line and finally perform the continuation. The Matsubara formalism selects the line $z = -\tau$ being real, i.e. time is considered as a complex temperature. The Matsubara function corresponding to the matrix element (8.78) is

$$G_{AB}^{\mathrm{Mat}}(\tau_x - \tau_y) = -\left\langle T\, A(\tau_x)\, B(\tau_y)\right\rangle^* . \tag{8.79}$$

The minus sign is conventional and T means that τ- ordered products must be taken. The operators are defined as $O(\tau) = \mathrm{e}^{\tau H} O \mathrm{e}^{-\tau H}$. The range of τ is restricted to the values: $-\beta^* \leq \tau \leq \beta^*$. This is because we must also include the possibility to evaluate

$$G_{AB}^{\mathrm{Mat}}(0 - \tau) = -\left\langle T\, A(0)\, B(-\tau)\right\rangle^* \tag{8.80}$$

and $B(-\tau) = \mathrm{e}^{-\tau H} O \mathrm{e}^{\tau H}$. The Matsubara function can be formally evaluated by inserting a complete set of eigenstates $H\,|n\rangle = E_n\,|n\rangle$. The resulting expression is easily shown to be periodic in the interval $[0, \beta^*]$ for bosons and anti-periodic for fermions, i.e. for $-\beta^* \leq \tau \leq 0$ we have

$$\begin{aligned} G_{AB}^{\mathrm{Mat}}(\tau) &= G_{AB}^{\mathrm{Mat}}(\tau + \beta^*)\,, && \text{bosons} \\ G_{AB}^{\mathrm{Mat}}(\tau) &= -G_{AB}^{\mathrm{Mat}}(\tau + \beta^*)\,, && \text{fermions.} \end{aligned} \tag{8.81}$$

This property enables a discrete Fourier expansion to be made

$$f(\tau) = \frac{1}{\beta^*} \sum_{l=-\infty}^{\infty} \tilde{f}(\mathrm{i}\omega_l) \mathrm{e}^{-\mathrm{i}\omega_l \tau} \tag{8.82}$$

$$\tilde{f}(\mathrm{i}\omega_l) = \int_0^{\beta^*} \mathrm{d}\tau\, f(\tau) \mathrm{e}^{\mathrm{i}\omega_l \tau} . \tag{8.83}$$

The *spectral representation* $S_{AB}(\omega)$ of the Matsubara function is defined as

$$\tilde{G}_{AB}^{\mathrm{Mat}}(\mathrm{i}\omega_l) = \int_{-\infty}^{\infty} \mathrm{d}\omega \frac{S_{AB}(\omega)}{\mathrm{i}\omega_l - \omega}\,, \qquad \omega_l = \frac{2\pi\, l}{\beta^*} . \tag{8.84}$$

It should be noted that the spectral function has the property that $S_{AB}(\omega)^* = -S_{AB}(-\omega)$. Plugging in a complete set of energy eigenstates directly into the kernel of equation (8.71), and comparing the result with the expression for the spectral function we obtain

$$K(A, B) = \frac{2}{\beta^*} \int_0^{\beta^*} \mathrm{d}\omega \left(\frac{\Re[S_{AB}(\omega)]}{\omega} \Re[\tilde{W}(\omega)] - \frac{\Im[S_{AB}(\omega)]}{\omega} \Im[\tilde{W}(\omega)] \right) , \tag{8.85}$$

where $\Re$ and $\Im$ stand for the real and imaginary parts. $\tilde{W}(\omega)$ is the Fourier transform of the integrated smearing function $W(t)$, namely

$$\tilde{W}(\omega) = -\frac{1}{\mathrm{i}\omega}\left(1 + \frac{1 - \mathrm{e}^{\mathrm{i}\omega s}}{\mathrm{i}\omega s}\right) . \tag{8.86}$$

If we had chosen the exponential smearing function of Zubarev [41, 42]

$$W(t) = \mathrm{e}^{-\varepsilon t} , \tag{8.87}$$

the Fourier transform would be

$$\tilde{W}(\omega) = \frac{1}{\varepsilon - \mathrm{i}\omega} . \tag{8.88}$$

Equation (8.85) is the sought for connection between the kernel and the Matsubara function. It should be noted that for most cases under consideration, in particular for auto-correlation functions, the Matsubara function is real and therefore the the second term in (8.85) is absent.

For s large or ε small, we consider the case where the details of the smearing function become irrelevant and $\tilde{W}(\omega) \to \pi\delta(\omega) + \mathrm{i}/\omega$. As a consequence

$$K(A, B) = \frac{\pi}{\beta^*} \lim_{\omega \to 0} \frac{\Re[S_{AB}(\omega)]}{\omega} . \tag{8.89}$$

In this equation we have no reference anymore to the use of advanced or retarded smearings. Various linear response methods are incorporated into this result. For instance, Zubarev's method is derived from treating the Liouville equation analogous to a scattering problem. His method builds in irreversibility by an infinitesimal term which is not time-reversal invariant. This term leads to transport coefficients with $W(t) = \mathrm{e}^{-\varepsilon t}$. Mori's theory performs advanced smearing. Now we have seen that all these details do not incorporate new physics. The static transport coefficients require some time averaging, of which the details are not visible anymore in the final result.

Exercise

8.1. Prove equation (8.39).

9. Wigner Distribution Functions

A popular method to include quantum effects into the calculation of the characteristics of submicron devices is based on the Wigner functions. The underlying idea is that quantum transport can be catched into generalized transport equations that are in the spirit of the Boltzmann transport equation, suitable extended with terms that represent quantum corrections.
In order to derive the so-called 'quantum-hydrodynamical' equations, it is convenient to consider the *Wigner distribution* functions. Wigner [43] constructed 'distribution' functions $f(\boldsymbol{x}, \boldsymbol{p})$ with the remarkable property that

$$P(\boldsymbol{x}) = \int \mathrm{d}^3 p \; f(\boldsymbol{x}, \boldsymbol{p}) \qquad \text{and} \qquad P(\boldsymbol{p}) = \int \mathrm{d}^3 x \; f(\boldsymbol{x}, \boldsymbol{p}) \, , \tag{9.1}$$

where $P(\boldsymbol{x})$ is the probability to find the particle at position $\boldsymbol{x}$ and $P(\boldsymbol{p})$ is the probability that the particle has momentum $P(\boldsymbol{p})$. Strictly speaking $f(\boldsymbol{x}, \boldsymbol{p})$ is not a distribution function since it does not represent a probability density. The function may take negative and complex values. Nevertheless, the equations (9.1) show a striking common feature with the Boltzmann distribution function $f(\boldsymbol{x}, \boldsymbol{p}, t)$. To construct the Wigner distribution function, i.e. the *quantum-mechanical* function $f(\boldsymbol{x}, \boldsymbol{p})$ that produces the probability densities $P(\boldsymbol{x})$ and $P(\boldsymbol{p})$, we rewrite the latter by using the density operator ϱ. Single-particle quantum mechanics implies that the probability of finding a particle at the position $\boldsymbol{x}$, while the particle is in the state $|\psi\rangle$, is given by

$$P(\boldsymbol{x}) = |\psi(\boldsymbol{x})|^2 = \psi^*(\boldsymbol{x})\psi(\boldsymbol{x}) = \langle \boldsymbol{x}|\psi\rangle \langle \psi|\boldsymbol{x}\rangle \; . \tag{9.2}$$

The 'matrix element' $\langle \boldsymbol{x}|\psi\rangle \langle \psi|\boldsymbol{x}\rangle$ is the diagonal term of the (continuous) matrix representation of the operator $|\psi\rangle \langle \psi|$ in the position representation. The generic matrix element of this operator in the position is $\langle \boldsymbol{x}|\psi\rangle \langle \psi|\boldsymbol{y}\rangle = \psi^*(\boldsymbol{y})\psi(\boldsymbol{x})$. If only statistical knowledge of the particle is available, then the single-particle state $|\psi\rangle$ must be replaced by an incoherent sum according to (7.28), i.e.

$$|\psi\rangle \langle \psi| \longrightarrow \sum_n p_n \; |\Phi\rangle \langle \Phi| \; = \; \varrho \, . \tag{9.3}$$

Therefore, the probability to find the particle at position $\boldsymbol{x}$ is

$$P(\boldsymbol{x}) = \sum_n p_n \langle \boldsymbol{x}|\Phi\rangle \langle\Phi|\boldsymbol{x}\rangle = \langle \boldsymbol{x}|\varrho|\boldsymbol{x}\rangle = \varrho(\boldsymbol{x},\boldsymbol{x}) . \tag{9.4}$$

In general, we have that

$$\varrho(\boldsymbol{x},\boldsymbol{y}) = \sum_n \mathrm{e}^{-\beta E_n} \langle \boldsymbol{y}|\Phi\rangle \langle\Phi|\boldsymbol{x}\rangle . \tag{9.5}$$

We may also apply the momentum representation. In a completely analogous manner we then obtain

$$P(\boldsymbol{p}) = \varrho(\boldsymbol{p},\boldsymbol{p}) \tag{9.6}$$

and in general

$$\varrho(\boldsymbol{p},\boldsymbol{q}) = \sum_n \mathrm{e}^{-\beta E_n} \langle \boldsymbol{q}|\Phi_n\rangle \langle\Phi_n|\boldsymbol{p}\rangle . \tag{9.7}$$

Since $P(\boldsymbol{x})$ is extracted from the density operator in the momentum representation and since $P(\boldsymbol{x})$ is extracted from the density operator in the position representation, one may guess that the function with the desired properties of (9.1), i.e. the Wigner function is related to the density operator in a 'mixed' representation: $\varrho \longrightarrow \langle \boldsymbol{p}|\varrho|\boldsymbol{x}\rangle \simeq f(\boldsymbol{x},\boldsymbol{p})$. This is indeed the case although a transformed set of coordinates is exploited. Starting from the momentum-probability distribution, we find

$$P(\boldsymbol{p}) = \varrho(\boldsymbol{p},\boldsymbol{p}) = \int \mathrm{d}^3x\mathrm{d}^3y\ \varrho(\boldsymbol{x},\boldsymbol{y})\ \exp\left(\frac{\mathrm{i}}{\hbar}\boldsymbol{p}\cdot(\boldsymbol{x}-\boldsymbol{y})\right) . \tag{9.8}$$

Setting $\boldsymbol{X} = (\boldsymbol{x}+\boldsymbol{y})/2$ and $\boldsymbol{\xi} = \boldsymbol{x}-\boldsymbol{y}$, we obtain that $\boldsymbol{x} = \boldsymbol{X}+\boldsymbol{\xi}/2$ and $\boldsymbol{y} = \boldsymbol{X}-\boldsymbol{\xi}/2$. Therefore,

$$\begin{aligned} P(\boldsymbol{p}) &= \int \mathrm{d}^3X\mathrm{d}^3\xi\ \varrho(\boldsymbol{X}+\boldsymbol{\xi}/2,\ \boldsymbol{X}-\boldsymbol{\xi}/2)\ \mathrm{e}^{-\frac{\mathrm{i}}{\hbar}\boldsymbol{p}\cdot\boldsymbol{\xi}} \\ &= \int \mathrm{d}^3X\ f(\boldsymbol{p},\boldsymbol{X}) , \end{aligned} \tag{9.9}$$

where

$$f(\boldsymbol{p},\boldsymbol{X}) = \int \mathrm{d}^3\xi\ \varrho(\boldsymbol{X}+\boldsymbol{\xi}/2,\ \boldsymbol{X}-\boldsymbol{\xi}/2)\ \mathrm{e}^{-\frac{\mathrm{i}}{\hbar}\boldsymbol{p}\cdot\boldsymbol{\xi}} . \tag{9.10}$$

This expression has the appearance of a mixed representation, although 'center-of-mass' coordinates are required.

We must now show that (9.10) for the Wigner function satisfies both relations of (9.1). Therefore we evaluate

$$\begin{aligned} \int \mathrm{d}^3p\ f(\boldsymbol{p},\boldsymbol{X}) &= \int \mathrm{d}^3\xi\mathrm{d}^3p\ \varrho(\boldsymbol{X}+\boldsymbol{\xi}/2,\ \boldsymbol{X}-\boldsymbol{\xi}/2)\ \mathrm{e}^{\frac{\mathrm{i}}{\hbar}\boldsymbol{p}\cdot\boldsymbol{\xi}} \\ &= \int \mathrm{d}^3\xi\ \delta(\boldsymbol{\xi})\ 2\pi\hbar\ \varrho(\boldsymbol{X}+\boldsymbol{\xi}/2,\ \boldsymbol{X}-\boldsymbol{\xi}/2) \\ &= 2\pi\hbar\ \varrho(\boldsymbol{X},\boldsymbol{X}) = 2\pi\hbar\ P(\boldsymbol{X}) . \end{aligned} \tag{9.11}$$

So we almost arrived at the desired result. The factor $2\pi\hbar$ can be easily absorbed into the defining relations (9.1). From now on we require that

$$P(\boldsymbol{x}) = \int \frac{\mathrm{d}^3 p}{2\pi\hbar} \, f(\boldsymbol{x}, \boldsymbol{p}) \quad \text{and} \quad P(\boldsymbol{p}) = \int \mathrm{d}^3 x \, f(\boldsymbol{x}, \boldsymbol{p}) \, . \tag{9.12}$$

such that the Wigner function satisfies the construction requirements. It should be emphasized that the Wigner function contains the same information as the density function. Given ϱ, we can construct f (as we have done), and given f, we can construct ϱ:

$$\begin{aligned} &\int \frac{\mathrm{d}^3 p}{2\pi\hbar} f\left(\boldsymbol{p}, \frac{\boldsymbol{x}+\boldsymbol{y}}{2}\right) \, \mathrm{e}^{\frac{\mathrm{i}}{\hbar}\boldsymbol{p}\cdot(\boldsymbol{x}-\boldsymbol{y})} \\ &\quad = \int \frac{\mathrm{d}^3 p}{2\pi\hbar} \int \mathrm{d}^3\xi \; \varrho(\boldsymbol{X}+\boldsymbol{\xi}/2, \; \boldsymbol{X}-\boldsymbol{\xi}/2) \, \mathrm{e}^{\frac{\mathrm{i}}{\hbar}\boldsymbol{p}\cdot(\boldsymbol{x}-\boldsymbol{y}) - \frac{\mathrm{i}}{\hbar}\boldsymbol{p}\cdot\boldsymbol{\xi}} \\ &\quad = \int \mathrm{d}^3\xi \; \varrho(\boldsymbol{X}+\boldsymbol{\xi}/2, \; \boldsymbol{X}-\boldsymbol{\xi}/2) \, \delta\left(\boldsymbol{\xi} - (\boldsymbol{x}-\boldsymbol{y})\right) = \varrho(\boldsymbol{x}, \boldsymbol{y}) \, . \end{aligned} \tag{9.13}$$

So far, we have been concerned with a single particle. The formalism can be straightforwardly extended for an arbitrary number of particles. In general, the density matrix takes the following form:

$$\begin{aligned} &\varrho(\boldsymbol{x}_1, \boldsymbol{x}_2, \ldots \boldsymbol{x}_N; \boldsymbol{y}_1, \boldsymbol{y}_2, \ldots \boldsymbol{y}_N) \\ &\quad = \sum_n \mathrm{e}^{-\beta E_n} \, \langle \boldsymbol{y}_1, \boldsymbol{y}_2, \ldots \boldsymbol{y}_N | \Phi \rangle \, \langle \Phi | \boldsymbol{x}_1, \boldsymbol{x}_2, \ldots \boldsymbol{x}_N \rangle \, . \end{aligned} \tag{9.14}$$

For each degree of freedom a 'center-of-mass' coordinate $\boldsymbol{X}_i = (\boldsymbol{x}_i + \boldsymbol{y}_i)/2$ is defined as well as a 'relative' coordinate $\boldsymbol{\xi}_i = \boldsymbol{x}_i - \boldsymbol{y}_i$.

There are various approaches to incorporate the time evolution into the Wigner functions. The methods differ in the intermediate steps. For instance, Hänsch [44] exploits non-equilibrium Green functions for the construction of Wigner–Boltzmann equation. At the heart of time evolution is of course the Schrödinger equation and this equation is exploited in one way or another to obtain the time evolution. Here we exploit the Liouville equation that for our purposes must now be formulated in 'center-of-mass' and 'relative' coordinates. The Liouville equation in the coordinate representation reads

$$\begin{aligned} \mathrm{i}\hbar\frac{\partial}{\partial t}\varrho(\boldsymbol{x}, \boldsymbol{y}) &= \langle \boldsymbol{x} | \, H\varrho - \varrho H \, | \boldsymbol{y} \rangle \\ &= \int \mathrm{d}^3 z \, \left(\langle \boldsymbol{x} | H | \boldsymbol{z} \rangle \, \langle \boldsymbol{z} | \varrho | \boldsymbol{y} \rangle \; - \; \langle \boldsymbol{x} | H | \boldsymbol{z} \rangle \, \langle \boldsymbol{z} | \varrho | \boldsymbol{y} \rangle \right) \, . \end{aligned} \tag{9.15}$$

Therefore, one finds that

$$\mathrm{i}\hbar\frac{\partial}{\partial t}\varrho(\boldsymbol{x}, \boldsymbol{y}) = H(\nabla_{\boldsymbol{x}}, \boldsymbol{x})\varrho(\boldsymbol{x}, \boldsymbol{y}) \; - \; H(\nabla_{\boldsymbol{y}}, \boldsymbol{y})\varrho(\boldsymbol{x}, \boldsymbol{y}) \tag{9.16}$$

After the transformation to the new coordinates, (9.15) becomes

$$\begin{aligned} i\hbar\frac{\partial}{\partial t}\varrho(\,\boldsymbol{X}+\boldsymbol{\xi},\,\boldsymbol{X}-\boldsymbol{\xi}\,) = &\; H\left(\nabla_{\boldsymbol{X}}+\frac{1}{2}\nabla_{\boldsymbol{\xi}}\,,\boldsymbol{X}+\frac{1}{2}\boldsymbol{\xi}\right)\,\varrho(\boldsymbol{X}+\boldsymbol{\xi},\,\boldsymbol{X}-\boldsymbol{\xi}) \\ & -H\left(\nabla_{\boldsymbol{X}}-\frac{1}{2}\nabla_{\boldsymbol{\xi}}\,,\boldsymbol{X}-\frac{1}{2}\boldsymbol{\xi}\right)\,\varrho(\boldsymbol{X}+\boldsymbol{\xi},\,\boldsymbol{X}-\boldsymbol{\xi})\,. \end{aligned} \tag{9.17}$$

Taking the Fourier transform with respect to the relative coordinate $\boldsymbol{\xi}$ results into the time-evolution equation for the Wigner function

$$\begin{aligned} i\hbar\frac{\partial}{\partial t}f(\,\boldsymbol{p},\boldsymbol{X}\,) = &\; H\left(\nabla_{\boldsymbol{X}}+\frac{i\boldsymbol{p}}{2\hbar}\,,\boldsymbol{X}+\frac{i\hbar}{2}\nabla_{\boldsymbol{p}}\right)\,f(\boldsymbol{p},\boldsymbol{X}) \\ & -H\left(\nabla_{\boldsymbol{X}}-\frac{i\boldsymbol{p}}{2\hbar}\,,\boldsymbol{X}-\frac{i\hbar}{2}\nabla_{\boldsymbol{p}}\right)\,f(\boldsymbol{p},\boldsymbol{X})\,. \end{aligned} \tag{9.18}$$

Equation (9.18) is the Wigner equation or Wigner–Boltzmann equation. The correspondence with the Boltzmann equation becomes manifest after expansion of the Hamiltonian. The ordering of the operators is non-trivial because quantum mechanics is sensitive to the order of the position and momentum operators. However, once an explicit form for the Hamiltonian is selected, progress can be made. The potential-energy term in the Hamiltonian, U, is usually approximated with a Taylor series. Also the method of moments may be applied to derive quantum-mechanically corrected transport equations ('quantum hydrodynamics') [45]. In order to gain some further insight, we consider the following Hamiltonian

$$H = -\sum_{n=1}^{N}\frac{\hbar^2}{2m}\nabla_n^2 + \sum_{n=1}^{N}U_1(\boldsymbol{r}_n) + \sum_{n<m}^{N}U_2(\boldsymbol{r}_n,\boldsymbol{r}_m)\,. \tag{9.19}$$

Then the Wigner equation takes the form

$$i\hbar\left(\frac{\partial}{\partial t}+\frac{\boldsymbol{p}}{m}\cdot\nabla_{\boldsymbol{X}}-\nabla_{\boldsymbol{X}}U\cdot\nabla_{\boldsymbol{p}}+\ldots\right)f(\boldsymbol{p},\boldsymbol{X},t) = i\hbar\frac{\partial f}{\partial t}[U_2]_C\,, \tag{9.20}$$

where the dots represent higher-order terms of the Taylor expansion of U_1 and are of $O(\hbar)$ and higher. The right-hand side describes a collision contribution that depends in this case on inter-particle scattering. Of particular interest is the density-gradient approximation [46] that results by retaining the lowest-order term of the sum in (9.20). Under a sufficient series of assumptions the quantum corrections to the Boltzmann transport equation can be casted in a form such that electron flux in the corresponding drift-diffusion model

$$\boldsymbol{J}_n = D_n\nabla n + \mu_n n\boldsymbol{E} - 2\mu_n b_n\nabla\left(\frac{\nabla^2\sqrt{n}}{\sqrt{n}}\right) \tag{9.21}$$

and $b_n = \hbar^2/(4dqm_n^*)$. The parameter d represents the dimensionality of the system, i.e. $d = 2$ for strongly confined carriers. In practice, d is a parameter that may be used to fit the quantum-corrected drift-diffusion model and therefore, d may be non-integer. The density gradient model is exploited for the construction of quantum-mechanically corrected device simulation programs [47].

10. Balance Equations

In this chapter a set of quantum mechanical balance equations will be derived, which can be further used as a starting point in order to establishing a workable scheme for concrete transport calculations. Contrary to earlier balance equation approaches [48, 49, 50, 51], no preliminary assumptions will be made regarding the nature of the external electromagnetic field supplying the driving force that is able to maintain the current flow. In particular, it is not necessary to restrict the derivation to uniform fields or fields that are slowly varying in space, which makes the formalism capable of studying – at least in principle – also charge transport and energy dissipation in mesoscopic structures and other layered structured where electric potentials may considerably fluctuate on a small distance scale. For the sake of clarity, all operators that are associated to physical variables (Hamiltonian, momentum, current, etc.) will be denoted by a hat, ˆ on top of the defining symbols throughout this chapter.

10.1 Basic Assumptions

1. The physical model that will be proposed to study the electric current flowing in a closed circuit, is an electron gas, confined in a multiply connected region Ω in three-dimensional configuration space. In order to simplify the algebra, we will explicitly assume that Ω contains only one 'hole'. In other words, the electric circuit takes the form of a torus (see Fig. 10.1) and the discussion is restricted to two-terminal devices connected to an external power supply.

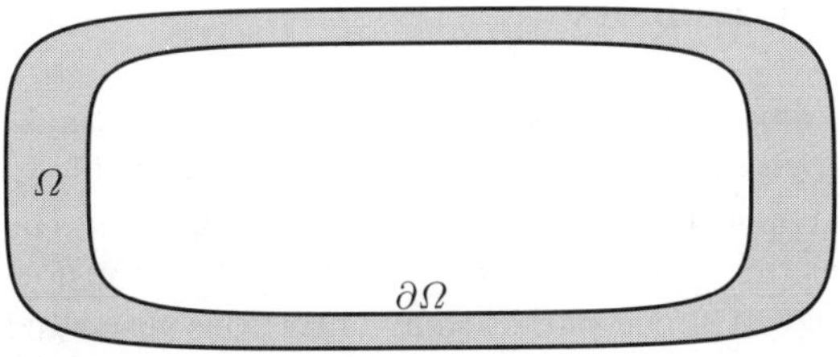

Fig. 10.1. Circuit topology

2. The outer surface $\partial\Omega$ bounding the circuit region Ω acts as an infinite potential well: electrons can therefore never reside on $\partial\Omega$ and all wave functions as well as the electron field operator ψ are bound to vanish on $\partial\Omega$:

$$\psi(\boldsymbol{r},t) = 0 \qquad \forall t,\ \forall \boldsymbol{r} \text{ on } \partial\Omega\,. \tag{10.1}$$

3. The electron gas is enabled to extract electrical power from the electromagnetic field. The total electromagnetic field governing the dynamics of the confined electron gas generally consists of an externally applied field and an induced field generated by the electric current and charge distributions. The basic variables describing the electromagnetic field are taken to be the scalar electric potential $V(\boldsymbol{r},t)$ and the vector potential $\boldsymbol{A}(\boldsymbol{r},t)$.
4. The electron gas is further interacting with other quantized or unquantized fields representing the scattering agents that are responsible for all decoherence and dissipation effects other than those caused by the interaction with a radiation field. We explicitly assume that the corresponding interaction Hamiltonian representing these scattering processes commutes with the electron density operator $\hat{\varrho} = -e\psi^\dagger\psi$.

The last assumption should not be seen as a restriction since various interactions, such as electron–phonon scattering are covered by a generic interaction of the form

$$\hat{H}' = \int_\Omega \mathrm{d}\tau\, \hat{\varrho}\, \chi\,, \tag{10.2}$$

where χ is a classical field or a quantized field commuting with $\hat{\varrho}$ so that $[\hat{\varrho}, \hat{H}'] = 0$ is an automatically satisfied.
The electric and magnetic field vectors characterizing the total electromagnetic field are denoted by $\boldsymbol{E}$ and $\boldsymbol{B}$ and may be extracted from the scalar potential V and the vector potential $\boldsymbol{A}$ according to the usual prescription:

$$\boldsymbol{E} = -\nabla V - \frac{\partial \boldsymbol{A}}{\partial t} \tag{10.3}$$

$$\boldsymbol{B} = \nabla \times \boldsymbol{A}\,. \tag{10.4}$$

10.2 Charge and Current Density

Electrical charge and current density are measurable quantities and must therefore not depend on a particular gauge of the electromagnetic field. In particular, the second-quantized operators defined in Chap. 6 to represent the electron charge and current densities must be invariant under gauge transformations. In analogy with the case of single-particle wave functions discussed in Chap. 5, we may infer that a gauge transformation

$$\begin{aligned} \boldsymbol{A}' &= \boldsymbol{A} + \nabla\theta \\ V' &= V - \frac{\partial\theta}{\partial t}\,, \end{aligned} \tag{10.5}$$

θ being an arbitrary function of $\boldsymbol{r}$ and t, gives rise to the following transformation of the electron field operators:

$$\psi' = \exp\left(-\frac{\mathrm{i}e\theta}{\hbar}\right)\psi; \quad \psi'^{\dagger} = \exp\left(\frac{\mathrm{i}e\theta}{\hbar}\right)\psi^{\dagger}\,. \tag{10.6}$$

Applying this transformation to the charge density operator, we obtain full gauge invariance:

$$\begin{aligned} \hat{\varrho}' &= -e\,\psi'^{\dagger}\psi' = -e\exp\left(\frac{\mathrm{i}e\theta}{\hbar}\right)\psi^{\dagger}\exp\left(-\frac{\mathrm{i}e\theta}{\hbar}\right)\psi \\ &= -e\,\psi^{\dagger}\psi = \hat{\varrho}\,. \end{aligned} \tag{10.7}$$

It is left as an exercise to show that gauge invariance is established also for the current density operator

$$\hat{\boldsymbol{J}} = -\frac{e}{2}\left[\psi^{\dagger}\,\boldsymbol{v}\psi + \left(\boldsymbol{v}^{*}\psi^{\dagger}\right)\psi\right] \tag{10.8}$$

that was introduced in Chap. 6.

10.3 Total Hamiltonian

The total Hamiltonian of the electric circuit can be written as follows:

$$\hat{H} = \hat{H}_{\mathrm{E}} + \hat{H}_{\mathrm{P}} + \hat{H}'\,. \tag{10.9}$$

$\hat{H}_{\mathrm{E}}$ describes the electron energy including the scalar and vector potential while $\hat{H}_{\mathrm{P}}$ and $\hat{H}'$ describe the free scattering agents and their interactions with the electron gas respectively. The precise form of $\hat{H}_{\mathrm{P}}$ and $\hat{H}'$ is left unspecified for the time being. On the other hand, $\hat{H}_{\mathrm{E}}$ can generally be written as follows:

$$\hat{H}_{\mathrm{E}} = \int_{\Omega}\mathrm{d}\tau\left[\frac{1}{2}m\left(\boldsymbol{v}^{*}\psi^{\dagger}\right)\cdot\boldsymbol{v}\psi + (U - eV)\,\psi^{\dagger}\psi\right]. \tag{10.10}$$

U describes any internal potential energy including built-in potentials and all kinds of potential wells, potential barriers etc.

10.4 Basic Equations of Motion

In the Heisenberg representation, the equations of motion satisfied by the field operators constitute a basic set of equations from which all dynamical equations may be derived. The electron field operators are evolving according to

$$i\hbar \frac{\partial \psi(\boldsymbol{r},t)}{\partial t} = [\psi(\boldsymbol{r},t),\, H] \tag{10.11}$$

$$i\hbar \frac{\partial \psi^\dagger(\boldsymbol{r},t)}{\partial t} = \left[\psi^\dagger(\boldsymbol{r},t),\, H\right] . \tag{10.12}$$

For the sake of notational simplicity, the explicit dependence of all quantities upon $\boldsymbol{r}$ and t will be dropped whenever possible. Evaluating the first two commutators in eq. (10.12) and (10.12), we arrive at the following basic equations:

$$i\hbar \frac{\partial \psi}{\partial t} = H_1 \psi \; - \; eV\, \psi + \left[\, \psi,\, \hat{H}' \,\right] \tag{10.13}$$

$$i\hbar \frac{\partial \psi^\dagger}{\partial t} = -H_1^* \psi^\dagger \, + eV\, \psi^\dagger + \left[\, \psi^\dagger,\, \hat{H}' \,\right] , \tag{10.14}$$

where H_1 is a short-hand notation for the single-particle Hamiltonian in the coordinate representation,

$$H_1 \equiv \frac{1}{2} m v^2 + U . \tag{10.15}$$

10.5 Continuity Equation

The well-known continuity equation can be shown to hold even on the operator level:

$$\frac{\partial \hat{\varrho}}{\partial t} + \nabla \cdot \hat{\boldsymbol{J}} = 0 .$$

The proof is straightforward. From (10.14) and (10.14) it follows respectively:

$$i\hbar\, \psi^\dagger \frac{\partial \psi}{\partial t} = \psi^\dagger H_1 \psi \; - \; eV\, \psi^\dagger \psi \, + \, \psi^\dagger \left[\psi,\, \hat{H}'\right] \tag{10.16}$$

$$i\hbar \frac{\partial \psi^\dagger}{\partial t} \psi = -\left(H_1^* \psi^\dagger\right) \psi \, + \, eV\, \psi^\dagger \psi \, + \, \left[\psi^\dagger,\, \hat{H}'\right] \psi . \tag{10.17}$$

Adding both sides of (10.17) and (10.17) we obtain

$$\begin{aligned} i\hbar \frac{\partial \hat{\varrho}}{\partial t} &= -e \left[\psi^\dagger H_1 \psi - \left(H_1^* \psi^\dagger\right) \psi\right] + [\hat{\varrho},\, \hat{H}'] \\ &= -\frac{em}{2} \left[\psi^\dagger v^2 \psi - \left(v^{*2} \psi^\dagger\right) \psi\right] + [\hat{\varrho},\, \hat{H}'] \\ &= -\frac{em}{2} \left[\psi^\dagger v^2 \psi - \left(v^{*2} \psi^\dagger\right) \psi\right] , \end{aligned} \tag{10.18}$$

where the latter equality follows from the fourth basic assumption.

On the other hand, the divergence of $\hat{\boldsymbol{J}}$ turns out to be

$$\begin{aligned} \nabla \cdot \hat{\boldsymbol{J}} = -\frac{e}{2} \Big[&\nabla \psi^\dagger \cdot \boldsymbol{v} \psi + \psi^\dagger \nabla \cdot \boldsymbol{v} \psi \\ &+ \left(\nabla \cdot \boldsymbol{v}^* \psi^\dagger\right) \psi + \left(\boldsymbol{v}^* \psi^\dagger\right) \cdot \nabla \psi \Big] . \end{aligned} \tag{10.19}$$

The easiest way to proceed is to make an appropriate choice between the two ways of representing ∇ in terms of $\boldsymbol{v}$ or $\boldsymbol{v}^*$:

$$\nabla = \frac{\mathrm{i}}{\hbar}(m\boldsymbol{v} - e\boldsymbol{A}) \tag{10.20}$$

$$\nabla = -\frac{\mathrm{i}}{\hbar}(m\boldsymbol{v}^* - e\boldsymbol{A}) \,. \tag{10.21}$$

Substituting (10.20) into the second and fourth term of the right-hand side of (10.19) and (10.21) into the remaining terms, we arrive at

$$\begin{aligned}\nabla \cdot \hat{\boldsymbol{J}} &= -\frac{\mathrm{i}e}{2\hbar}\left\{-\left[(m\boldsymbol{v}^* - e\boldsymbol{A})\psi^\dagger\right]\cdot\boldsymbol{v}\psi + \psi^\dagger\left[(m\boldsymbol{v} - e\boldsymbol{A})\cdot\boldsymbol{v}\psi\right]\right.\\ &\quad \left. -\left[(m\boldsymbol{v}^* - e\boldsymbol{A})\cdot\boldsymbol{v}^*\psi^\dagger\right]\psi + (\boldsymbol{v}^*\psi^\dagger)\cdot(m\boldsymbol{v} - e\boldsymbol{A})\psi\right\}\\ &= -\frac{\mathrm{i}em}{2\hbar}\left[\psi^\dagger v^2\psi - \left(v^{*2}\psi^\dagger\right)\psi\right] \,.\end{aligned} \tag{10.22}$$

From (10.18) and (10.22) we conclude:

$$\mathrm{i}\hbar\frac{\partial\hat{\varrho}}{\partial t} = -\frac{em}{2}\frac{2\mathrm{i}\hbar}{em}\nabla\cdot\hat{\boldsymbol{J}}$$

or finally,

$$\frac{\partial\hat{\varrho}}{\partial t} = -\nabla\cdot\hat{\boldsymbol{J}} \,. \tag{10.23}$$

□

10.6 Energy Balance Equation

Introducing the electron energy density operator W as

$$W = \frac{1}{2}m\boldsymbol{v}^*\psi^\dagger\cdot\boldsymbol{v}\psi + U\psi^\dagger\psi \,, \tag{10.24}$$

we may expect to derive an energy balance equation of the form:

$$\frac{\partial W}{\partial t} = -\nabla\cdot\boldsymbol{K} + \boldsymbol{E}\cdot\hat{\boldsymbol{J}} - \frac{\mathrm{i}}{\hbar}\left[W, \hat{H}'\right] \,. \tag{10.25}$$

The first term of the right-hand side of eq. (10.25) represents the divergence of some energy flux vector $\boldsymbol{K}$ whereas the second and the third terms respectively refer to the power gained from the field and released to the environment.

Proceeding in the same way as in the previous section, we first calculate the time derivative of W starting from the basic equation (10.14). If we let act the single-particle operator $\boldsymbol{v}$ onto both sides of (10.14), we obtain:

$$\mathrm{i}\hbar\,\boldsymbol{v}\frac{\partial\psi}{\partial t} = \boldsymbol{v}H_1\psi - e\boldsymbol{v}V\,\psi + \left[\boldsymbol{v}\psi, \hat{H}'\right] \,. \tag{10.26}$$

Since the vector potential may be explicitly depend on time, we cannot just commute $\boldsymbol{v}$ with $\frac{\partial}{\partial t}$. Instead, we should write

$$\frac{\partial(\boldsymbol{v}\psi)}{\partial t} = \frac{\partial \boldsymbol{v}}{\partial t}\psi + \boldsymbol{v}\frac{\partial \psi}{\partial t} = \frac{e}{m}\frac{\partial \boldsymbol{A}}{\partial t}\psi + \boldsymbol{v}\frac{\partial \psi}{\partial t}\,. \tag{10.27}$$

Similarly, we have

$$\boldsymbol{v}\,(V\psi) = \frac{1}{m}\left[(-\mathrm{i}\hbar\nabla V)\,\psi + V\,(-\mathrm{i}\hbar\nabla + e\boldsymbol{A})\,\psi\right] = -\frac{\mathrm{i}\hbar}{m}(\nabla V)\,\psi + V\boldsymbol{v}\psi\,. \tag{10.28}$$

Substituting (10.27) and (10.28) into (10.26), we obtain

$$\mathrm{i}\hbar\frac{\partial(\boldsymbol{v}\psi)}{\partial t} = \boldsymbol{v}H_1\psi + \frac{\mathrm{i}e\hbar}{m}\left(\nabla V + \frac{\partial \boldsymbol{A}}{\partial t}\right)\psi - eV\boldsymbol{v}\psi + \left[\boldsymbol{v}\psi,\,\hat{H}'\right] = \boldsymbol{v}H_1\psi - \frac{\mathrm{i}e\hbar}{m}\psi\boldsymbol{E} - eV\boldsymbol{v}\psi + \left[\boldsymbol{v}\psi,\,\hat{H}'\right]. \tag{10.29}$$

Multiplying both sides from the left with $\boldsymbol{v}^*\psi^\dagger$ yields

$$\mathrm{i}\hbar\left(\boldsymbol{v}^*\psi^\dagger\right)\cdot\frac{\partial(\boldsymbol{v}\psi)}{\partial t} = \left(\boldsymbol{v}^*\psi^\dagger\right)\cdot\boldsymbol{v}H_1\psi - \frac{\mathrm{i}e\hbar}{m}\left(\boldsymbol{v}^*\psi^\dagger\right)\psi\cdot\boldsymbol{E} - eV\left(\boldsymbol{v}^*\psi^\dagger\right)\cdot\boldsymbol{v}\psi + \left(\boldsymbol{v}^*\psi^\dagger\right)\cdot\left[\boldsymbol{v}\psi,\,\hat{H}'\right]. \tag{10.30}$$

The Hermitian conjugate equation reads:

$$-\mathrm{i}\hbar\left[\frac{\partial\left(\boldsymbol{v}^*\psi^\dagger\right)}{\partial t}\right]\cdot(\boldsymbol{v}\psi) = -\left(\boldsymbol{v}^*H_1^*\psi^\dagger\right)\cdot(\boldsymbol{v}\psi) + \frac{\mathrm{i}e\hbar}{m}\boldsymbol{E}\cdot\left(\psi^\dagger\boldsymbol{v}\psi\right) - eV\left(\boldsymbol{v}^*\psi^\dagger\right)\cdot\boldsymbol{v}\psi - \left[\boldsymbol{v}^*\psi^\dagger,\,\hat{H}'\right]\cdot(\boldsymbol{v}\psi)\,. \tag{10.31}$$

Combining (10.30) and (10.31) with (10.24) we obtain

$$\frac{\partial W}{\partial t} = \frac{1}{2}m\frac{\partial}{\partial t}\left(\left(\boldsymbol{v}^*\psi^\dagger\right)\cdot\boldsymbol{v}\psi\right) - \frac{U}{e}\frac{\partial\hat{\varrho}}{\partial t} = -\frac{\mathrm{i}m}{2\hbar}\left[\left(\boldsymbol{v}^*\psi^\dagger\right)\cdot(\boldsymbol{v}H_1\psi) - \left(\boldsymbol{v}^*H_1^*\psi^\dagger\right)\cdot(\boldsymbol{v}\psi)\right] - \frac{e}{2}\left[\left(\boldsymbol{v}^*\psi^\dagger\right)\psi + \psi^\dagger\boldsymbol{v}\psi\right]\cdot\boldsymbol{E} - \frac{U}{e}\frac{\partial\hat{\varrho}}{\partial t} - \frac{\mathrm{i}}{\hbar}\left[W,\,\hat{H}'\right] \tag{10.32}$$

or

$$\frac{\partial W}{\partial t} = -\frac{\mathrm{i}m}{2\hbar}\left[\left(\boldsymbol{v}^*\psi^\dagger\right)\cdot(\boldsymbol{v}H_1\psi) - \left(\boldsymbol{v}^*H_1^*\psi^\dagger\right)\cdot(\boldsymbol{v}\psi)\right] + \hat{\boldsymbol{J}}\cdot\boldsymbol{E} - \frac{U}{e}\frac{\partial\hat{\varrho}}{\partial t} - \frac{\mathrm{i}}{\hbar}\left[W,\,\hat{H}'\right]. \tag{10.33}$$

We will now relate the expression between square brackets to the divergence of an energy flux vector $\boldsymbol{K}$. A simple, appropriate and gauge invariant definition is given by:

$$\boldsymbol{K} = \frac{1}{2}\left[\left(\boldsymbol{v}^*\psi^\dagger\right) H_1\psi + \left(H_1^*\psi^\dagger\right)\boldsymbol{v}\psi\right] . \tag{10.34}$$

Indeed, taking the divergence of $\boldsymbol{K}$, we find that

$$\begin{aligned}\nabla\cdot\boldsymbol{K} = \frac{1}{2}&\left[\left(\boldsymbol{v}^*\psi^\dagger\right)\cdot\nabla H_1\psi + \left(\nabla H_1^*\psi^\dagger\right)\cdot\boldsymbol{v}\psi + \left(\nabla\cdot\boldsymbol{v}^*\psi^\dagger\right)H_1\psi\right.\\ &\left.+\left(H_1^*\psi^\dagger\right)\nabla\cdot\boldsymbol{v}\psi\right] .\end{aligned} \tag{10.35}$$

Inserting (10.20) and (10.21) respectively into the second and the first term, we get:

$$\begin{aligned}\nabla\cdot\boldsymbol{K} = \frac{1}{2}&\left\{\frac{im}{\hbar}\left[\left(\boldsymbol{v}^*\psi^\dagger\right)\cdot\boldsymbol{v}H_1\psi - \left(\boldsymbol{v}^*H_1^*\psi^\dagger\right)\cdot\boldsymbol{v}\psi\right]\right.\\ &-\frac{\mathrm{i}e}{\hbar}\boldsymbol{A}\cdot\left[\left(\boldsymbol{v}^*\psi^\dagger\right)\boldsymbol{v}H_1\psi - \left(H_1^*\psi^\dagger\right)\boldsymbol{v}\psi\right]\\ &\left.+\left(\nabla\cdot\boldsymbol{v}^*\psi^\dagger\right)H_1\psi + \left(H_1^*\psi^\dagger\right)\nabla\cdot\boldsymbol{v}\psi\right\} .\end{aligned} \tag{10.36}$$

Recollecting the terms in $\boldsymbol{A}$ and ∇, we arrive at

$$\begin{aligned}\nabla\cdot\boldsymbol{K} &= \frac{1}{2}\left\{\frac{\mathrm{i}m}{\hbar}\left[\left(\boldsymbol{v}^*\psi^\dagger\right)\cdot\boldsymbol{v}H_1\psi - \left(\boldsymbol{v}^*H_1^*\psi^\dagger\right)\cdot\boldsymbol{v}\psi\right]\right.\\ &\quad\left.-\frac{\mathrm{i}m}{\hbar}\left[\left(v^{*2}\psi^\dagger\right)H_1\psi - \left(H_1^*\psi^\dagger\right)v^2\psi\right]\right\}\\ &= \frac{\mathrm{i}m}{2\hbar}\left\{\left(\boldsymbol{v}^*\psi^\dagger\right)\cdot\boldsymbol{v}H_1\psi - \left(\boldsymbol{v}^*H_1^*\psi^\dagger\right)\cdot\boldsymbol{v}\psi\right.\\ &\quad\left.-U\left[\left(v^{*2}\psi^\dagger\right)\psi - \psi^\dagger v^2\psi\right]\right\}\\ &= \frac{\mathrm{i}m}{2\hbar}\left[\left(\boldsymbol{v}^*\psi^\dagger\right)\cdot\boldsymbol{v}H_1\psi - \left(\boldsymbol{v}^*H_1^*\psi^\dagger\right)\cdot\boldsymbol{v}\psi\right] - \frac{U}{e}\nabla\cdot\hat{\boldsymbol{J}}\end{aligned} \tag{10.37}$$

where the last equality follows from (10.22).

Inserting (10.37) into (10.33) yields

$$\frac{\partial W}{\partial t} = -\nabla\cdot\boldsymbol{K} + \hat{\boldsymbol{J}}\cdot\boldsymbol{E} - \frac{U}{e}\left(\nabla\cdot\hat{\boldsymbol{J}} + \frac{\partial\hat{\varrho}}{\partial t}\right) - \frac{\mathrm{i}}{\hbar}\left[W,\hat{H}'\right] . \tag{10.38}$$

Finally, we may invoke the continuity equation $\nabla\cdot\hat{\boldsymbol{J}} + \partial\hat{\varrho}/\partial t = 0$ to write the rate equation for W as follows:

$$\frac{\partial W}{\partial t} = -\nabla\cdot\boldsymbol{K} + \hat{\boldsymbol{J}}\cdot\boldsymbol{E} - \frac{\mathrm{i}}{\hbar}\left[W,\hat{H}'\right] . \tag{10.39}$$

□

This is the local energy balance equation balance which, in principle, allows us to study the spatial distribution of the quantum mechanical energy density of the electron gas, and therefore also the local properties of energy dissipation.

As a corollary, we may derive the global energy balance equation governing the time evolution of the Heisenberg operator $\hat{H}_\mathrm{E}(t)$ which is nothing but the integral of $W\boldsymbol{r},t)$ over the circuit region:

$$\hat{H}_{\mathrm{E}}(t) = \int_{\Omega} \mathrm{d}\tau\, W(\boldsymbol{r}, t) \,. \tag{10.40}$$

Indeed, remembering that ψ and $\psi^{\dagger}$ are vanishing at the outer surface $\partial\Omega$, we may integrate the the local energy balance equation over the circuit region to obtain

$$\frac{\mathrm{d}\hat{H}_{\mathrm{E}}(t)}{\mathrm{d}t} = \int_{\Omega} \mathrm{d}\tau \hat{\boldsymbol{J}}(\boldsymbol{r}, t) \cdot \boldsymbol{E}(\boldsymbol{r}, t) - \frac{\mathrm{i}}{\hbar}\left[\hat{H}_{\mathrm{E}}(t),\, \hat{H}'(t)\right] \,. \tag{10.41}$$

□

The above derived continuity equation as well as the local and global energy balance equations are operator equation established in the Heisenberg picture. As such they should be transformed into tractable balance equations governing the time evolution of the physical observables represented by the operators entering the equations. This can be accomplished formally by taking appropriate ensemble averages describing generally the non-equilibrium state of the electron gas. Since we have adopted the Heisenberg picture in which the time dependence is shifted entirely onto the operators, we merely need to know the initial statistical operator (density matrix) ϱ_0 in order to construct the full time-dependent expectation value $X(t)$ of an arbitrary observable represented by a Heisenberg operator $\hat{X}(t)$:

$$X(t) = \mathrm{Tr}[\varrho_0 \hat{X}(t)] \equiv \langle \hat{X}(t) \rangle_0 \,. \tag{10.42}$$

For instance, taking the ensemble average of (10.41), we may recast the global energy balance equation as follows:

$$\frac{\mathrm{d}H_{\mathrm{E}}(t)}{\mathrm{d}t} = \int_{\Omega} \mathrm{d}\tau\, \boldsymbol{J}(\boldsymbol{r}, t) \cdot \boldsymbol{E}(\boldsymbol{r}, t) - \frac{\mathrm{i}}{\hbar}\left\langle [\hat{H}_{\mathrm{E}}(t),\, \hat{H}'(t)] \right\rangle_0 \tag{10.43}$$

with $H_{\mathrm{E}}(t) = \langle \hat{H}_{\mathrm{E}}(t) \rangle_0$, $\boldsymbol{J}(\boldsymbol{r}, t) = \langle \hat{\boldsymbol{J}}(\boldsymbol{r}, t) \rangle_0$, $\varrho(\boldsymbol{r}, t) = \langle \hat{\varrho}(\boldsymbol{r}, t) \rangle_0$, etc.

In particular, when the circuit carries DC current, a steady state will usually be established after a very long time ($t \to \infty$) while all energy supplied by the external electric field is relaxed to the environment. Such a steady state may be characterized by a steady value of the total electron energy:

$$\lim_{t \to \infty} \frac{\mathrm{d}H_{\mathrm{E}}(t)}{\mathrm{d}t} = 0 \tag{10.44}$$

and therefore

$$\int_{\Omega} \mathrm{d}\tau\, \boldsymbol{J}(\boldsymbol{r}) \cdot \boldsymbol{E}(\boldsymbol{r}) = \frac{\mathrm{i}}{\hbar} \lim_{t \to \infty} \left\langle [\hat{H}_{\mathrm{E}}(t),\, \hat{H}'(t)] \right\rangle_0 \,, \tag{10.45}$$

where $\boldsymbol{J}(\boldsymbol{r}) = \lim_{t\to\infty} \langle \hat{\boldsymbol{J}}(\boldsymbol{r}, t) \rangle_0$ and the electric field, though being *non*-conservative, does not depend on time. Representing the total power extracted from the electric field, the left-hand side of (10.45) may formally be rewritten as the product of the total DC current I and the applied electromotive force V_{ε}. Indeed, since the continuity equation reduces to $\nabla \cdot \boldsymbol{J} = 0$ for a DC current distribution and $\nabla \times \boldsymbol{E} = \boldsymbol{0}$ for a DC field we may apply a new integral theorem [52] disentangling the integrated product of solenoidal and irrotational vector fields defined on a multiply connected region Ω:

$$\int_{\Omega} \mathrm{d}\tau\, \boldsymbol{J}\cdot\boldsymbol{E} = \left[\int_{\Sigma} \boldsymbol{J}\cdot \mathrm{d}\boldsymbol{S}\right] \times \left[\oint_{\Gamma} \boldsymbol{E}\cdot \mathrm{d}\boldsymbol{r}\right] , \tag{10.46}$$

where Γ is an arbitrary, simple closed curve lying inside Ω encircling the 'hole' of Ωi once and only once, and Σ is an arbitrary, single cross section of Ω. Clearly, it is well-known from the theory of electromagnetism that $V_{\varepsilon} = \oint_{\Gamma} \boldsymbol{E}\cdot \mathrm{d}\boldsymbol{r}$ represents the electromotive force while the total DC current may be written as the integral of the current density $\boldsymbol{J}(\boldsymbol{r})$ over the cross-section Σ, i.e. $I = \int_{\Sigma} \boldsymbol{J}\cdot \mathrm{d}\boldsymbol{S}$. Consequently, the energy balance equation takes the appealing form

$$IV_{\varepsilon} = \frac{\mathrm{i}}{\hbar} \lim_{t\to\infty} \left\langle [\hat{H}_{\mathrm{E}}(t),\, \hat{H}'(t)] \right\rangle_0 . \tag{10.47}$$

10.7 Linear and Angular Momentum Balance Equation

Although the algebra is a lot more cumbersome, it is possible to set up also balance equations for the linear and angular momentum in a similar way.

The gauge invariant operators $\hat{\boldsymbol{P}}$ and $\hat{\boldsymbol{L}}$ describing the total linear and angular momentum of a the electron gas may be obtained by applying the prescriptions of second quantization discussed in Chap. 6:

$$\begin{aligned}
\hat{\boldsymbol{P}} &= \int_{\Omega} \mathrm{d}\tau\, \psi^{\dagger} m\boldsymbol{v}\psi \\
&= \frac{1}{2} m \int_{\Omega} \mathrm{d}\tau\, \left[\psi^{\dagger}\boldsymbol{v}\psi + \left(\boldsymbol{v}^{*}\psi^{\dagger}\right)\right] = -\frac{m}{e}\int_{\Omega} \mathrm{d}\tau \hat{\boldsymbol{J}} \\
\hat{\boldsymbol{L}} &= \int_{\Omega} \mathrm{d}\tau\, \psi^{\dagger} \left(\boldsymbol{r}\times m\boldsymbol{v}\right)\psi \\
&= \frac{1}{2} m \int_{\Omega} \mathrm{d}\tau\, \boldsymbol{r}\times \left[\psi^{\dagger}\boldsymbol{v}\psi + \left(\boldsymbol{v}^{*}\psi^{\dagger}\right)\right] = -\frac{m}{e}\int_{\Omega} \mathrm{d}\tau\, \boldsymbol{r}\times\hat{\boldsymbol{J}} .
\end{aligned} \tag{10.48}$$

The intimate connection between the total momentum and the current density operator encourages us to derive also a rate equation for the current density, as was done for the energy density:

$$\frac{\partial\hat{\boldsymbol{J}}(\boldsymbol{r},t)}{\partial t} = \left[\hat{\boldsymbol{J}}(\boldsymbol{r},t), H\right] + \frac{e}{m}\,\hat{\varrho}(\boldsymbol{r},t)\,\frac{\partial\boldsymbol{A}(\boldsymbol{r},t)}{\partial t} . \tag{10.49}$$

Inspired by the analogy with the energy balance equation, we are tempted to construct an operator similar to the energy flux density, the divergence of which we expect to find in the right-hand side of the rate equation. Having the dimensions of a force per unit area, such an operator corresponds to the quantum-mechanical version of the 'hydrodynamic pressure tensor **S**' and its explicit expression reads in rectangular coordinates:

$$S_{jr} = \frac{1}{4}m\left[v_j^* v_r^* \psi^\dagger \psi + v_j^* \psi^\dagger v_r \psi + v_r^* \psi^\dagger v_j \psi + \psi^\dagger v_j v_r \psi\right] . \tag{10.50}$$

A lengthy calculation finally yields the rate equation for the current density – and therefore for the local momentum:

$$\frac{\partial \hat{\boldsymbol{J}}}{\partial t} = -\nabla \cdot \mathbf{S} - \frac{e}{m}\left[\hat{\varrho}\left(\boldsymbol{E} - \frac{1}{e}\nabla U\right) + \hat{\boldsymbol{J}} \times \boldsymbol{B}\right] - \frac{\mathrm{i}}{\hbar}\left[\hat{\boldsymbol{J}},\, \hat{H}'\right] . \tag{10.51}$$

The corresponding momentum balance equation is obtained by integrating (10.51) over the volume Ω whence the divergence term vanishes:

$$\frac{\mathrm{d}\boldsymbol{P}}{\mathrm{d}t} = \int_\Omega \mathrm{d}\tau \left[\hat{\varrho}\left(\boldsymbol{E} - \frac{1}{e}\nabla U\right) + \hat{\boldsymbol{J}} \times \boldsymbol{B}\right] - \frac{\mathrm{i}}{\hbar}\left[\boldsymbol{P},\, \hat{H}'\right] . \tag{10.52}$$

The first two operators at the right-hand side of (10.52) correspond to the force executed by the total electric field and the well-known Lorentz force respectively. The last term represents the frictional force counter-balancing the driving forces, and its explicit expression depends on the details of the scattering mechanisms. Similarly, the global balance equation for the angular momentum reads:

$$\frac{\mathrm{d}\hat{\boldsymbol{L}}}{\mathrm{d}t} = \int_\Omega \mathrm{d}\tau\, \boldsymbol{r} \times \left[\hat{\varrho}\left(\boldsymbol{E} - \frac{1}{e}\nabla U\right) + \hat{\boldsymbol{J}} \times \boldsymbol{B}\right] - \frac{\mathrm{i}}{\hbar}\left[\hat{\boldsymbol{L}},\, \hat{H}'\right] , \tag{10.53}$$

where the integral at the right-hand side represents the torque exerted by the total electric field, the built-in field and the Lorentz force.

10.8 Calculation of the DC Current

The balance equation of the previous section needs to be evaluated in more detail in order to obtain the current as a function of the applied bias. So far, no approximations have been made concerning the microscopic equations of motion and the basic assumptions formulated at the beginning of this chapter. However, apart from rewriting the basic equations of motions as a set of intuitively appealing balance equations, we still need to perform a real, quantitative evaluation of the current flowing through the circuit region Ω, which is generally a giant task. A brute-force solution would not only require us to determine exactly the time dependence of the Heisenberg operators, but also to calculate the initial density matrix ϱ_0 in the presence of all scattering mechanisms described by the interaction term $\hat{H}'$. In the electron gas were in thermal equilibrium at $t = 0$ thereby carrying no current in the absence of any bias voltage, such a calculation should in principle enable us to obtain explicitly an irreversible evolution for all relevant macroscopic quantities, in spite of the reversible microscopic equations of motion, mainly by exploiting the infinite number of scattering modes to which electrons may relax their energy and momentum.

10.8.1 Gedankenexperiment

Unfortunately, the above described exact solution of the Heisenberg equations is only available for a limited number of interacting systems for which the algebra of Heisenberg operators is closed and the set of resulting equations can be solved. A non-trivial example of such a system is discussed in the next chapter. In general however, the hierarchy of equation cannot be terminated exactly and one needs to rely on approximate solutions. One particular scheme that will be generalized is the balance equations approach introduced two decades ago by Devreese and Peeters [48] and Lei and Ting [49, 50, 51]. Being designed originally to study steady-state hot-electron transport in uniform electric fields, the formalism may be extended to the more general case in which the total electric field may be inhomogeneous in space thereby varying significantly within the small-sized active parts of the circuit.

The starting point is the 'Gedankenexperiment' [49, 50, 51] in which the non-equilibrium steady state of the electron gas is considered. Such a steady state is a complicated interacting many-particle state as it is capable of maintaining a momentum and energy balance between the power supply on one hand and the elastic and dissipative scattering mechanisms on the other hand. Suppose now that we would be able to turn off simultaneously and adiabatically the driving field and all scattering mechanisms. Having scaled down both counteracting effects in a uniform way, we would end up with a non-interacting electron gas that has preserved the electric current and the average energy of the interacting ensemble, while it would be entirely decoupled from all scattering agents. The density matrix describing such a decoupled state would differ considerably from that of the exact, interacting steady state but it would at least give rise to the same electric current and energy distribution. In this light, we may decide to take the initial density matrix ϱ_0 to be the density matrix representing the decoupled but current carrying state rather than the equilibrium state because then the difference between ϱ_0 and the true steady-state density matrix $\varrho_\infty = \lim_{t\to\infty} \varrho(t)$ could be studied in a perturbation scheme as far as the current and the average energy are concerned.

10.8.2 Equilibrium Currents and Broken Time Reversal Symmetry

Adopting the perturbation scheme introduced in the previous section, the first step will be the construction of an appropriate 'boosted' density matrix for a non-interacting electron gas in the absence of an electromagnetic field. Having solved the single-particle Schrödinger equation for the energy eigenstates $\{\phi_k(\boldsymbol{r})\}$ and eigenvalues $\{\varepsilon_k\}$ of a single electron residing in the circuit region Ω, we may set up the non-interacting Hamiltonian

$$\hat{H}_{\mathrm{E}}^{(0)} = \sum_{k\sigma} \varepsilon_k c_{k\sigma}^\dagger c_{k\sigma} \,. \tag{10.54}$$

Time reversal symmetry requires that for each eigenfunction $\phi_k(\boldsymbol{r})$ there exist a time-reversed solution $\phi_{\tilde{k}}(\boldsymbol{r})$ with the same energy ε_k. Since the eigensolutions $\{\phi_k(\boldsymbol{r})\}$ constitute a set of stationary states, it is clear that for each k we must have

$$\phi_{\tilde{k}}(\boldsymbol{r}) = \phi_k^*(\boldsymbol{r}) \tag{10.55}$$

and

$$\varepsilon_{\tilde{k}} = \varepsilon_k \,. \tag{10.56}$$

Moreover, if the single-particle spectrum is doubly degenerated, the eigenstates are carrying electric current and the current carried by a state $|\phi_k\rangle$ is exactly the opposite of the current of the time-reversed state $|\phi_{\tilde{k}}\rangle$. This can easily be seen by considering the current operator and its matrix elements introduced in Chap. 6:

$$\hat{\boldsymbol{J}}(\boldsymbol{r}) = \sum_{kk'\sigma} \boldsymbol{J}_{k'k}(\boldsymbol{r}) c_{k'\sigma}^\dagger c_{k\sigma} \tag{10.57}$$

with

$$\boldsymbol{J}_{k'k}(\boldsymbol{r}) = \frac{\mathrm{i}e\hbar}{2m}\left[\phi_{k'}^*(\boldsymbol{r})\nabla\phi_k(\boldsymbol{r}) - \left(\nabla\phi_{k'}^*(\boldsymbol{r})\right)\phi_k(\boldsymbol{r})\right] \,. \tag{10.58}$$

From $\phi_{\tilde{k}}(\boldsymbol{r}) = \phi_k^*(\boldsymbol{r})$, we immediately infer that

$$\boldsymbol{J}_{\tilde{k}\tilde{k}}(\boldsymbol{r}) = -\boldsymbol{J}_{kk}(\boldsymbol{r}) \tag{10.59}$$

Moreover, invoking the Schrödinger equation

$$-\frac{\hbar^2}{2m}\nabla^2\phi_k(\boldsymbol{r}) + \left(U_{\mathrm{conf}}(\boldsymbol{r}) - \varepsilon_k\right)\phi_k(\boldsymbol{r}) = 0 \,, \tag{10.60}$$

where $U_{\mathrm{conf}}(\boldsymbol{r})$ denotes the confining potential, we may conclude

$$\nabla \cdot \boldsymbol{J}_{k'k}(\boldsymbol{r}) = \frac{\mathrm{i}e}{\hbar}\left(\varepsilon_{k'} - \varepsilon_k\right)\phi_{k'}^*(\boldsymbol{r})\phi_k(\boldsymbol{r}) \tag{10.61}$$

and, in particular,

$$\nabla \cdot \boldsymbol{J}_{kk}(\boldsymbol{r}) = 0 \,. \tag{10.62}$$

As a consequence, the total current carried by the state $|\phi_k\rangle$, i.e.

$$I_k = \int_\Sigma \boldsymbol{J}_{kk} \cdot \mathrm{d}\boldsymbol{S} \tag{10.63}$$

does not depend on a particular choice of the circuit cross section Σ and is therefore unequivocally determined by the wave function $\phi_k(\boldsymbol{r})$. Combining (10.62) and (10.63) finally yields

$$I_{\tilde{k}} = -I_k \,. \tag{10.64}$$

As an extremely simple example, we could study electron transport on a finite one-dimensional ring with radius R. If θ denotes the polar angle, then the spectrum would be given by

$$\begin{aligned} \phi_k(\theta) &= \frac{1}{\sqrt{2\pi}}\,\mathrm{e}^{\mathrm{i}k\theta} \\ \varepsilon_k &= \frac{\hbar^2 k^2}{2mR^2}, \qquad k = 0, \pm 1, \pm 2, \dots \,. \end{aligned} \tag{10.65}$$

Clearly, the energy eigenstates are also diagonalizing the z-component of the angular momentum operator $\boldsymbol{r}\times\boldsymbol{p}$ and time reversal symmetry invokes a one-to-one correspondence between each left rotating state and its right rotating partner at the same energy.

Next, we will artificially break the time reversal symmetry of the non-interacting electron gas by adding a current dependent term to the electron energies:

$$\varepsilon_k \to \varepsilon_k - \gamma_0 I_k \,, \tag{10.66}$$

where γ_0 is left unspecified for the time being.

The latter obviously amounts to the substitution

$$\hat{H}_{\mathrm{E}}^{(0)} \to \hat{H}_{\mathrm{E}}^{(B)} = \sum_{k\sigma} (\varepsilon_k - \gamma_0 I_k)\, c_{k\sigma}^\dagger c_{k\sigma} \,. \tag{10.67}$$

Bearing in mind that the 'boosted' Hamiltonian $\hat{H}_{\mathrm{E}}^{(\mathrm{B})}$ commutes with $\hat{H}_{\mathrm{P}}$, the Hamiltonian of the free scatterers, we may now construct the following density matrix describing the free scatterers in equilibrium and the non-interacting electron gas:

$$\begin{aligned} \varrho_{\mathrm{B}} &= \frac{1}{Z_{\mathrm{B}}} \exp\left[-\beta \hat{H}_{\mathrm{P}}\right] \exp\left[-\beta_{\mathrm{e}}\left(\hat{H}_{\mathrm{E}}^{(\mathrm{B})} - \mu \hat{N}\right)\right] \\ &= \frac{1}{Z_{\mathrm{B}}} \exp\left[-\beta \hat{H}_{\mathrm{P}}\right] \prod_{k\sigma} \exp\left[-\beta_{\mathrm{e}} (\varepsilon_k - \gamma_0 I_k - \mu)\, c_{k\sigma}^\dagger c_{k\sigma}\right] \,, \end{aligned} \tag{10.68}$$

where $\hat{N} = \sum_{k\sigma} c_{k\sigma}^\dagger c_{k\sigma}$ is the electron number operator, $T_{\mathrm{e}} = 1/(k_{\mathrm{B}}\beta_{\mathrm{e}})$ represents the electron temperature characterizing the average electron energy in the boosted equilibrium state and Z_{B} is the partition function for the 'boosted' ensemble. Hence, choosing the initial density matrix as the boosted one, $\varrho_0 = \varrho_{\mathrm{B}}$ and taking the average of $\boldsymbol{J}(\boldsymbol{r})$ with ϱ_0, we obtain straightaway:

$$\begin{aligned} \langle \hat{\boldsymbol{J}}(\boldsymbol{r}) \rangle &= \mathrm{Tr}\left(\varrho_0 \hat{\boldsymbol{J}}(\boldsymbol{r})\right) \\ &= \sum_{k\sigma} F_{\mathrm{e}}(\varepsilon_k - \gamma_0 I_k) \boldsymbol{J}_{kk}(\boldsymbol{r}) \end{aligned}$$

with

$$F_e(E) = \frac{1}{1 + e^{\beta_e(E-\mu)}} \tag{10.69}$$

and μ denotes the chemical potential of the electron gas.

Integrating over the cross section Σ, we finally obtain the equilibrium current as being parameterized by γ_0 and T_e:

$$I = \sum_{k\sigma} F_e(\varepsilon_k - \gamma_0 I_k) I_k \; . \tag{10.70}$$

Similarly, taking the average of $\hat{H}_E^{(0)}$, we may as well parameterize the average electron energy:

$$H_E^{(0)} = \sum_{k\sigma} F_e(\varepsilon_k - \gamma_0 I_k) \varepsilon_k \; . \tag{10.71}$$

If γ_0 tends to zero, then we conclude from (10.55) (10.56) and (10.63) that $I = 0$, while $H_E^{(0)}$ reduces to its thermal equilibrium value, as we should expect. However, if $\gamma_0 \neq 0$ and T_e differs from the ambient temperature T, we may in principle reach any desired value of I and $\hat{H}_E^{(0)}$ by an appropriate choice of γ_0 and T_e.

Although both parameters are introduced *ad hoc* in order to create an artificial equilibrium state carrying a non-zero current and and an energy excess, they may be interpreted as the conjugated variables of the electron Hamiltonian and the total current operator. As such they might appear as a set of Lagrange multipliers so as to maximize the system entropy under the constraints $\langle \hat{I} \rangle = I$ and $\langle \hat{H}_E^{(0)} \rangle = H_E^{(0)}$.

10.8.3 Perturbative Solution Scheme

The second step, the most difficult one, is to integrate the boosted density matrix in an appropriate perturbation scheme that should enable us to determine self-consistently the values of γ_0 and T_e – and hence the current and the average energy – corresponding to a given set of elastic and inelastic scattering mechanisms.

In view of the previous section, it becomes obvious to choose the unperturbed Hamiltonian $\hat{H}_0$ to be the sum of the boosted free electron Hamiltonian and the Hamiltonian of the free scatterers

$$\hat{H}_0 = \hat{H}_E^{(B)} + \hat{H}_P \; . \tag{10.72}$$

Consequently, the perturbation Hamiltonian will be given by the difference between the full Hamiltonian and the unperturbed one:

$$\hat{H}_{\text{int,t}} = \hat{H}_t - \hat{H}_E^{(B)} - \hat{H}_P \; , \tag{10.73}$$

where the subscript t in $\hat{H}_t$ – and therefore in $\hat{H}_{\text{int,t}}$ – reflects any possible explicit time dependence due to external fields. We are now in a position to define an interaction representation that will be helpful to calculate the time dependence of the Heisenberg operators entering the balance equations up to first order in $\hat{H}_{\text{int,t}}$. Formally, we can represent the time evolution of an arbitrary Heisenberg operator $\hat{X}(t)$ by a unitary transformation generated by the evolution operator $U(t)$:

$$\hat{X}(t) = U^\dagger(t)\, \hat{X}\, U(t) \,, \tag{10.74}$$

where $\hat{X} \equiv \hat{X}(t=0)$, $U(t)$ is a unitary operator solving the initial-value problem

$$\begin{aligned} \mathrm{i}\hbar \frac{\mathrm{d}U(t)}{\mathrm{d}t} &= \hat{H}_\mathrm{t}\, U(t) \\ U(0) &= \mathbf{1} \end{aligned} \tag{10.75}$$

and the physical dependency on time is governed by the total Hamiltonian $\hat{H}_\mathrm{t} = \hat{H}_0 + \hat{H}_{\text{int,t}}$. If the interaction were turned off, the Heisenberg representation would reduce to the so-called interaction representation in which all time dependence is extracted from the unperturbed Hamiltonian $\hat{H}_0$ that usually does not have any explicit time dependence:

$$\hat{X}^{(0)}(t) = \exp\left(\frac{\mathrm{i}\hat{H}_0 t}{\hbar}\right) \hat{X} \exp\left(-\frac{\mathrm{i}\hat{H}_0 t}{\hbar}\right) , \tag{10.76}$$

where the superscript $^{(0)}$ refers to the interaction representation. This equation may look as complicated as (10.74) but in practice one may always write down explicitly the time dependent operators in the interaction representation and we will therefore assume that this has been accomplished for all relevant operators. For instance, it can easily be shown that the time evolution of the electron operator $c_{k\sigma}$ is given by the following expression in the interaction representation:

$$c^{(0)}_{\mathrm{k}\sigma}(t) = \exp\left(-\frac{\mathrm{i}\varepsilon_k t}{\hbar}\right) c_{k\sigma} \,. \tag{10.77}$$

The aim of the time-dependent perturbation theory is now to express the time dependence of the Heisenberg operators in the framework of the interaction picture up to a given order of $\hat{H}_{\text{int}}$ – first order in our case. The link between the Heisenberg picture and the interaction picture can be established by another unitary transformation

$$\hat{X}(t) = S^\dagger(t)\hat{X}^{(0)}(t)S(t) \,, \tag{10.78}$$

where $S(t)$ needs to be calculated up to first order in $\hat{H}_{\text{int}}$. Comparing the three unitary transformations (10.74), (10.76) and (10.78) with each other for an arbitrary observable X, we obtain

$$S(t) = \exp\left(\frac{\mathrm{i}\hat{H}_0 t}{\hbar}\right) U(t) \,. \tag{10.79}$$

Differentiating both sides w.r.t. time, yields a differential equation for $S(t)$:

$$\begin{aligned} \mathrm{i}\hbar \frac{\mathrm{d}S(t)}{\mathrm{d}t} &= -\hat{H}_0 \exp\left(\frac{\mathrm{i}\hat{H}_0 t}{\hbar}\right) U(t) + \exp\left(\frac{\mathrm{i}\hat{H}_0 t}{\hbar}\right) H_t\, U(t) \\ &= -\exp\left(\frac{\mathrm{i}\hat{H}_0 t}{\hbar}\right) \hat{H}_0 U(t) + \exp\left(\frac{\mathrm{i}\hat{H}_0 t}{\hbar}\right) \left(\hat{H}_0 + \hat{H}_{\mathrm{int,t}}\right) U(t) \\ &= \exp\left(\frac{\mathrm{i}\hat{H}_0 t}{\hbar}\right) \hat{H}_{\mathrm{int,t}} U(t) \\ &= \exp\left(\frac{\mathrm{i}\hat{H}_0 t}{\hbar}\right) \hat{H}_{\mathrm{int,t}} \exp\left(\frac{-\mathrm{i}\hat{H}_0 t}{\hbar}\right) \exp\left(\frac{\mathrm{i}\hat{H}_0 t}{\hbar}\right) \hat{H}_{\mathrm{int,t}} \\ &= \hat{H}^{(0)}_{\mathrm{int,t}}(t)\, S(t)\,, \end{aligned} \tag{10.80}$$

which, together with the initial condition $S(0) = \mathbf{1}$, can be recast in the form of an integral equation

$$S(t) = \mathbf{1} - \frac{\mathrm{i}}{\hbar} \int_0^t \mathrm{d}t'\, \hat{H}^{(0)}_{\mathrm{int,t'}}(t')\, S(t')\,. \tag{10.81}$$

The explicit time dependence due to external fields may be present even for DC fields. Up to first order in $\hat{H}_{\mathrm{int,t}}$, its solution reads

$$S(t) = \mathbf{1} - \frac{\mathrm{i}}{\hbar} \int_0^t \mathrm{d}t'\, \hat{H}^{(0)}_{\mathrm{int,t'}}(t')\,. \tag{10.82}$$

Correspondingly, the Heisenberg operator $X(t)$ can be obtained approximately by substituting (10.82) into (10.78) thereby neglecting the second order term:

$$\begin{aligned} \hat{X}(t) &= S^{\dagger}(t)\, \hat{X}^0(t)\, S(t) \\ &= \hat{X}^0(t) - \frac{\mathrm{i}}{\hbar} \int_0^t \mathrm{d}t' \left[\hat{X}^0(t), \hat{H}^{(0)}_{\mathrm{int,t'}}(t')\right]\,. \end{aligned} \tag{10.83}$$

Finally, the time dependent expectation value of $\hat{X}$ may be calculated by taking the average of (10.83) with ϱ_0 as the initial density matrix:

$$\begin{aligned} X(t) \equiv \langle \hat{X}(t) \rangle_0 &= \mathrm{Tr}[\hat{X}(t)\varrho_0] \\ &= \langle \hat{X}^0(t) \rangle_0 - \frac{\mathrm{i}}{\hbar} \int_0^t \mathrm{d}t' \left\langle \left[\hat{X}^0(t), \hat{H}^{(0)}_{\mathrm{int,t'}}(t')\right] \right\rangle_0\,. \end{aligned} \tag{10.84}$$

Being the outcome of a first-order perturbation calculation, (10.84) is valid only if the action of the perturbation $\hat{H}^{(0)}_{\mathrm{int,t}}$ can be considered small. The latter, in turn, is strongly depends on the features of the density matrix ϱ_0, chosen to represent the initial state of the electron gas. If the driving field is relatively small, we may be inclined to assume that the electron gas will be in the 'ohmic regime' where the current response would be proportional to the applied voltage and the steady state would accordingly close to the thermal equilibrium state. In that case, we may consider to take ϱ_0 equal to the thermal equilibrium density matrix which would make (10.84) equivalent to Kubo's linear response result [11]. However, if the electric field can reach as high values as 10 kV/cm (and higher) in various semiconductor devices, it becomes clear that the 'boosted' density matrix (10.68) is a much better candidate since it would share by construction at least two ensemble averages (current and average energy) with the true density matrix describing the interacting steady state.

The above argument does not provide a formal justification for the use of the proposed balance equation approach, but merely intends to make the latter plausible from intuitive arguments while the real justification lies in the assessment of the results obtained. Thus, taking $\varrho_0 = \exp[-\beta\hat{H}_{\rm P}]\exp[-\beta_{\rm e}(\hat{H}_{\rm E}^{\rm (B)} - \mu\hat{N})]/Z_{\rm B}$ we may express the time-dependent ensemble average of $\hat{X}(t)$ in terms of retarded Green functions [53]:

$$\langle\hat{X}(t)\rangle_0 = \langle\hat{X}^{(0)}(t)\rangle_0 + \int_0^\infty {\rm d}t' \,\langle\langle\,\hat{X}^{(0)}(t)\,;\,\hat{H}^{(0)}_{\rm int,t'}(t')\,\rangle\rangle\,, \tag{10.85}$$

where

$$\langle\langle\,\hat{A}^{(0)}(t)\,;\,\hat{B}^{(0)}(t')\,\rangle\rangle = -\frac{\rm i}{\hbar}\theta(t-t')\left\langle\left[\hat{A}^{(0)}(t),\,\hat{B}^{(0)}(t')\right]\right\rangle_0 \tag{10.86}$$

defines the *r*etarded Green function associated with a set of operators $\hat{A}$ and $\hat{B}$.

Since the expectation values $\langle\ldots\rangle_0$ are formally calculated as the 'boosted equilibrium averages' related to H_0 that is also governing the time dependence of the interaction picture, the Green functions depend on time only through the difference of the two time arguments:

$$\langle\langle\,\hat{A}^{(0)}(t)\,;\,\hat{B}^{(0)}(t')\,\rangle\rangle = \langle\langle\,\hat{A}^{(0)}(t-t')\,;\,\hat{B}^{(0)}(0)\,\rangle\rangle \tag{10.87}$$

$$= \langle\langle\,\hat{A}^{(0)}(0)\,;\,\hat{B}^{(0)}(t'-t)\,\rangle\rangle\,. \tag{10.88}$$

We may therefore define Fourier transforms with respect to the difference $t - t'$ according to

$$\langle\langle\,\hat{A}^{(0)}(t)\,;\,\hat{B}^{(0)}(t')\,\rangle\rangle = \int_{-\infty}^{\infty} {\rm d}E\,{\rm e}^{-{\rm i}E(t-t')/\hbar}\,\langle\langle\,\hat{A}\,;\,\hat{B}\,\rangle\rangle_E\,,\ \Im(E) \geq 0 \tag{10.89}$$

Similarly, averages such as $\langle\hat{X}(t)\rangle_0$, containing only one time argument t, turn out to be independent on t and therefore we may write:

$$\begin{aligned}\langle\hat{X}(t)\rangle_0 &= \langle\hat{X}\rangle_0 + \int_0^\infty {\rm d}t'\,\langle\langle\,\hat{X}^{(0)}(t-t')\,;\,\hat{H}^{(0)}_{\rm int,t'}(t')\,\rangle\rangle \\ &= \langle\hat{X}\rangle_0 + \int_0^t {\rm d}t'\,\langle\langle\,\hat{X}^{(0)}(t')\,;\,\hat{H}^{(0)}_{\rm int}\,\rangle\rangle\end{aligned} \tag{10.90}$$

so that, finally,

$$\langle \hat{X} \rangle_{\mathrm{S}} = \langle \hat{X} \rangle_0 + 2\pi\hbar \lim_{\eta \to 0^+} \langle\langle \hat{X} \, ; \, \hat{H}^{(0)}_{\mathrm{int}} \rangle\rangle_{\mathrm{i}\eta} \, , \tag{10.91}$$

where the subscript S denotes the final steady-state average, i.e. $\langle \hat{X} \rangle_{\mathrm{S}} \equiv \lim_{t \to \infty} \langle \hat{X}(t) \rangle_0$.

In practice, $\hat{X}$ essentially stands for elementary operator products of the form $c^{\dagger}_{k\sigma} c_{k\sigma}$ and some more complicated combinations from which all relevant physical quantities may be constructed. The detailed calculation of the corresponding Green functions entering (10.91) directly depends on the scattering interactions that are relevant to the problem under investigation, and will therefore be postponed to Chaps. 11 and 17. However, the general idea of the balance equation approach is to apply the above perturbation analysis to the operators arising in the steady-state momentum and energy balance equations and to write the latter eventually in the generic form

$$\int_{\Omega} \mathrm{d}\tau \left[\langle \hat{\varrho} \rangle_S (\gamma_0, T_{\mathrm{e}}) \left(\boldsymbol{E} - \frac{1}{e} \nabla U \right) + \langle \hat{\boldsymbol{J}} \rangle_S (\gamma_0, T_{\mathrm{e}}) \times \boldsymbol{B} \right] + \boldsymbol{F}(\gamma_0, T_{\mathrm{e}}) = \boldsymbol{0} \tag{10.92}$$

$$I V_{\varepsilon} + P(\gamma_0, T_{\mathrm{e}}) = 0 \tag{10.93}$$

$\boldsymbol{F}(\gamma_0, T_{\mathrm{e}})$ and $P(\gamma_0, T_{\mathrm{e}})$ respectively denote the frictional force and the dissipated energy describing the momentum and energy relaxation of the electron gas, both terms being parameterized by the Lagrange multipliers γ_0 and T_{e}.

Equations (10.92) and (10.93) may now be combined with (10.70) and (10.71) to set up a numerical algorithm in order to evaluate the current I and, as a byproduct, the two Lagrange multipliers and the average energy of the electron gas.

Exercise

10.1. Prove (10.88) by expanding ϱ_0 in the eigenstates of $\hat{H}_0$.

Part II

Applications

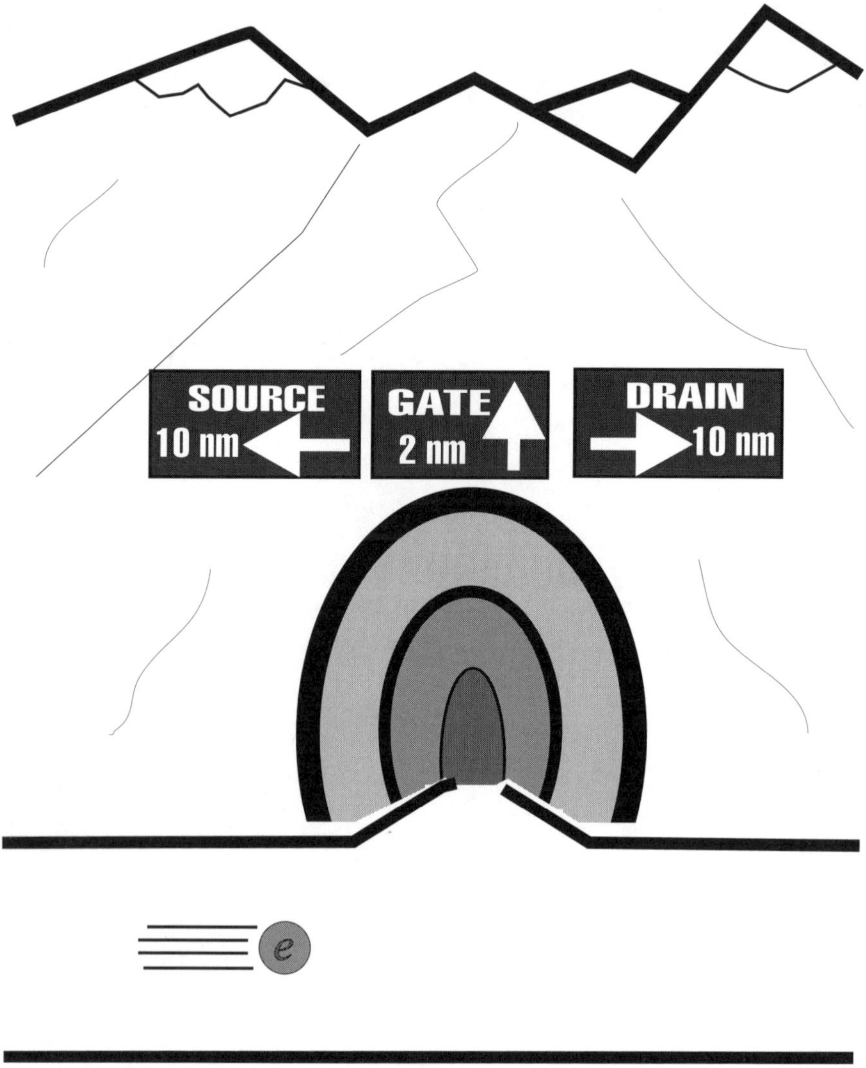

SOURCE
10 nm
GATE
2 nm
DRAIN
10 nm
e

11. Velocity–Field Characteristics of a Silicon MOSFET

11.1 Momentum and Energy Balance Equations for a MOSFET Channel

In this chapter we will study the motion of a quasi-two-dimensional electron gas residing in the conduction channel established in the inversion layer of a p-type silicon (Si) MOSFET due to a positive gate voltage (see Fig. 11.1). When an positive voltage is applied between source and drain, all electrons in the channel acquire a drift velocity due to the lateral, driving electric field. The major goal is now to derive the relation between the electron drift velocity and the electric field. For the sake of simplicity, we will make three approximations.

First, we ignore the source and drain contacts in the calculation assuming that the channel is long enough to invoke full translational invariance in the two directions parallel with the Si/SiO_2 interface. Secondly, the lateral electric field parallel to the source–drain direction is taken to be uniform along the channel. Finally, we assume that the insulating oxide layer, separating the gate electrode from the inversion layer, is sufficiently thick to suppress any noticeable leakage currents flowing from the channel to the gate electrode. This approximation will be relaxed in the section on gate leakage currents where a tunneling through ultra-thin oxide barriers will be treated. Although the first two approximation will generally break down when deep

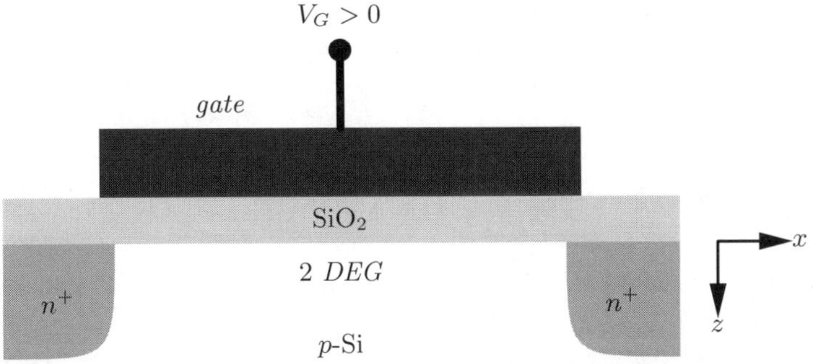

Fig. 11.1. Schematic view of a p-type MOSFET in inversion mode

submicron transistors are envisaged, reliable predictions of the low- and high-field electron mobilities may still be obtained for relatively large gate lengths exceeding, say, 0.5 μm.

Furthermore, choosing the (x, y)-plane at the Si/SiO_2 interface, we will account for the subbands arising from the electron confinement in the z-direction which is determined by the conduction band profile $E_c(z)$.

Due to the translational invariance in the x- and y-directions, at least in the absence of the lateral field $E_L\, \boldsymbol{e}_x$, we may write the one-electron wave functions $\{\Psi_{\alpha l \boldsymbol{k}}(\boldsymbol{r}, z)\}$ as plane waves modulated by envelope subband wave functions $\{\phi_{\alpha l}(z)\}$:

$$\Psi_{\alpha l \boldsymbol{k}}(\boldsymbol{r}, z) = \frac{1}{\sqrt{L_x L_y}} \mathrm{e}^{\mathrm{i}\boldsymbol{k}\cdot\boldsymbol{r}} \mathrm{e}^{\mathrm{i}k_{\alpha z} z} \phi_{\alpha l}(z) \tag{11.1}$$

and the energies as

$$\varepsilon_{\alpha l \boldsymbol{k}} = \hbar^2 \left[\frac{(k_x - k_{\alpha x})^2}{2m_{\alpha x}} + \frac{(k_y - k_{\alpha y})^2}{2m_{\alpha y}} \right] + W_{\alpha l} \,. \tag{11.2}$$

Here, α, l, $\boldsymbol{k}$, $\boldsymbol{r}$, L_x and L_y are the valley and subband indices, two-dimensional wave vectors and position vectors and channel dimensions in the x- and y-directions respectively. $m_{\alpha x}$, $m_{\alpha y}$ and $m_{\alpha z}$ are the effective masses along the principal directions of the αth valley. The conduction band minimum of the latter is denoted by $\boldsymbol{k}_\alpha$. The minima of the six valleys are located near one of the six X-points defining the zone boundaries of the first Brillouin zone:

$$|\boldsymbol{k}_\alpha| = 0.86 \, \frac{2\pi}{a} \,. \tag{11.3}$$

The subband energies $\{W_{\alpha l}\}$ and the corresponding wavefunctions $\{\phi_{\alpha l}(z)\}$ are the solutions of the effective Schrödinger equation

$$-\frac{\hbar^2}{2m_{\alpha z}} \frac{\mathrm{d}^2 \phi_{\alpha l}(z)}{\mathrm{d}z^2} + E_c(z)\, \phi_{\alpha l}(z) = W_{\alpha l}\, \phi_{\alpha l}(z) \,. \tag{11.4}$$

Introducing creation and annihilation operators for electrons along the lines of previous section, we may now express the total momentum of the electron gas arising in the scattering term of the momentum balance equation:

$$P_x = \sum_{\alpha l \boldsymbol{k} \sigma} \hbar k_x \, c^\dagger_{\alpha l \boldsymbol{k} \sigma} c_{\alpha l \boldsymbol{k} \sigma} \tag{11.5}$$

$$P_y = \sum_{\alpha l \boldsymbol{k} \sigma} \hbar k_y \, c^\dagger_{\alpha l \boldsymbol{k} \sigma} c_{\alpha l \boldsymbol{k} \sigma} \tag{11.6}$$

$$P_z = -\mathrm{i}\hbar \sum_{\alpha l l' \boldsymbol{k} \sigma} \int_0^\infty \mathrm{d}z \, \phi_{\alpha l'}(z) \frac{\mathrm{d}\phi_{\alpha l}(z)}{\mathrm{d}z} \, c^\dagger_{\alpha l' \boldsymbol{k} \sigma} c_{\alpha l \boldsymbol{k} \sigma} \,. \tag{11.7}$$

It should be noted that the term containing the vector potential $\boldsymbol{A} = -\boldsymbol{E}_\mathrm{L}\, t$ has been omitted in (11.5) since it is proportional to the density operator

$\varrho(t)$ and would therefore give a vanishing contribution to scattering term due to basic assumption $[\varrho, \hat{H}'] = 0$. The derivation is left as an exercise.

Starting from the momentum balance equation (10.52) with a uniform driving field $\boldsymbol{E} = -E_{\mathrm{L}} \boldsymbol{e}_x$, we obtain for the steady state:

$$-E_{\mathrm{L}} \lim_{t\to\infty} \int_{\Omega} \mathrm{d}\tau \, \langle \varrho(\boldsymbol{r}, z, t) \rangle_0 - \frac{\mathrm{i}}{\hbar} \lim_{t\to\infty} \left\langle \left[P_x(t), \, \hat{H}'(t) \right] \right\rangle_0 = 0 \, . \qquad (11.8)$$

The first integral is nothing but the total charge of all free electrons by the electron gas and therefore reduces to $-Ne$, where N is the number of electrons present, yielding

$$NeE_{\mathrm{L}} - \mathrm{i} \sum_{\alpha l \boldsymbol{k} \sigma} k_x \lim_{t\to\infty} \left\langle \left[c^{\dagger}_{\alpha l \boldsymbol{k} \sigma}(t) c_{\alpha l \boldsymbol{k} \sigma}(t), \, \hat{H}'(t) \right] \right\rangle_0 = 0 \qquad (11.9)$$

with $E_{\mathrm{L}} = |V_{\mathrm{DS}}|/L_{\mathrm{gate}}$.

The difficult part however is in the evaluation of the scattering terms which requires an explicit representation of H' in terms of $\{c_{\alpha l \boldsymbol{k} \sigma}\}$ and the operators referring to the scattering agents. In the case of silicon, the most important ones are:

- ionized impurity scattering (e.g. acceptors for an inversion layer, donors for an accumulation layer in an the case of p-type substrates)
- scattering by acoustic phonons
- inter-valley scattering
- surface roughness scattering

For the sake of simplicity we have only accounted for acoustic phonons in the present calculation. After all, acoustic phonons contributing to the electron–phonon Hamiltonian $\hat{H}'$ are causing a great deal of energy dissipation hampering charge transport through the conduction channel and the details of the corresponding inelastic scattering processes can easily be modified to treat also other phonon scattering events.

For a silicon inversion layer, the appropriate expression for the electron–phonon Hamiltonian can be obtained from standard solid-state physics textbooks: in the case of acoustic phonons, the electrons interact with the (acoustic) lattice vibrations through the dilation $D(\boldsymbol{r}, z)$ of the displacement field:

$$\hat{H}' = \int_{\Omega} \mathrm{d}\tau \, \psi^{\dagger}(\boldsymbol{r}, z) \psi(\boldsymbol{r}, z) D(\boldsymbol{r}, z) \, , \qquad (11.10)$$

where the displacement field is quantized by expanding it into normal modes and imposing the usual Bose–Einstein commutation rules to the expansion coefficients $\{a_{\boldsymbol{Q}}, a^{\dagger}_{\boldsymbol{Q}}\}$:

$$D(\boldsymbol{r}, z) = \sum_{\boldsymbol{Q}} \left[M(\boldsymbol{Q}) \, a_{\boldsymbol{Q}} \mathrm{e}^{\mathrm{i}(\boldsymbol{k}\cdot\boldsymbol{r} + q_z z)} + M^*(\boldsymbol{Q}) \, a^{\dagger}_{\boldsymbol{Q}} \mathrm{e}^{-\mathrm{i}(\boldsymbol{k}\cdot\boldsymbol{r} + q_z z)} \right] . \quad (11.11)$$

Throughout this section capital letters will be used to denote 3D vectors, e.g. a 3D phonon wave vector $\boldsymbol{Q}, \ldots$ represents $\boldsymbol{q} + q_z \boldsymbol{e}_z$.

For longitudinal acoustic phonons the coupling constant $M(\boldsymbol{Q})$ can be expressed in terms of the sound velocity v_{S}, the mass density ϱ_{Si} and the deformation potential D_{AC} as follows:

$$M(\boldsymbol{Q}) = \mathrm{i}D_{\mathrm{AC}}\sqrt{\frac{\hbar|\boldsymbol{Q}|}{2\varrho_{\mathrm{Si}}\Omega v_{\mathrm{S}}}}\,, \tag{11.12}$$

while the dispersion relation is borrowed from the Debye approximation

$$\omega_{\boldsymbol{Q}} = v_S|\boldsymbol{Q}| \quad \text{for } |\boldsymbol{Q}| \le Q_{\mathrm{D}}\,, \tag{11.13}$$

where Q_{D} denotes the Debye wave vector [54].

The above expansion of the displacement field in a set of plane waves representing the normal modes silently implies full translational invariance in all three directions. Strictly speaking, this does not hold for a MOSFET or any other structure in which there is an interface between two different materials. We will however ignore the lack of lattice periodicity in the z-direction and assume that the corresponding surface effects are duly accounted for by the electronic subband structure that does reflect the interface features. Expanding the electron field operators, we obtain

$$\begin{aligned}\hat{H}' = \frac{1}{L_x L_y}&\sum_{\alpha\sigma}\sum_{ll'}\sum_{\boldsymbol{k}\boldsymbol{k}'}\sum_{\boldsymbol{q},q_z} c^{\dagger}_{\alpha l'\boldsymbol{k}+\boldsymbol{q}\sigma}c_{\alpha l\boldsymbol{k}\sigma}\\ &\times\left[M(\boldsymbol{q},q_z)\int \mathrm{d}^2r\,\mathrm{e}^{\mathrm{i}(\boldsymbol{k}-\boldsymbol{k}'+\boldsymbol{q})\cdot\boldsymbol{r}}\int_0^\infty \mathrm{d}z\,\phi_{\alpha l}(z)\phi_{\alpha l'}(z)\mathrm{e}^{\mathrm{i}q_z z}a_{\boldsymbol{Q}}\right.\\ &\left.+\,M^*(\boldsymbol{q},q_z)\int \mathrm{d}^2r\,\mathrm{e}^{\mathrm{i}(\boldsymbol{k}-\boldsymbol{k}'-\boldsymbol{q})\cdot\boldsymbol{r}}\int_0^\infty \mathrm{d}z\,\phi_{\alpha l}(z)\phi_{\alpha l'}(z)\mathrm{e}^{-\mathrm{i}q_z z}a^{\dagger}_{\boldsymbol{Q}}\right].\end{aligned} \tag{11.14}$$

Continuing the evaluation, we obtain

$$\hat{H}' = \sum_{\alpha\sigma ll'}\sum_{\boldsymbol{k}\boldsymbol{q}q_z} M_{\alpha ll'}(\boldsymbol{q},q_z)\,c^{\dagger}_{\alpha l'\boldsymbol{k}+\boldsymbol{q}\sigma}c_{\alpha l\boldsymbol{k}\sigma}\,(a_{\boldsymbol{Q}} + a^{\dagger}_{-\boldsymbol{Q}})\,, \tag{11.15}$$

where the effective coupling constants are to be extracted from their 'bulk' counterparts by simple multiplication with the subband structure factors $I_{\alpha ll'}(q_z)$:

$$\begin{aligned}M_{\alpha ll'}(\boldsymbol{q},q_z) &= I_{\alpha ll'}(q_z)M(\boldsymbol{q},q_z),\\ I_{\alpha ll'}(q_z) &= \int_0^\infty \mathrm{d}z\,\phi_{\alpha l}(z)\,\phi_{\alpha l'}(z)\,\mathrm{e}^{\mathrm{i}q_z z}\,.\end{aligned} \tag{11.16}$$

Going from (11.14) to (11.15) we have explicitly made use of

$$M(-\boldsymbol{q},-q_z) = M^*(\boldsymbol{q},q_z)\,, \tag{11.17}$$

in order to rewrite the sum over $\boldsymbol{q}$ and q_z in the second term of the right-hand side of (11.14) as a sum over $-\boldsymbol{q}$ and $-q_z$. Moreover, the translational symmetry in the x- and y-directions enabled us to reduce the two-dimensional integrals to the Kronecker delta's $\delta_{\boldsymbol{k}',\boldsymbol{k}+\boldsymbol{q}}$ and $\delta_{\boldsymbol{k}',\boldsymbol{k}-\boldsymbol{q}}$ respectively. They account for momentum conservation in the directions parallel to the interface, whereas the subband structure factors reflect the absence of translational invariance in the z-direction and the lack of the corresponding momentum conservation.

It is clear from (11.15) that emission and absorption of one phonon are the two basic processes leading to a net momentum and energy loss of the electron gas, where electrons can make individual transitions from one subband to another as well as loosing or gaining 2D momentum. Note that not only the spin index σ but also the valley index α is left unchanged in these processes since intervalley scattering requires a momentum transfer of the order of the reciprocal lattice vectors and these large momenta cannot be supplied by the long wavelength part of the acoustic phonon spectrum which provides the basis of the continuum description adopted in this discussion. On the other hand, the short-wavelength edges of both the acoustic and optical phonon branches which are inducing the intervalley transfer processes, may equally be incorporated in the corresponding intervalley frictional force but are skipped from the the present discussion.

According to (11.9) the frictional force due to acoustic phonon scattering is given by

$$F_{\mathrm{AC}x} = -\mathrm{i}\sum_{\alpha l\boldsymbol{k}\sigma} k_x \lim_{t\to\infty}\left\langle\left[c^\dagger_{\alpha l\boldsymbol{k}\sigma}(t)c_{\alpha l\boldsymbol{k}\sigma}(t),\, \hat{H}'(t)\right]\right\rangle_0 , \tag{11.18}$$

where the evaluation of the frictional force operator between the brackets is straightforward on using the elementary commutator relation

$$[c_{\alpha l\boldsymbol{k}\sigma}, \hat{H}'] = \sum_{\alpha l'\sigma}\sum_{\boldsymbol{q}q_z} M_{\alpha l'l}(\boldsymbol{q}, q_z)c_{\alpha l'\boldsymbol{k}-\boldsymbol{q}\sigma}(a_{\boldsymbol{Q}} + a^\dagger_{-\boldsymbol{Q}}) \tag{11.19}$$

yielding

$$\begin{aligned}[c^\dagger_{\alpha l\boldsymbol{k}\sigma}c_{\alpha l\boldsymbol{k}\sigma}, \hat{H}'] = \sum_{\alpha l'\sigma}\sum_{\boldsymbol{q}q_z} M_{\alpha l'l}(\boldsymbol{q}, q_z)\left[c^\dagger_{\alpha l\boldsymbol{k}\sigma}c_{\alpha l'\boldsymbol{k}-\boldsymbol{q}\sigma} - c^\dagger_{\alpha l'\boldsymbol{k}+\boldsymbol{q}\sigma}c_{\alpha l\boldsymbol{k}\sigma}\right]\\ \times(a_{\boldsymbol{Q}} + a^\dagger_{-\boldsymbol{Q}}) .\end{aligned} \tag{11.20}$$

Replacing one sum over $\boldsymbol{q}$ with a sum over $\boldsymbol{k}+\boldsymbol{q}$ and using the symmetry of the coupling constants w.r.t. the subband indexes, we finally obtain

$$\begin{aligned}F_{\mathrm{AC}x} = -\mathrm{i}\sum_{\alpha ll'\sigma}\sum_{\boldsymbol{k}\boldsymbol{q}q_z} q_x\, M_{\alpha l'l}(\boldsymbol{q}, q_z)\left[\lim_{t\to\infty}\left\langle c^\dagger_{\alpha l'\boldsymbol{k}+\boldsymbol{q}\sigma}(t)c_{\alpha l\boldsymbol{k}\sigma}(t)a_{\boldsymbol{Q}}(t)\right\rangle_0\right.\\ \left. + \lim_{t\to\infty}\left\langle c^\dagger_{\alpha l'\boldsymbol{k}+\boldsymbol{q}\sigma}(t)c_{\alpha l\boldsymbol{k}\sigma}(t)a^\dagger_{-\boldsymbol{Q}}(t)\right\rangle_0\right] .\end{aligned} \tag{11.21}$$

We will now apply the Green function technique described in chapter (10) to calculate the two expectation values in (11.21) up to first order in the perturbation $\hat{H}_{\mathrm{int}}$. The zeroth order parts of the expectation values reduce to zero because they depend linearly on the Bose operators $a_{\boldsymbol{Q}}(t)$ and $a^\dagger_{-\boldsymbol{Q}}(t)$

while the latter appear in a bilinear combination entering the boosted density matrix (10.68). Indeed, the unperturbed Hamiltonian H_0 clearly commutes with the total number of both electrons and phonons. In particular, it follows that within the relevant Green functions all *unperturbed* expectation values, computed w.r.t. H_0 and containing an *odd* number of phonon operators are identically vanishing. Furthermore, for the same reason we may skip in the interaction Hamiltonian $\hat{H}_{\text{int}}$ all terms that do not contain any phonon operators. Consequently, we only need to calculate Green functions with $c^{\dagger}_{\alpha l'\boldsymbol{k}+\boldsymbol{q}\sigma}c_{\alpha l\boldsymbol{k}\sigma}a_{\boldsymbol{Q}}$ respectively $c^{\dagger}_{\alpha l'\boldsymbol{k}+\boldsymbol{q}\sigma}c_{\alpha l\boldsymbol{k}\sigma}a^{\dagger}_{-\boldsymbol{Q}}$ as the left operator and the bare electron–phonon Hamiltonian $\hat{H}'$ as the operator on the right:

$$\begin{aligned}\langle\langle\, c^{\dagger}_{\alpha l'\boldsymbol{k}+\boldsymbol{q}\sigma}c_{\alpha l\boldsymbol{k}\sigma}a_{\boldsymbol{Q}}\,;\,\hat{H}_{\text{int}}\,\rangle\rangle_E &= \langle\langle\, c^{\dagger}_{\alpha l'\boldsymbol{k}+\boldsymbol{q}\sigma}c_{\alpha l\boldsymbol{k}\sigma}a_{\boldsymbol{Q}}\,;\,\hat{H}'\,\rangle\rangle_E\\ \langle\langle\, c^{\dagger}_{\alpha l'\boldsymbol{k}+\boldsymbol{q}\sigma}c_{\alpha l\boldsymbol{k}\sigma}a^{\dagger}_{-\boldsymbol{Q}}\,;\,\hat{H}_{\text{int}}\,\rangle\rangle_E &= \langle\langle\, c^{\dagger}_{\alpha l'\boldsymbol{k}+\boldsymbol{q}\sigma}c_{\alpha l\boldsymbol{k}\sigma}a^{\dagger}_{-\boldsymbol{Q}}\,;\,\hat{H}'\,\rangle\rangle_E\,. \end{aligned}\tag{11.22}$$

The subscript 'E' was defined in (10.89). Analogously, we have to retain in $\hat{H}'$ only those terms that are pairing one phonon creation operator at the left with one annihilation operator to the right or vice versa:

$$\begin{aligned}&\langle\langle\, c^{\dagger}_{\alpha l'\boldsymbol{k}+\boldsymbol{q}\sigma}c_{\alpha l\boldsymbol{k}\sigma}a_{\boldsymbol{Q}}\,;\,\hat{H}'\,\rangle\rangle_E\\ &=\sum_{l_1l_2}\sum_{\boldsymbol{k}_1\boldsymbol{Q}_1}M_{\alpha l_1l_2}(\boldsymbol{Q}_1)\,\langle\langle\, c^{\dagger}_{\alpha l'\boldsymbol{k}+\boldsymbol{q}\sigma}c_{\alpha l\boldsymbol{k}\sigma}a_{\boldsymbol{Q}}\,;\,c^{\dagger}_{\alpha l_2\boldsymbol{k}_1+\boldsymbol{q}_1\sigma}c_{\alpha l_1\boldsymbol{k}_1\sigma}a^{\dagger}_{-\boldsymbol{Q}_1}\,\rangle\rangle_E\\ &\langle\langle\, c^{\dagger}_{\alpha l'\boldsymbol{k}+\boldsymbol{q}\sigma}c_{\alpha l\boldsymbol{k}\sigma}a^{\dagger}_{-\boldsymbol{Q}}\,;\,\hat{H}'\,\rangle\rangle_E\\ &=\sum_{l_1l_2}\sum_{\boldsymbol{k}_1\boldsymbol{Q}_1}M_{\alpha l_1l_2}(\boldsymbol{Q}_1)\,\langle\langle\, c^{\dagger}_{\alpha l'\boldsymbol{k}+\boldsymbol{q}\sigma}c_{\alpha l\boldsymbol{k}\sigma}a^{\dagger}_{-\boldsymbol{Q}}\,;\,c^{\dagger}_{\alpha l_2\boldsymbol{k}_1+\boldsymbol{q}_1\sigma}c_{\alpha l_1\boldsymbol{k}_1\sigma}a_{\boldsymbol{Q}_1}\,\rangle\rangle_E\,. \end{aligned}\tag{11.23}$$

11.2 Calculation of the Elementary Green Functions

From the equations of motion in the interaction picture we obtain straightaway

$$\begin{aligned}&\mathrm{i}\hbar\frac{\mathrm{d}}{\mathrm{d}t}\langle\langle\, c^{(0)\dagger}_{\alpha l'\boldsymbol{k}+\boldsymbol{q}\sigma}(t)c^{(0)}_{\alpha l\boldsymbol{k}\sigma}(t)a^{(0)}_{\boldsymbol{Q}}(t)\,;\,c^{\dagger}_{\alpha l_2\boldsymbol{k}_1+\boldsymbol{q}_1\sigma}c_{\alpha l_1\boldsymbol{k}_1\sigma}a^{\dagger}_{-\boldsymbol{Q}_1}\,\rangle\rangle=\delta(t)\\ &\quad\times\left\langle\left[c^{\dagger}_{\alpha l'\boldsymbol{k}+\boldsymbol{q}\sigma}c_{\alpha l\boldsymbol{k}\sigma}a_{\boldsymbol{Q}},c^{\dagger}_{\alpha l_2\boldsymbol{k}_1+\boldsymbol{q}_1\sigma}c_{\alpha l_1\boldsymbol{k}_1\sigma}a^{\dagger}_{-\boldsymbol{Q}_1}\right]\right\rangle_0+(-\varepsilon_{\alpha l'\boldsymbol{q}+\boldsymbol{k}}+\varepsilon_{\alpha l\boldsymbol{k}}\\ &\quad+\hbar\omega_{\boldsymbol{Q}})\,\langle\langle\, c^{(0)\dagger}_{\alpha l'\boldsymbol{k}+\boldsymbol{q}\sigma}(t)c^{(0)}_{\alpha l\boldsymbol{k}\sigma}(t)a^{(0)}_{\boldsymbol{Q}}(t)\,;\,c^{\dagger}_{\alpha l\boldsymbol{k}\sigma}c_{\alpha l_1\boldsymbol{k}_1\sigma}a^{\dagger}_{-\boldsymbol{Q}_1}\,\rangle\rangle \end{aligned}\tag{11.24}$$

and, after Fourier transformation,

$$\begin{aligned}&E\,\langle\langle\, c^{\dagger}_{\alpha l'\boldsymbol{k}+\boldsymbol{q}\sigma}c_{\alpha l\boldsymbol{k}\sigma}a_{\boldsymbol{Q}}\,;\,c^{\dagger}_{\alpha l_2\boldsymbol{k}_1+\boldsymbol{q}_1\sigma}c_{\alpha l_1\boldsymbol{k}_1\sigma}a^{\dagger}_{-\boldsymbol{Q}_1}\,\rangle\rangle_E\\ &=\frac{1}{2\pi}\left\langle\left[c^{\dagger}_{\alpha l'\boldsymbol{k}+\boldsymbol{q}\sigma}c_{\alpha l\boldsymbol{k}\sigma}a_{\boldsymbol{Q}},c^{\dagger}_{\alpha l_2\boldsymbol{k}_1+\boldsymbol{q}_1\sigma}c_{\alpha l_1\boldsymbol{k}_1\sigma}a^{\dagger}_{-\boldsymbol{Q}_1}\right]\right\rangle_0+(-\varepsilon_{\alpha l'\boldsymbol{k}+\boldsymbol{q}}+\varepsilon_{\alpha l\boldsymbol{k}}\\ &\quad+\hbar\omega_{\boldsymbol{Q}})\,\langle\langle\, c^{\dagger}_{\alpha l'\boldsymbol{k}+\boldsymbol{q}\sigma}c_{\alpha l\boldsymbol{k}\sigma}a_{\boldsymbol{Q}}\,;\,c^{\dagger}_{\alpha l_2\boldsymbol{k}_1+\boldsymbol{q}_1\sigma}c_{\alpha l_1\boldsymbol{k}_1\sigma}a^{\dagger}_{-\boldsymbol{Q}_1}\,\rangle\rangle_E\,. \end{aligned}\tag{11.25}$$

whence

$$\langle\langle\, c^{\dagger}_{\alpha l' \boldsymbol{k}+\boldsymbol{q}\sigma} c_{\alpha l \boldsymbol{k}\sigma} a_{\boldsymbol{Q}}\,;\, c^{\dagger}_{\alpha l_2 \boldsymbol{k}_1+\boldsymbol{q}_1\sigma} c_{\alpha l_1 \boldsymbol{k}_1\sigma} a^{\dagger}_{-\boldsymbol{Q}_1}\,\rangle\rangle_E = \frac{1}{2\pi}\,\frac{\left\langle \left[c^{\dagger}_{\alpha l' \boldsymbol{k}+\boldsymbol{q}\sigma} c_{\alpha l \boldsymbol{k}\sigma} a_{\boldsymbol{Q}}\,,\, c^{\dagger}_{\alpha l_2 \boldsymbol{k}_1+\boldsymbol{q}_1\sigma} c_{\alpha l_1 \boldsymbol{k}_1\sigma} a^{\dagger}_{-\boldsymbol{Q}_1} \right] \right\rangle_0}{E + \varepsilon_{\alpha l' \boldsymbol{k}+\boldsymbol{q}} - \varepsilon_{\alpha l \boldsymbol{k}} - \hbar\omega_{\boldsymbol{Q}}}\,. \tag{11.26}$$

The commutator appearing in the last equation can be evaluated as follows:

$$\begin{aligned}
\left[c_1^{\dagger} c_2 a_3,\, c_4^{\dagger} c_5 a_6^{\dagger}\right] &= c_1^{\dagger} c_2 \left[a_3,\, c_4^{\dagger} c_5 a_6^{\dagger}\right] + \left[c_1^{\dagger} c_2,\, c_4^{\dagger} c_5 a_6^{\dagger}\right] a_3 \\
&= c_1^{\dagger} c_2 c_4^{\dagger} c_5 \left[a_3,\, a_6^{\dagger}\right] + \left[c_1^{\dagger} c_2,\, c_4^{\dagger} c_5\right] a_6^{\dagger} a_3 \\
&= c_1^{\dagger} c_2 c_4^{\dagger} c_5 \left[a_3,\, a_6^{\dagger}\right] + c_1^{\dagger} \left\{c_2,\, c_4^{\dagger}\right\} c_5 a_6^{\dagger} a_3 \\
&\quad - c_4^{\dagger} \left\{c_1^{\dagger},\, c_5\right\} c_2 a_6^{\dagger} a_3 \\
&= \delta_{36}\, c_1^{\dagger} c_2 c_4^{\dagger} c_5 + \delta_{24}\, c_1^{\dagger} c_5 a_6^{\dagger} a_3 - \delta_{15}\, c_4^{\dagger} c_2 a_6^{\dagger} a_3 \\
&= \delta_{36}\, c_1^{\dagger} c_2 c_4^{\dagger} c_5 + \left(\delta_{24}\, c_1^{\dagger} c_5 - \delta_{15}\, c_4^{\dagger} c_2\right) a_6^{\dagger} a_3\,.
\end{aligned} \tag{11.27}$$

The expectation value w.r.t. $\hat{H}_0$ then reads

$$\begin{aligned}
\left\langle \left[c_1^{\dagger} c_2 a_3,\, c_4^{\dagger} c_5 a_6^{\dagger}\right] \right\rangle_0 &= \delta_{36} \left\langle c_1^{\dagger} c_2 c_4^{\dagger} c_5 \right\rangle_0 + \left(\delta_{24} \left\langle c_1^{\dagger} c_5 \right\rangle_0 - \delta_{15} \left\langle c_4^{\dagger} c_2 \right\rangle_0\right) \\
&\quad \times \left\langle a_6^{\dagger} a_3 \right\rangle_0\,,
\end{aligned} \tag{11.28}$$

Since we are dealing with unperturbed expectation values we may 'disentangle' all four-point averages like $\left\langle c_1^{\dagger} c_2 c_4^{\dagger} c_5 \right\rangle_0$ and express them in terms of Fermi–Dirac distribution functions:

$$\left\langle c_1^{\dagger} c_2 c_4^{\dagger} c_5 \right\rangle_0 = \delta_{12}\, \delta_{45}\, F_{\mathrm{e}}(\varepsilon_1) F_{\mathrm{e}}(\varepsilon_4) + \delta_{15}\, \delta_{24}\, F_{\mathrm{e}}(\varepsilon_1)\, [(1 - F_{\mathrm{e}}(\varepsilon_2)]\,. \tag{11.29}$$

At this point it is convenient to introduce the ratio between the lattice temperature and the electron temperature:

$$\gamma = \frac{T}{T_{\mathrm{e}}}\,. \tag{11.30}$$

Combining now (11.28) and (11.29) with the identity

$$F_{\mathrm{e}}(E)\,[1 - F_{\mathrm{e}}(E')] = [F_{\mathrm{e}}(E') - F_{\mathrm{e}}(E)]\; \nu(\gamma(E - E')) \tag{11.31}$$

the original expectation value may be simplified to be

$$\begin{aligned}
\left\langle c_1^{\dagger} c_2 a_3 c_4^{\dagger} c_5 a_6^{\dagger} \right\rangle_0 &= \delta_{15}\, \delta_{24}\, \delta_{36}\; [F_{\mathrm{e}}(\varepsilon_1) - F_{\mathrm{e}}(\varepsilon_2)]\; [\nu(\hbar\omega_3) - \nu(\gamma(\varepsilon_1 - \varepsilon_2))] \\
&\quad + \delta_{12}\, \delta_{36}\, \delta_{45}\, F_{\mathrm{e}}(\varepsilon_1) F_{\mathrm{e}}(\varepsilon_4)
\end{aligned} \tag{11.32}$$

where $\nu(E)$ denotes the Bose–Einstein distribution function:

$$\nu(E) = \frac{1}{\mathrm{e}^{\beta E} - 1} . \tag{11.33}$$

Inserting this result into (11.26), we obtain the full expression for the first Green function:

$$\begin{aligned}
&\langle\langle\, c^{\dagger}_{\alpha l' \boldsymbol{k}+\boldsymbol{q}\sigma} c_{\alpha l \boldsymbol{k}\sigma} a_{\boldsymbol{Q}} \,;\, c^{\dagger}_{\alpha l_2 \boldsymbol{k}_1+\boldsymbol{q}_1\sigma} c_{\alpha l_1 \boldsymbol{k}_1\sigma} a^{\dagger}_{-\boldsymbol{Q}_1} \rangle\rangle_E \\
&= \frac{1}{2\pi} \frac{1}{E + \varepsilon_{\alpha l' \boldsymbol{k}+\boldsymbol{q}} - \varepsilon_{\alpha l \boldsymbol{k}} - \hbar\omega_{\boldsymbol{Q}}} \times \{\delta_{l_1 l'}\, \delta_{l_2, l}\, \delta_{\boldsymbol{k}_1, \boldsymbol{k}+\boldsymbol{q}}\, \delta_{\boldsymbol{k}_2, \boldsymbol{k}}\, \delta_{\boldsymbol{Q}_1, -\boldsymbol{Q}} \\
&\quad \times [F_{\mathrm{e}}(\varepsilon_{\alpha l' \boldsymbol{k}+\boldsymbol{q}}) - F_{\mathrm{e}}(\varepsilon_{\alpha l \boldsymbol{k}})]\, [\nu(\hbar\omega_{\boldsymbol{Q}}) - \nu(\gamma\varepsilon_{\alpha l' \boldsymbol{k}+\boldsymbol{q}} - \gamma\varepsilon_{\alpha l \boldsymbol{k}})] \\
&\quad + \delta_{l l'}\, \delta_{l_2 l_1}\, \delta_{\boldsymbol{q}\boldsymbol{0}}\, \delta_{\boldsymbol{k}_2 \boldsymbol{k}_1}\, \delta_{\boldsymbol{Q}_1, -\boldsymbol{Q}}\, F_{\mathrm{e}}(\varepsilon_{\alpha l \boldsymbol{k}}) F_{\mathrm{e}}(\varepsilon_{\alpha l_1 \boldsymbol{k}_1})\} .
\end{aligned} \tag{11.34}$$

Similarly, the second Green function reads

$$\begin{aligned}
&\langle\langle\, c^{\dagger}_{\alpha l' \boldsymbol{k}+\boldsymbol{q}\sigma} c_{\alpha l \boldsymbol{k}\sigma} a^{\dagger}_{-\boldsymbol{Q}} \,;\, c^{\dagger}_{\alpha l_2 \boldsymbol{k}_1+\boldsymbol{q}_1\sigma} c_{\alpha l_1 \boldsymbol{k}_1\sigma} a_{\boldsymbol{Q}_1} \rangle\rangle_E \\
&= \frac{1}{2\pi} \frac{1}{E + \varepsilon_{\alpha l' \boldsymbol{k}+\boldsymbol{q}} - \varepsilon_{\alpha l \boldsymbol{k}} - \hbar\omega_{\boldsymbol{Q}}} \times \{\delta_{l_1 l'}\, \delta_{l_2, l}\, \delta_{\boldsymbol{k}_1, \boldsymbol{k}+\boldsymbol{q}}\, \delta_{\boldsymbol{k}_2, \boldsymbol{k}}\, \delta_{\boldsymbol{Q}_1, -\boldsymbol{Q}} \\
&\quad \times [F_{\mathrm{e}}(\varepsilon_{\alpha l' \boldsymbol{k}+\boldsymbol{q}}) - F_{\mathrm{e}}(\varepsilon_{\alpha l \boldsymbol{k}})]\, [1 + \nu(\hbar\omega_{\boldsymbol{Q}}) - \nu(\gamma\varepsilon_{\alpha l' \boldsymbol{k}+\boldsymbol{q}} - \gamma\varepsilon_{\alpha l \boldsymbol{k}})] \\
&\quad + \delta_{l l'}\, \delta_{l_2 l_1}\, \delta_{\boldsymbol{q}\boldsymbol{0}}\, \delta_{\boldsymbol{k}_2 \boldsymbol{k}_1}\, \delta_{\boldsymbol{Q}_1, -\boldsymbol{Q}}\, F_{\mathrm{e}}(\varepsilon_{\alpha l \boldsymbol{k}}) F_{\mathrm{e}}(\varepsilon_{\alpha l_1 \boldsymbol{k}_1})\} .
\end{aligned} \tag{11.35}$$

Finally, we may accomplish the calculation of the frictional force by inserting the above Green functions into expression (11.21) for the frictional force and applying (10.91) with

$$\left\langle c^{(0)\dagger}_{\alpha l' \boldsymbol{k}+\boldsymbol{q}\sigma}(t) c^{(0)}_{\alpha l \boldsymbol{k}\sigma}(t) a^{(0)}_{\boldsymbol{Q}}(t) \right\rangle_0 = \left\langle c^{(0)\dagger}_{\alpha l' \boldsymbol{k}+\boldsymbol{q}\sigma}(t) c^{(0)}_{\alpha l \boldsymbol{k}\sigma}(t) a^{(0)\dagger}_{-\boldsymbol{Q}}(t) \right\rangle_0 = 0 . \tag{11.36}$$

The algebra is tedious but straightforward and therefore only the final result will be quoted here:

$$\begin{aligned}
F_{\mathrm{AC}x} = -2 \sum_{\alpha l l'} \sum_{\boldsymbol{k} q_z} q_x\, |M_{\alpha l l'}(\boldsymbol{q}, q_z)|^2\, [\nu(\hbar\omega_{\boldsymbol{Q}}) - \nu(\gamma\hbar\omega_{\boldsymbol{Q}} + \gamma\hbar q_x v_{\mathrm{D}})] \\
\times \Gamma_{\alpha l l'}(\boldsymbol{q}, \omega_{\boldsymbol{Q}} + q_x v_{\mathrm{D}}) ,
\end{aligned} \tag{11.37}$$

where $\Gamma_{\alpha l l'}(\boldsymbol{q}, \omega)$ is the so-called electron–electron density correlation function. The latter should in principle account for the electron–electron interactions in the inversion layer. However, in the present, simplified treatment, the electron–electron interaction is treated in the Hartree approximation, which includes the average electrostatic field felt by one electron due to the presence of all others. As such the electron–electron interaction is embedded in the electrostatic potential $V(z)$ rather than in a two-particle interaction Hamiltonian that would be quartic in the electron operators. Consequently, the electron–electron density correlation function takes the relatively simple form

$$\Gamma_{\alpha ll'}(\boldsymbol{q},\omega)=\frac{2\pi}{N}\sum_{\boldsymbol{k}}\left[F_{\mathrm{e}}(\varepsilon_{\alpha l\boldsymbol{k}}-\hbar\omega)-F_{\mathrm{e}}(\varepsilon_{\alpha l\boldsymbol{k}})\right]\delta\left(\varepsilon_{\alpha l'\boldsymbol{k}+\boldsymbol{q}}-\varepsilon_{\alpha l\boldsymbol{k}}+\hbar\omega\right)\ , \tag{11.38}$$

where N denotes the number of electrons.

Quite similarly, the expression for the dissipated power may be obtained:

$$P_{\mathrm{AC}}=-2\sum_{\alpha ll'}\sum_{\boldsymbol{k}q_z}\omega_{\boldsymbol{Q}}\left|M_{\alpha ll'}(\boldsymbol{q},q_z)\right|^2\left[\nu(\hbar\omega_Q)\ -\ \nu(\gamma\hbar\omega_{\boldsymbol{Q}}\ +\ \gamma\hbar q_x v_{\mathrm{D}})\right]$$
$$\times\Gamma_{\alpha ll'}(\boldsymbol{q},\omega_{\boldsymbol{Q}}\ +\ q_x v_{\mathrm{D}})\ . \tag{11.39}$$

Although the expressions in (11.37), (11.38) and (11.39) are not yet in their most appropriate form for numerical processing, it's worth taking a closer look since they take the generic form of all frictional forces/dissipation rates due to electron–phonon interactions. The summand consists of four factors:

- The first one represents the momentum (energy) exchanged during an elementary scattering process. i.e. the absorption or emission of one phonon.
- The next factor is proportional to the strength of the electron–phonon interaction for a particular mode $\boldsymbol{Q}$ and differs from its bulk value through the valley and subband dependent structure factors $\{I_{\alpha ll'}(q_z)\}$ reflecting the inversion layer geometry.
- The third factor is measures the balance between absorbed and emitted phonons at a particular $(v_{\mathrm{D}}, E_{\mathrm{L}})$ point.
- Finally, the last term is to account for the electron–electron interaction as far as it is included as an explicit two-particle interaction. However, as stated before, within a Hartree approximation adopted here no correlations are available to 'smear out' the energy conservation of the electron–phonon interaction and a typical delta function is observed.

Note that the energy conservation imposed by the delta function relates to the relative motion of the electrons as can be seen from the Doppler-like term $q_x v_{\mathrm{D}}$ appearing in (11.38). Moreover, both the third and fourth factors are depending on v_{D} in a highly non-linear way, which illustrates the capability of the balance equation approach to investigate transport properties beyond the so-called ohmic regime where the electric fields are low enough to ensure the validity of a linear v_{D}–E_{L} relationship.

For completeness, we will terminate this discussion with some expressions for $F_{\mathrm{AC}x}$ and P_{AC} which are suitable for numerical implementation.

In most cases of interest, it turns out that the geometrical structure factors are sharply peaked around $q_z = 0$ and therefore it is a very good approximation to replace all other factors with their value at $q_z = 0$. This way, the summation (or integration) on q_z can be done analytically and we are left with the following simplified set of equations:

$$F_{\mathrm{AC}x} = -2L_z \sum_{\alpha ll'} \Lambda_{\alpha ll'} \sum_{\boldsymbol{q}} q_x \, |M(\boldsymbol{q})|^2 \, [\nu(\hbar\omega_{\boldsymbol{q}}) - \nu(\gamma\hbar\omega_{\boldsymbol{q}} + \gamma\hbar q_x v_{\mathrm{D}})] \times \Gamma_{\alpha ll'}(\boldsymbol{q}, \omega_{\boldsymbol{q}} + q_x v_{\mathrm{D}}) \tag{11.40}$$

$$P_{\mathrm{AC}} = -2L_z \sum_{\alpha ll'} \Lambda_{\alpha ll'} \sum_{\boldsymbol{q}} \omega_{\boldsymbol{q}} \, |M(\boldsymbol{q})|^2 \, [\nu(\hbar\omega_{\boldsymbol{q}}) - \nu(\gamma\hbar\omega_{\boldsymbol{q}} + \gamma\hbar q_x v_{\mathrm{D}})] \times \Gamma_{\alpha ll'}(\boldsymbol{q}, \omega_{\boldsymbol{q}} + q_x v_{\mathrm{D}}) \tag{11.41}$$

$$\Lambda_{\alpha ll'} = \int_0^\infty \mathrm{d}z \, \phi_{\alpha l}^2(z) \, \phi_{\alpha l'}^2(z) \, . \tag{11.42}$$

Converting all sums over wave vectors into integrals, denoting the number of electrons per unit area by N_s, introducing polar coordinates in the (q_x, q_y)-plane according to

$$\begin{aligned} q_x &= q\cos\phi \\ q_y &= q\sin\phi \\ 0 \le q &\le Q_D; \; 0 \le \phi \le 2\pi \end{aligned} \tag{11.43}$$

and inserting all material parameters, we finally obtain:

$$F_{\mathrm{AC}x} = -\frac{D_{\mathrm{AC}}^2}{2\pi^2\hbar\varrho_{\mathrm{Si}}N_{\mathrm{s}}} \frac{1}{v_{\mathrm{S}}} \sum_{\alpha ll'} \Lambda_{\alpha ll'} \sqrt{m_{\alpha x} m_{\alpha y}} \int_0^{Q_{\mathrm{D}}} \mathrm{d}q \, q^2 \int_0^\pi \mathrm{d}\phi \frac{\cos\phi}{\Lambda_\alpha(\phi)} \times [\nu(\hbar v_{\mathrm{S}} q) - \nu(\gamma\hbar\omega)] \, G_{\alpha ll'}(q,\phi) \tag{11.44}$$

$$P_{\mathrm{AC}} = -\frac{D_{\mathrm{AC}}^2}{2\pi^2\hbar\varrho_{\mathrm{Si}}N_{\mathrm{s}}} \sum_{\alpha ll'} \Lambda_{\alpha ll'} \sqrt{m_{\alpha x} m_{\alpha y}} \int_0^{Q_{\mathrm{D}}} \mathrm{d}q \, q^2 \int_0^\pi \mathrm{d}\phi \frac{1}{\Lambda_\alpha(\phi)} \times [\nu(\hbar v_{\mathrm{S}} q) - \nu(\gamma\hbar\omega)] \, G_{\alpha ll'}(q,\phi) \tag{11.45}$$

$$G_{\alpha ll'}(q,\phi) = \frac{\pi\hbar^2 N_{\mathrm{s}} q \Lambda_\alpha(\phi)}{\sqrt{m_{\alpha x} m_{\alpha y}}} \Gamma_{\alpha ll'}(q\cos\phi, q\sin\phi, 0)$$

$$\omega \equiv v_{\mathrm{S}} q + q v_{\mathrm{D}} \cos\phi$$

$$\Lambda_\alpha(\phi) \equiv \sqrt{\left(\frac{\cos\phi}{m_{\alpha x}}\right)^2 + \left(\frac{\sin\phi}{m_{\alpha y}}\right)^2} \, m_0 \, . \tag{11.46}$$

All other kinds of frictional forces and dissipation rates may evaluated according to the same recipe. Putting everything together we would end up with a set of nonlinear balance equations of the type

$$-eE_{\mathrm{L}} = F_x(T_{\mathrm{e}}, v_{\mathrm{D}}) \tag{11.47}$$

$$-eE_{\mathrm{L}} v_{\mathrm{D}} = P(T_{\mathrm{e}}, v_{\mathrm{D}}) \, . \tag{11.48}$$

As a typical example we show in Fig. 11.2 the (v_D, E_L)-characteristics of a silicon inversion layer in the presence of scattering by acoustic phonons, acoustic and optic intervalley phonons and uniform acceptor doping for different values of the areal electron density N_s. Similarly, Fig. 11.3 displays the electron temperature as a function of the electric field which is pointing in the (100) direction.

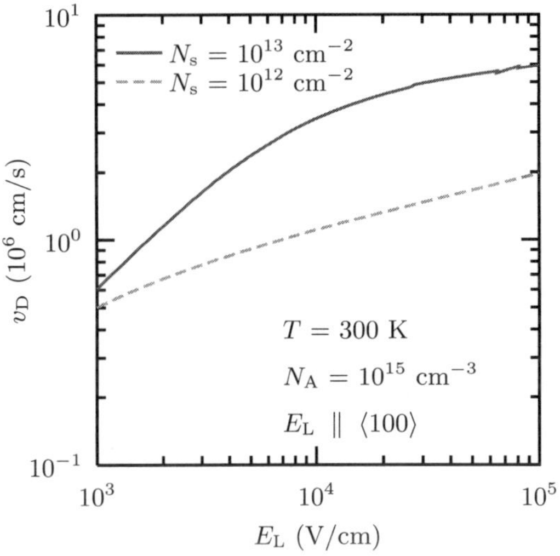

Fig. 11.2. Drift velocity versus electric field in a MOSFET channel

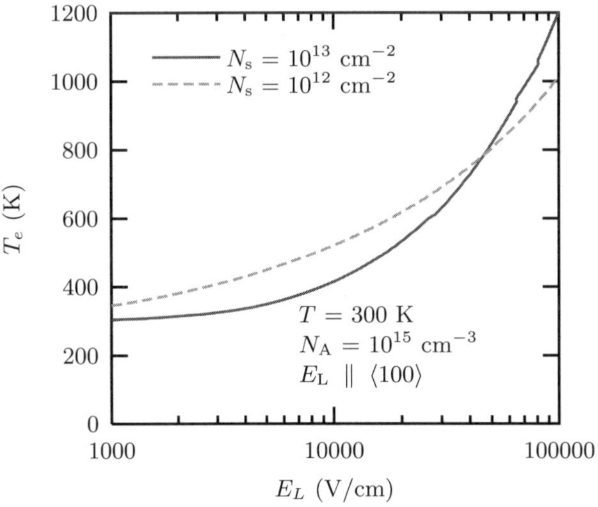

Fig. 11.3. Electron temperature versus electric field in a MOSFET channel

12. Gate Leakage Currents

We will present a method for the evaluation of the charge distribution and quantum-mechanical leakage currents in ultra-thin metal-insulator-semiconductor (MIS) gate stacks that may be composed of several layers of materials. Also the charge distribution due to the finite penetration depth inside the insulating material stack is obtained. The method successfully applies the Breit–Wigner theory of nuclear decay to the confined carrier states in inversion layers and provides an alternative approach for the evaluation of the gate currents to that based on the Wentzel–Kramers–Brillouin (WKB) approximation or Bardeen's perturbative method. A comparison between experimental and simulated current–voltage characteristics has been carried out. The methods of this section were published in [55].

The recent size reduction of metal-oxide-semiconductor field-effect transistors in the sub 0.1 μm regime has led to devices and structures with relatively high doping values and thin insulating layers separating the channel region from the gate layer. Since the thickness of the insulating layers may become of the order of 20 Å or less, the semiconductor region may not be regarded as being uncoupled from the gate. Consequently, a conceptually correct quantum mechanical model predicting both the charge distribution in the channel region and the leakage current flowing between the gate and the channel due to tunneling should treat the gate, the insulator and the semiconductor substrate as a single spatial region accessible to all free charge carriers. The model presented here provides such a coherent treatment for a silicon MIS capacitor with a p-type inversion layer. We will show that the discrete subband states characterizing the energy spectrum of electrons in the inversion layer are emerging as virtual bound states i.e. as sharply peaked resonances of the continuous energy spectrum governing the *entire* MIS structure. It follows that the penetration of the subband wave functions into the insulator region strongly depends on the resonance widths, as is reflected in the formula for the spatial electron distribution. The widths of the bound state resonances are inversely proportional to the lifetimes of the subband states, thereby providing the key quantity for the leakage current formula. The latter is then derived in the framework of the non-interacting particle approximation including however the non-equilibrium features of quantum transport. Furthermore, numerical simulations are presented and discussed for a selected set of MIS capacitors with different gate layer stacks. The theoretical predictions are compared with experimental data.

12.1 Subband States and Resonances

A planar p-type silicon MIS capacitor consisting of a gate electrode, a gate stack and a silicon substrate is considered. The gate stack has a thickness t_{ox} ranging from 15 to 40 Å and contains N_{ox} layers of insulating material such as SiO_2, Si_3Ni_4, Ta_2O_5 etc. When a positive gate voltage V_{G} is applied to the gate electrode, the electrons residing in the electron inversion layer formed near the Si/insulator interface, are coupled to both the gate and the gate stack through non-vanishing tunneling amplitudes. As a result, measurable tunneling currents are observed that involve a net migration of electrons from the leaky inversion layer to the gate electrode.

The z-axis is chosen to be perpendicular to the silicon/oxide interface that is taken to be the (x, y)-plane. The gate, gate stack and semiconductor region are defined by $-\infty \le z < -t_{\text{ox}}$, $-t_{\text{ox}} \le z < 0$ and $0 \le z \le +\infty$ respectively, as depicted in Fig. 12.1. All electron energies including the chemical potential, are measured with respect to the edge of the conduction band at the silicon/insulator interface. The potential energy takes a uniform value in the gate region whereas it approaches the limit U_{sub} in the bulk substrate.

Unlike the Bardeen approach to model tunneling currents [56], we consider the whole MIS capacitor as a single quantum mechanical entity for which the Schrödinger equation needs to be solved. Adopting the effective mass approximation for the electrons in the different valleys, and the Hartree approximation to describe the electron–electron interaction in the inversion layer, the 3D time independent Schrödinger equation for the semiconductor region takes the form:

$$
\begin{aligned}
&-\frac{\hbar^2}{2}\left(\frac{1}{m_{\alpha x}}\frac{\partial^2}{\partial x^2}+\frac{1}{m_{\alpha y}}\frac{\partial^2}{\partial y^2}+\frac{1}{m_{\alpha z}}\frac{\partial^2}{\partial z^2}\right)\psi_\alpha(\boldsymbol{r},z)\\
&+\left[U(z)-E\right]\psi_\alpha(\boldsymbol{r},z) = 0\,,
\end{aligned}
\tag{12.1}
$$

where $\boldsymbol{r} = (x, y)$, α is a valley index and $m_{\alpha x}$, $m_{\alpha y}$ and $m_{\alpha z}$ denote the effective masses along the principle directions of the silicon valleys. The same equation applies to the other regions upon insertion of appropriate effective masses.

Assuming translation invariance in the lateral directions, one may write each one-electron wave function as a plane wave modulated by a one-dimensional envelope wave function $\phi_\alpha(W, z)$ and the corresponding one-electron energy $E_{\alpha\boldsymbol{k}}(W)$ as follows

$$
\psi_{\alpha\boldsymbol{k}}(W,\boldsymbol{r},z)=\frac{1}{\sqrt{L_x L_y}}\,\mathrm{e}^{\,\mathrm{i}\boldsymbol{k}\cdot\boldsymbol{r}}\,\phi_\alpha(W,z)
$$

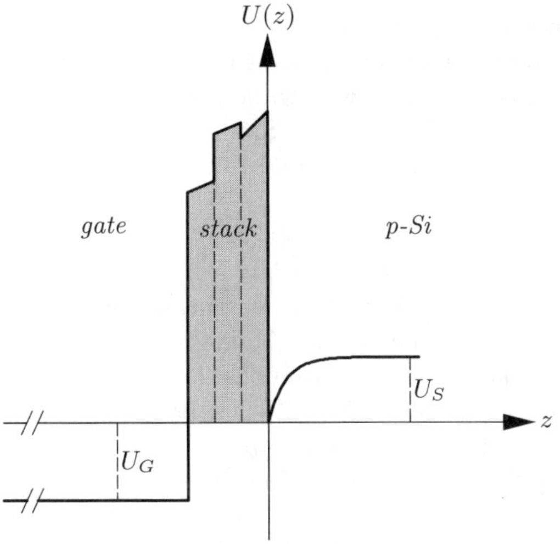

Fig. 12.1. Conduction band profile of a MIS capacitor for $V_G > 0$. t_{ox} is the thickness of the gate stack whereas U_g and U_s respectively denote the potential energy in the gate region and the bulk substrate

$$E_{\alpha\boldsymbol{k}}(W)=\frac{\hbar^2}{2}\left(\frac{k_x^2}{m_{\alpha x}}+\frac{k_y^2}{m_{\alpha y}}\right)+W\;, \tag{12.2}$$

where $\boldsymbol{k} = (k_x, k_y)$ and $\phi_\alpha(W, z)$ is an eigenfunction of the 1D Schrödinger equation

$$-\frac{\hbar^2}{2m_{\alpha z}}\frac{\mathrm{d}^2\phi_\alpha(W,z)}{\mathrm{d}z^2} + [U(z)-W]\,\phi_\alpha(W,z) = 0 \tag{12.3}$$

corresponding to the eigenvalue W.

Since the size of the whole system is assumed to be large in all directions, the energy spectrum will be dense and in particular the eigenvalues W can take all real values exceeding $-U_g$. Moreover, the complete set of wave functions solving (12.3) constitutes an orthogonal, continuous basis for which a proper delta normalization is invoked

$$\langle\phi_\alpha(W')|\phi_\alpha(W)\rangle \equiv \int_{-\infty}^{\infty}\mathrm{d}z\;\phi_\alpha^*(W,z)\phi_\alpha(W,z) = \delta(W'-W)\;. \tag{12.4}$$

Although the insulating layers are relatively thin, the energy barriers separating the inversion layer from the gate electrode are generally high enough to prevent a flood of electrons leaking away into the gate. In other words, in most cases of interest the potential well, hosting the majority of inversion layer electrons, will be coupled only weakly to the gate region. It follows from ordinary quantum mechanics [25] that the relative probability of finding an electron in the inversion layer well should exhibit sharply peaked maxima for a discrete set of W-values. The latter are the resonant energies corresponding to a set of virtually bound states, also called quasi-bound states, that may be regarded as the subband states of the coupled system. This becomes intuitively clear when the thickness of the barrier region is arbitrarily increased so that the coupling between the gate electrode and the semiconductor region vanishes. In this limiting case, the resonant energies will coincide with the true subband energies of the isolated potential well while the resonant wave functions drop to zero at the interface plane $z = 0$. Similarly, the spectral widths of the resonant wave functions tend to zero and the resonance peaks turn into genuine delta functions of W.

The above picture provides an alternative way to investigate the subband structure of an inversion layer. In contrast to the more conventional approaches where the subband wave functions and energies are extracted from the numerical diagonalization of the one-electron Hamiltonian or from a proper variational method, we calculate the full, continuous wave functions by applying a transfer matrix approach to a piecewise constant potential profile and tracing the maxima of the squared wave function amplitudes as a function of W. Once the sequence of resonant subband energies $\{W_{\alpha l} \mid l = 1, 2, \ldots\}$ and the corresponding wave functions are found, we analytically determine the spectral widths that are directly related to the second derivative of the wave functions, with respect to W, evaluated at the resonant energies.

Within the Hartree approximation invoked here, the potential energy profile $U(z)$ needs to be determined by solving self-consistently the above mentioned Schrödinger equation (12.3) and the one-dimensional Poisson equation

$$\frac{\mathrm{d}^2 U(z)}{\mathrm{d}z^2} = -\frac{e^2}{\varepsilon_{\mathrm{S}}} \left[n(z) - p(z) + N_{\mathrm{A}}^{-}(z) \right] , \tag{12.5}$$

where $n(z)$, $p(z)$, $N_{\mathrm{A}}^{-}(z)$ and ε_{S} denote respectively the electron, hole and acceptor concentrations and the permittivity in the silicon part of the structure. Here we do not treat the occurrence of free charges in the gate and the gate stack. On the other hand, charges trapped by interface states are incorporated through a surface charge density D_{it}.

As mentioned above, the potential energy is modeled by a piecewise constant profile defined on a 1D mesh reflecting the gate stack layers and a user-defined number of substrate layers. In this light the self-consistent link between $n(z)$ and $U(z)$ is not provided for each point in the inversion layer but rather for their averages over the subsequent cells of the mesh. This approach is adequate whenever the number of cells is sufficiently large and it has been successfully employed in the past [57, 58, 59]. In the following however, we focus on the procedure to extract the resonant energies and spectral widths.

Since the continuous energy variable W can take any value above, below or equal to the potential energy constant assigned to a particular cell, it proves convenient to define a generic set of basis functions as follows

$$u_1(k,z) = \begin{Bmatrix} \cos kz \\ 1 \\ \cosh kz \end{Bmatrix}, \quad u_2(k,z) = \begin{Bmatrix} \dfrac{\sin kz}{k} \\ z \\ \dfrac{\sinh kz}{k} \end{Bmatrix} \quad \text{for} \quad \begin{Bmatrix} W > U \\ W = U \\ W < U \end{Bmatrix} , \tag{12.6}$$

where m and U respectively denote the effective mass along the z-direction and the potential energy associated with a generic cell, and k given by

$$k = \frac{\sqrt{2m|W-U|}}{\hbar} \, . \tag{12.7}$$

The solutions to the Schrödinger equation for the layered structure can now compactly be written as linear combinations of u_1 and u_2. Moreover, the regular behavior of the basis functions for $k \to 0$ ensures numerical stability on evaluating the transfer matrices for the subsequent layers.

Considering the case of a gate with thickness $t_{\mathrm{g}}(\to \infty)$, a gate stack containing N_{ox} layers with thicknesses $t_1, \ldots, t_{N_{\mathrm{ox}}}$, an inversion layer potential well covered by N_{w} cells with thicknesses $t_{1,\mathrm{w}}, \ldots, t_{N,\mathrm{w}}$ and a substrate region, the linear combinations take the form:

$$\phi_\alpha(W,z) =$$

$$\begin{cases} A_{\mathrm{g},\alpha}\, u_1(k_{\mathrm{g},\alpha}, z+t_{\mathrm{g}}+t_{\mathrm{ox}}) & +B_{\mathrm{g},\alpha}\, u_2(k_{\mathrm{g},\alpha}, z+t_{\mathrm{g}}+t_{\mathrm{ox}}) \\ A_{1,\mathrm{ox},\alpha}\, u_1(k_{1,\mathrm{ox},\alpha}, z+t_{\mathrm{ox}}) & +B_{1,\mathrm{ox},\alpha}\, u_2(k_{1,\mathrm{ox},\alpha}, z+t_{\mathrm{ox}}) \\ \ldots & \\ A_{N_{\mathrm{ox}},\mathrm{ox},\alpha}\, u_1(k_{N_{\mathrm{ox}},ox,\alpha}, z+t_{N_{\mathrm{ox}}}) & +B_{N_{\mathrm{ox}},\mathrm{ox},\alpha}\, u_2(k_{N_{\mathrm{ox}},ox,\alpha}, z+t_{N_{\mathrm{ox}}}) \\ A_{1,\mathrm{w},\alpha}\, u_1(k_{1,\mathrm{w},\alpha}, z) & +B_{1,\mathrm{w},\alpha}\, u_2(k_{1,\mathrm{w},\alpha}, z) \\ \ldots & \\ A_{N_{\mathrm{ox}},\mathrm{w},\alpha}\, u_1(k_{N_{ox},\mathrm{w},\alpha}, z-a+t_{N,\mathrm{w}}) & +B_{N_{\mathrm{ox}},\mathrm{w},\alpha} u_2(k_{N_{ox},\mathrm{w},\alpha}, z-a+t_{N,\mathrm{w}}) \\ A_{\mathrm{s},\alpha}\, u_1(k_{\mathrm{s},\alpha}, z-a) & +B_{\mathrm{s},\alpha}\, u_2(k_{\mathrm{s},\alpha}, z-a) \end{cases} \tag{12.8}$$

where

$$a = \sum_{j=1}^{N_{\mathrm{w}}} t_{j,\mathrm{w}} \tag{12.9}$$

and the coefficients $A_{\mathrm{g},\alpha}, B_{\mathrm{g},\alpha}, \ldots, A_{\mathrm{s},\alpha}, B_{\mathrm{s},\alpha}$ depend on the energy W through the wave numbers $k_{\mathrm{g},\alpha}, \ldots, k_{\mathrm{s},\alpha}$. The subscripts $\mathrm{g}, \mathrm{ox}, \mathrm{w}, \mathrm{s}$ refer to the subsequent layers where the wave functions are evaluated, i.e. the gate, insulator (oxide), well and substrate regions. The connection rules reflecting the layer boundaries enable us to write the total transfer matrix $T_\alpha(W)$ straightforwardly as a product of $N_{\mathrm{ox}} + N_{\mathrm{w}} + 1$ transfer matrices of size 2×2

$$T_\alpha(W) \;=\; T_{\mathrm{g},\alpha}(W) \prod_{j=1}^{N_{\mathrm{ox}}} T_{j,\mathrm{ox},\alpha}(W) \prod_{j=1}^{N_{\mathrm{w}}} T_{j,\mathrm{w},\alpha}(W) \, , \tag{12.10}$$

where the generic transfer matrix connecting two adjacent regions 1 and 2 is given by

$$T_{1\to 2} = \begin{pmatrix} \frac{\partial u_2}{\partial z}(k_1,d_1) - \eta\, u_2(k_1,d_1) \\ \\ -\frac{\partial u_1}{\partial z}(k_1,d_1) \quad \eta\, u_1(k_1,d_1) \end{pmatrix} . \tag{12.11}$$

The variable d_1 denotes the thickness of region 1 and η is the effective mass ratio

$$\eta = \frac{m_1}{m_2} . \tag{12.12}$$

Taking the limit $t_{\mathrm{g}} \to \infty$ and invoking delta normalization according to (12.4) for the gate region, we obtain the following relation between the coefficients $A_{\mathrm{g},\alpha}$ and $B_{\mathrm{g},\alpha}$

$$|A_{\mathrm{g},\alpha}|^2 + \frac{|B_{\mathrm{g},\alpha}|^2}{k_{\mathrm{g},\alpha}^2} = \frac{2m_{\mathrm{g}\alpha z}}{\pi\hbar^2 k_{\mathrm{g},\alpha}} . \tag{12.13}$$

On the the other hand, in the case of an inversion layer only states with energies below U_{s} are occupied and therefore the corresponding wave functions vanish asymptotically in the substrate ($z \to \infty$). Consequently, the coefficient of the increasing exponential arising in (12.8) for $z > a$ is required to be zero, whence

$$B_{\mathrm{s},\alpha} = -k_{\mathrm{s},\alpha}\, A_{\mathrm{s},\alpha} . \tag{12.14}$$

Next, in order to trace the resonance peaks and spectral widths, a numerically stable probability function scanning the presence of an electron in the inversion layer as a function of W, needs to be determined. Rewriting the gate and substrate wave functions as

$$\phi_\alpha(W,z) = \begin{cases} C_{\mathrm{g},\alpha} \sin(k_{\mathrm{g},\alpha}(z + t_{\mathrm{ox}}) + \theta_\alpha) & \text{for} \quad z < -t_{\mathrm{ox}} \\ C_{\mathrm{s},\alpha} \exp(-k_{\mathrm{s},\alpha}(z - a)) & \text{for} \quad z > a \end{cases} \tag{12.15}$$

we obtain the relative probability of an electron for being in the inversion layer

$$P_\alpha(W) \equiv \left| \frac{C_{\mathrm{s},\alpha}(W)}{C_{\mathrm{g},\alpha}(W)} \right|^2 = \frac{1/4\,(A_{\mathrm{s},\alpha}(W) - B_{\mathrm{s},\alpha}(W)/k_{\mathrm{s},\alpha})^2}{A_{\mathrm{g},\alpha}(W)^2 + B_{\mathrm{g},\alpha}^2(W)/k_{\mathrm{g},\alpha}^2} . \tag{12.16}$$

From (12.13, 12.14), it follows that

$$P_\alpha(W) = \frac{\pi\hbar^2 k_{\mathrm{g},\alpha}(W)}{2m_{\mathrm{g}\alpha z}} |A_{\mathrm{s},\alpha}(W)|^2 . \tag{12.17}$$

Emerging as resonance energies in the continuous energy spectrum, the subband energies $W_{\alpha l}$ correspond to distinct and sharply peaked maxima of the $P_\alpha(W)$, or well defined minima of $P_\alpha^{-1}(W)$, even for oxide thicknesses as low as 1 nm. A typical plot of the function $P_\alpha(W)$ is shown in Fig. 12.2. As a consequence, expanding $P_\alpha^{-1}(W)$ in a Taylor series around $W = W_{\alpha l}$, we may replace $P_\alpha(W)$ by a sum of Lorentz-shaped functions

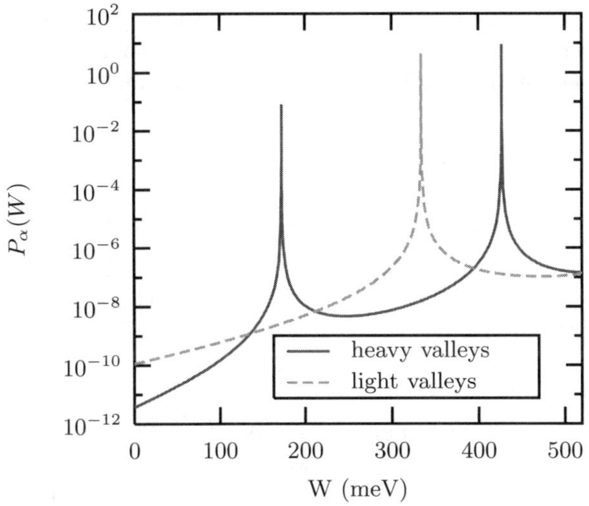

Fig. 12.2. Relative probability to find an electron in the inversion layer versus energy

$$P_\alpha(W) \;\rightarrow\; \sum_l P_\alpha(W_{\alpha l}) \frac{\Gamma_{\alpha l}^2}{(W \;-\; W_{\alpha l})^2 \;+\; \Gamma_{\alpha l}^2}\,, \tag{12.18}$$

where the resonance widths $\Gamma_{\alpha l}^2$ are related to the second derivative of $P_\alpha^{-1}(W)$ through

$$\Gamma_{\alpha l}^2 \;=\; 2P_\alpha^{-1}(W_{\alpha l}) \left[\frac{\partial^2 P_\alpha^{-1}}{\partial W^2}(W_{\alpha l})\right]^{-1} \tag{12.19}$$

and can be directly extracted from the transmission matrices and their derivatives, evaluated at $W = W_{\alpha l}$.

Finally, the electron density in the inversion layer channel can be evaluated for a given temperature T from the following formula including spin degeneracy

$$n(\boldsymbol{r}, z) \;=\; 2\sum_{\alpha \boldsymbol{k}} \int_0^\infty \mathrm{d}W \; |\psi_{\alpha \boldsymbol{k}}(W, \boldsymbol{r}, z)|^2 F(E_{\alpha \boldsymbol{k}}(W), E_\mathrm{F})\,, \tag{12.20}$$

where $\beta \;=\; 1/k_\mathrm{B}T$ and E_F denotes the Fermi energy associated with the inversion layer electrons and

$$F(E, \mu) \;=\; \frac{1}{1 + \exp\left(\beta(E - \mu)\right)} \tag{12.21}$$

represents the Fermi–Dirac distribution function. Converting $(1/L_x Ly)\sum_{\boldsymbol{k}}$ into $(1/4\pi^2)\int \mathrm{d}^2k$ and carrying out the two-dimensional integral, we obtain

$$n(z) = \frac{1}{\pi\hbar^2\beta} \sum_\alpha \sqrt{m_{\alpha x} m_{\alpha y}} \int_0^\infty \mathrm{d}W \;\phi_\alpha^2(W, z) \times \log\left[1 \;+\; \exp(\beta(E_\mathrm{F} - W))\right]\,. \tag{12.22}$$

In analogy with (12.18), the wave functions are essentially vanishing within the inversion layer unless W is close to one of the resonances and may adequately be replaced with their Lorentzian approximations

$$\phi_\alpha^2(W,z) \to \sum_l \phi_{\alpha l}^2(z) \frac{\Gamma_{\alpha l}^2}{(W - W_{\alpha l})^2 + \Gamma_{\alpha l}^2} . \tag{12.23}$$

Substitution of (12.23) into (12.22) allows us to evaluate analytically the energy integral to arrive at the final result

$$n(z)=\frac{1}{\pi \hbar^2 \beta} \sum_\alpha \sqrt{m_{\alpha x} m_{\alpha y}} \sum_l \phi_{\alpha l}^2(z)\, \Gamma_{\alpha l} \left(\frac{\pi}{2} + \mathrm{Arctan}\, \frac{W_{\alpha l}}{\Gamma_{\alpha l}} \right)$$
$$\times \log \left[1 + \exp(\beta(E_\mathrm{F} - W_{\alpha l}))\right] . \tag{12.24}$$

This formula was numerically implemented to calculate the electron density in the channel and the gate stack, using the appropriate effective masses as well as the areal electron density $N_S = \int_0^\infty n(z)\, \mathrm{d}z$. It should be noted that the arc tangent arising in (12.24) hardly differs from $\pi/2$ in most relevant cases, i.e. when $\Gamma_{\alpha l} << W_{\alpha l}$. A typical density profile is shown in Figs. 12.3 and 12.4. The logarithmic plot shows both the penetration of electrons in the oxide and the occurrence of oscillations arising from the sine-like shape of the wave functions in the metallic gate. The latter is not contradicting the observation that in a metal $n(z)$ should be huge and exhibit no oscillations at all, because we have not incorporated the continuous part of the spectrum that has no further relevance for a study of the inversion layer.

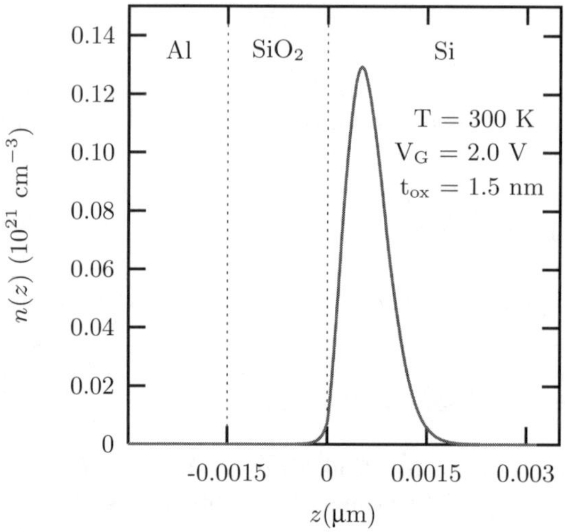

Fig. 12.3. Electron density profile in a leaky MIS capacitor versus distance. The Si/SiO_2interface is the plane $z = 0$

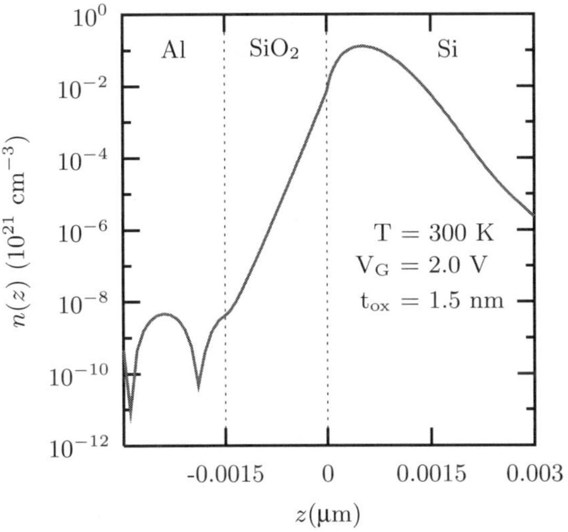

Fig. 12.4. Same electron density profile on a logarithmic scale. The oscillations reflect the sine-shape of the wave functions in the gate region

12.2 Tunneling Gate Currents

The subband structure of a p-type inversion layer channel may be seen to emerge from an enumerable set of sharp resonances appearing in the continuous energy spectrum of the composed system consisting of the gate contact, the gate stack (insulating layers), the inversion layer and the substrate contact. In particular, the discreteness of the subband states is intimately connected with the presence of energy barriers in the gate stack that restrict the coupling between the channel and the gate regions and therefore the amplitude for tunneling of electrons through the barriers (see Fig. 12.1). Clearly, the smallness of the above mentioned coupling is reflected in the size of the resonance width – or equivalently, the resonance lifetime $\tau_{\alpha l} = \hbar/2\Gamma_{\alpha l}$ – as compared to the resonance energy.

It is tempting to identify the gate leakage current as electrons originating from decaying subband states. However, before such a link can be established, a conceptual problem should be resolved. Although intuition obviously suggests that an electron residing in a particular subband $|\alpha l\rangle$ should contribute an amount $-e/\tau_{\alpha l}$ to the gate current, this is apparently contradicted by the observation that the current density corresponding to each individual subband wave function identically vanishes. The latter is due to the nature of the resonant states. Contrary to the case of the doubly degenerate running states having energies above the bottom of the conduction band in the substrate, the inversion layer resonances states are non-degenerate and virtually bound, and the wave functions are rapidly decaying into the substrate area. As a consequence, all wave functions are real (up to an irrelevant phase factor) and the diagonal matrix elements of the current density operator vanishes. The vanishing of the current for the envelope wave functions was also noted in

[60, 61]. Therefore, we need to establish a sound physical model (workaround) resolving the current paradox and connecting the resonance lifetimes to the gate current. Since we do not adopt a plane-wave hypothesis for the inversion layer electrons in the perpendicular direction, our resolution of the paradox differs from the one that is proposed in [60].

The paradox can be resolved by noting that the resonant states, though diagonalizing the electron Hamiltonian in the presence of the gate bias, are constituting a *non-equilibrium* state of the whole system which is not necessarily described by a Gibbs-like statistical operator, even not when the steady state is reached. There are at least two alternatives to solve the problem in practice.

The most rigorous approach aims at solving the full time dependent problem starting from a MIS capacitor that is in thermal equilibrium ($V_G = 0$) until some initial time $t = 0$. Before $t = 0$, the potential profile is essentially determined by the gate stack barriers and, due to the absence of an appreciable inversion layer potential well, all eigensolutions of the time independent Schrödinger equation are linear combinations of transmitted and reflected waves. In other words, almost all states are carrying current, although the thermal average is of course zero (equilibrium). However, it should be possible to calculate the time evolution of the creation and annihilation operators related to the unperturbed states. The perturbed resonant states, defining the subband structure for $V_G > 0$, would serve as a set of intermediate states participating in all transitions between the unperturbed states caused by the applied gate voltage. Although such an approach is conceptually straightforward, it is probably rather cumbersome to be carried out in practice.

We propose a strategy that is borrowed from the theory of nuclear decay [20, 21]. The resulting model leads to a concise calculation scheme for the gate current. Under the assumption[1] that the resonance widths of the virtual bound states are much smaller than their energies, the corresponding real wave functions can be extended to the complex plane if the resonance energies and the corresponding resonant widths are combined to form complex energy eigenvalues of the Schrödinger equation. Such an extension enables us to mimic both the supply (creation) and the decay (disintegration) of particles in a resonant bound state by studying the wave functions in those regions of space where the real, i.e. non-complex, wave functions would be standing waves either asymptotically or exactly.

In the case of a leaky inversion layer covering the interval $[0, +\infty]$, the above mentioned region would be the gate area, i.e. the interval $[-\infty, -t_{\mathrm{ox}}]$ that is characterized by a constant potential energy U_{g}. By introducing complex wave numbers k and complex energies W, we may generally write the time-independent envelope wave functions as follows

[1] This assumption is valid if the gate stack has sufficiently large or thick barriers in order to keep the coupling between gate and semiconductor areas relatively small.

$$\phi_\alpha(W,z) = C_\alpha(W)\,\mathrm{e}^{\,\mathrm{i}k(z+t_{\mathrm{ox}})} + C^*_\alpha(W)\,\mathrm{e}^{-\mathrm{i}k(z+t_{\mathrm{ox}})}\,, \tag{12.25}$$

where $k_1 = \Re(k) > 0$ and both the sign of $k_2 = \Im(k)$ and the relationship between k and W are determined by the class to which the states belong. $C^*_\alpha(W)$ is the complex conjugate function to C_α, taken however at the same, complex argument W. On the other hand, if W were chosen to be real, the real wave functions would be recovered.

Decaying states describe the process of inversion layer electrons leaking away into the gate region through gate tunneling. Anticipating the exponential decay as a function of time, we write the complex energy eigenvalues in the form:

$$W = W_{\alpha l} - \mathrm{i}\Gamma_{\alpha l}\,. \tag{12.26}$$

Since the inversion layer is emptied into the gate region, we have to look for complex eigensolutions describing transport in the negative z-direction. In other words, the complex eigenvalues of (12.26) should be the solutions to

$$C_\alpha(W_{\alpha l} - \mathrm{i}\Gamma_{\alpha l}) = 0\,. \tag{12.27}$$

For energy values W close to $W_{\alpha l} - \mathrm{i}\Gamma_{\alpha l}$, we may expand $C_\alpha(W)$ in a Taylor series around $W = W_{\alpha l} - \mathrm{i}\Gamma_{\alpha l}$

$$\begin{aligned} C_\alpha(W) &= C_\alpha(W_{\alpha l} - \mathrm{i}\Gamma_{\alpha l}) + D_{\alpha l}\,(W - W_{\alpha l} + \mathrm{i}\Gamma_{\alpha l}) + \ldots \\ &= D_{\alpha l}\cdot(W - W_{\alpha l} + \mathrm{i}\Gamma_{\alpha l})\,. \end{aligned} \tag{12.28}$$

The first term vanishes because our present interest concerns decaying states for which $C^*_\alpha(W) = 0$. The resonance approach allows us to truncate the Taylor series after the one term, since only values close to $W_{\alpha l} - \mathrm{i}\Gamma_{\alpha l}$ are contributing.

Correspondingly, $C^*_\alpha(W)$ reads

$$C^*_\alpha(W) = D^*_{\alpha l}\cdot(W - W_{\alpha l} - \mathrm{i}\Gamma_{\alpha l})\,. \tag{12.29}$$

In particular, the envelope wave function corresponding to $W = W_{\alpha l} - \mathrm{i}\Gamma_{\alpha l}$ is given by

$$\begin{aligned} \phi_{\alpha l}(z) \equiv \phi_\alpha(W_{\alpha l} - \mathrm{i}\Gamma_{\alpha l}, z) &= C^*_\alpha(W_{\alpha l} - \mathrm{i}\Gamma_{\alpha l})\,\mathrm{e}^{-\mathrm{i}k(z+t_{\mathrm{ox}})} \\ &= -2\mathrm{i}\,\Gamma_{\alpha l}D^*_{\alpha l}\,\mathrm{e}^{-\mathrm{i}k(z+t_{\mathrm{ox}})} \\ &= -2\mathrm{i}\,\Gamma_{\alpha l}D^*_{\alpha l}\,\mathrm{e}^{-\mathrm{i}k_1(z+t_{\mathrm{ox}})}\,\mathrm{e}^{\,k_2(z+t_{\mathrm{ox}})}\,. \end{aligned} \tag{12.30}$$

Using the relation between the complex energy eigenvalue $W_{\alpha l} - \mathrm{i}\Gamma_{\alpha l}$ and its associated complex wave number $k = k_1 + \mathrm{i}k_2$ and substituting (12.30) into the Schrödinger equation (for the gate region) we obtain

$$\frac{\hbar^2(k_1^2 - k_2^2)}{2m_{\mathrm{g}\alpha z}} = W_{\alpha l} - U_{\mathrm{g}} \tag{12.31}$$

$$\frac{\hbar^2 k_1 k_2}{m_{\mathrm{g}\alpha z}} = -\Gamma_{\alpha l} \tag{12.32}$$

From the latter and $k_1 > 0$, $\Gamma_{\alpha l} > 0$, it becomes clear that $k_2 < 0$ and

$$\phi_{\alpha l}(z) = -2\mathrm{i}\,\Gamma_{\alpha l} D^*_{\alpha l}\, \mathrm{e}^{-\mathrm{i}k_1(z+t_{\mathrm{ox}})}\, \mathrm{e}^{-|k_2|(z+t_{\mathrm{ox}})} \,. \tag{12.33}$$

Inversion layer confinement imposes the boundary condition

$$\lim_{z\to+\infty} \phi_{\alpha l}(z) = 0 \,. \tag{12.34}$$

Furthermore, substitution of (12.32) into (12.31) and invoking the restriction $k_1^2 > 0$ yields

$$k_1^2 = \frac{m_{\mathrm{g}\alpha z}}{\hbar^2}\left(W_{\alpha l} - U_{\mathrm{g}} + \sqrt{(W_{\alpha l} - U_{\mathrm{g}})^2 + \Gamma^2_{\alpha l}}\right) , \tag{12.35}$$

which, together with $\Gamma_{\alpha l} << W_{\alpha l}$, shows that k_1 is mainly determined by the resonant energy $W_{\alpha l}$.

Finally, since we assumed that $\Gamma_{\alpha l} << W_{\alpha l}$, we may determine the explicit value of $D_{\alpha l}$ by estimating the current carried by a particular, total wave function as follows. Denoting respectively by $\boldsymbol{k}$ and $\boldsymbol{r}$ the total wave vector and position vector in the (x,y)-plane, we may write the total, time dependent wave function of a particular decaying eigenstate $\alpha l\boldsymbol{k}$ as

$$\psi_{\alpha l\boldsymbol{k}}(\boldsymbol{r},z,t) = \chi_{\alpha\boldsymbol{k}}(\boldsymbol{r})\,\phi_{\alpha l}(z)\,\mathrm{e}^{-\mathrm{i}(E_{\alpha\boldsymbol{k}} + W_{\alpha l})t/\hbar}\,\mathrm{e}^{-\Gamma_{\alpha l}t/\hbar} \tag{12.36}$$

with

$$E_{\alpha\boldsymbol{k}} = \frac{\hbar^2}{2}\left(\frac{k_x^2}{m_{\mathrm{g}\alpha x}} + \frac{k_y^2}{m_{\mathrm{g}\alpha y}}\right) \tag{12.37}$$

$$\chi_{\alpha\boldsymbol{k}}(\boldsymbol{r}) = \frac{1}{\sqrt{L_x L_y}}\,\mathrm{e}^{\mathrm{i}\boldsymbol{k}\cdot\boldsymbol{r}} , \tag{12.38}$$

where L_x and L_y denote the lateral dimensions of the MIS capacitor. Being reduced to its z-component, the current density carried by the eigenstate $\alpha l\boldsymbol{k}$ reads

$$\begin{aligned} J^z_{\alpha l\boldsymbol{k}}(\boldsymbol{r},z,t) &= \frac{\mathrm{i}e\hbar}{2m_{\mathrm{g}\alpha z}}\left[\psi^*_{\alpha l\boldsymbol{k}}(\boldsymbol{r},z,t)\frac{\partial\psi_{\alpha l\boldsymbol{k}}(\boldsymbol{r},z,t)}{\partial z} - \mathrm{c.c.}\right]\mathrm{e}^{-2\Gamma_{\alpha l}t/\hbar} \\ &= \frac{\mathrm{i}e\hbar}{2m_{\mathrm{g}\alpha z}}|\chi_{\alpha\boldsymbol{k}}(\boldsymbol{r})|^2\left[\phi^*_{\alpha l}(z)\frac{\partial\phi_{\alpha l}(z)}{\partial z} - \mathrm{c.c.}\right]\mathrm{e}^{-2\Gamma_{\alpha l}t/\hbar} \,. \end{aligned} \tag{12.39}$$

Combining (12.33), (12.25) and (12.39), we arrive at

$$J^z_{\alpha l\boldsymbol{k}}(\boldsymbol{r},z,t) = 4e\frac{\hbar k_1}{m_{\mathrm{g}\alpha z}}|\chi_{\alpha\boldsymbol{k}}(\boldsymbol{r})|^2\,\Gamma^2_{\alpha l}|D_{\alpha l}|^2\,\mathrm{e}^{-2\left(|k_2|(z+t_{\mathrm{ox}}) + \Gamma_{\alpha l}t/\hbar\right)} \tag{12.40}$$

while the total current $I_{\alpha l}(z,t) = \int \mathrm{d}^2 r J^z_{\alpha l\boldsymbol{k}}(\boldsymbol{r},z,t)$ turns out to be (due to $\int \mathrm{d}^2 r|\chi_{\alpha\boldsymbol{k}}(\boldsymbol{r})|^2 = 1$)

$$
\begin{aligned}
I_{\alpha l}(z,t) &= \int \mathrm{d}^2 r\, J^z_{\alpha l \boldsymbol{k}}(\boldsymbol{r},z,t) \\
&= 4e \frac{\hbar k_1}{m_{\mathrm{g}\alpha z}}\, \Gamma^2_{\alpha l} |D_{\alpha l}|^2\, \mathrm{e}^{-2\left(|k_2|(z+t_{\mathrm{ox}}) + \Gamma_{\alpha l} t/\hbar\right)} \, .
\end{aligned} \tag{12.41}
$$

In the neighborhood of the insulating layer, this current should not appreciably differ from the current leaving the inversion layer at $z = 0$, i.e. no charges are piling up in the insulator

$$
I_{\alpha l}(-t_{\mathrm{ox}}, t) \approx I_{\alpha l}(0,t) \, . \tag{12.42}
$$

The latter can be easily connected to the instantaneous electron charge $Q_{\alpha l}(t)$ leaking away from the inversion layer through the continuity equation which is a direct consequence of the time dependent Schrödinger equation

$$
\begin{aligned}
\frac{\mathrm{d}Q_{\alpha l}(t)}{\mathrm{d}t} &= -\int \mathrm{d}^2 r \int_0^\infty \mathrm{d}z\, \nabla \cdot \boldsymbol{J}_{\alpha l \boldsymbol{k}}(\boldsymbol{r},z,t) \\
&= -\int_0^\infty \mathrm{d}z\, \frac{\partial I_{\alpha l}(z,t)}{\partial z} = I_{\alpha l}(0,t) \, ,
\end{aligned} \tag{12.43}
$$

where $Q_{\alpha l}(t)$ is defined by

$$
Q_{\alpha l}(t) = -e \int \mathrm{d}^2 r \int_0^\infty \mathrm{d}z\, |\psi_{\alpha l \boldsymbol{k}}(\boldsymbol{r},z,t)|^2 \, . \tag{12.44}
$$

On the other hand, the electron charge can obviously be extracted from the wave function for the semiconductor region:

$$
Q_{\alpha l}(t) = -e\, \mathrm{e}^{-2\Gamma_{\alpha l} t/\hbar} \int_0^\infty \mathrm{d}z\, |\phi_{\alpha l}(z)|^2 \, . \tag{12.45}
$$

Imposing the condition that an electron occupying the eigenstate $\alpha l \boldsymbol{k}$ be located in the inversion layer at some arbitrary instant $t = 0$, i.e.

$$
Q_{\alpha l}(0) = -e \tag{12.46}
$$

we find

$$
I_{\alpha l}(0,t) = \frac{\mathrm{d}Q_{\alpha l}(t)}{\mathrm{d}t} = \frac{2e\Gamma_{\alpha l}}{\hbar}\, \mathrm{e}^{-2\Gamma_{\alpha l} t/\hbar} \, . \tag{12.47}
$$

From (12.41), (12.42) and (12.47) it now becomes possible to fix $|D_{\alpha l}|$

$$
|D_{\alpha l}|^2 = \frac{m_{\mathrm{g}\alpha z}}{2\hbar^2 k_1 \Gamma_{\alpha l}} \, . \tag{12.48}
$$

Insertion into (12.40) finally yields the anticipated result

$$
J^z_{\alpha l \boldsymbol{k}}(\boldsymbol{r},z,t) = J^z_{\alpha l}(z,t) = \frac{1}{L_x L_y} \frac{e}{\tau_{\alpha l}}\, \mathrm{e}^{-2\left(|k_2|(z+t_{\mathrm{ox}}) + \Gamma_{\alpha l} t/\hbar\right)} \, , \tag{12.49}
$$

where $\tau_{\alpha l} = \hbar/2\Gamma_{\alpha l}$ is the lifetime of the eigenstate $\alpha l \boldsymbol{k}$.

Equation (12.49) describes a current that decreases exponentially. This result corresponds to the situation in which the electron states can decay only once and is analogous to the Breit–Wigner theory for describing nuclear decay [62]. However, in contradistinction to nuclear decay where each nucleus decays only once, the resonant states are filled by electrons from the external electromotive force or, for floating regions, from the Schottky–Read–Hall generation mechanism in junctions. We will follow here the heuristic approach that was advocated in [60] and assume that any decayed state is instantaneously refilled. In a more refined approach, a trade-off between the decay-time constant and the refill-time constant may be studied. The instantaneous refilling however allows us to ignore the time dependence from the exponential part. The resulting steady-state situation is also reflected also in the particular form of the statistical operator that is proposed at the end of this section.

Furthermore, assuming a uniform gate area and relying on the basic assumption $\Gamma_{\alpha l} << W_{\alpha l}$, we may assign the following constant current density to each decaying subband state $\alpha l \boldsymbol{k}$:

$$J_{\alpha l}^{z} = \frac{1}{L_x L_y} \frac{e}{\tau_{\alpha l}} . \tag{12.50}$$

A similar result has been proposed by Lo et al. [63, 64] who exploited the analogy with inhomogeneously filled waveguides to calculate the resonant energies and the leakage current flowing through a single oxide layer. Although above expression is extremely condensed, the evaluation of the lifetimes requires all the detailed knowledge of the gate stack and cannot be done without using numerical methods.

The global approach of the present theory implies a remarkable symmetry. Just as inversion-layer electrons may decay into the gate, it is also possible that gate electrons 'decay' into the inversion layer. This process will be referred to as state loading in distinction to state emptying that was mentioned above. Unlike the latter, state loading is described by a right traveling wave $\mathrm{e}^{\mathrm{i}k_1 z}\,\mathrm{e}^{-k_2 z}$ in the expansion (12.25). Clearly, since the wave functions of all states in the range $0 < W_{\alpha l} < U_{\mathrm{s}}$ are dying out in the bulk of the substrate, k_2 should be positive. Consequently, the complex energy eigenvalues must be of the form:

$$W = W_{\alpha l} + \mathrm{i}\Gamma_{\alpha l} . \tag{12.51}$$

Analogously, the complex energy eigenvalues are now zeros of the coefficient $C_{\alpha}^{*}(W)$ in (12.25):

$$C_{\alpha}^{*}(W_{\alpha l} + \mathrm{i}\Gamma_{\alpha l}) = 0 . \tag{12.52}$$

The corresponding Taylor expansion of $C_{\alpha}^{*}(W)$ around $W = W_{\alpha l} + \mathrm{i}\Gamma_{\alpha l}$ reads

$$C^*_\alpha(W) = C^*_\alpha(W_{\alpha l} + \mathrm{i}\Gamma_{\alpha l}) + D^*_{\alpha l}\,(W - W_{\alpha l} - \mathrm{i}\Gamma_{\alpha l}) + \dots$$
$$= D^*_{\alpha l} \cdot (W - W_{\alpha l} - \mathrm{i}\Gamma_{\alpha l})\,. \tag{12.53}$$

The derivation of the asymptotic form of the subband wave functions and the corresponding current densities can be evaluated in the same way as before. The subsequent results are quoted for completeness below.

$$C_\alpha(W) = D_{\alpha l} \cdot (W - W_{\alpha l} + \mathrm{i}\Gamma_{\alpha l}) \tag{12.54}$$

$$\phi_{\alpha l}(z) = 2\mathrm{i}\,\Gamma_{\alpha l} D_{\alpha l}\, \mathrm{e}^{\,\mathrm{i}k_1(z + t_{\mathrm{ox}})}\, \mathrm{e}^{-k_2(z + t_{\mathrm{ox}})} \tag{12.55}$$

$$\psi_{\alpha l \boldsymbol{k}}(\boldsymbol{r}, z, t) = \chi_{\alpha \boldsymbol{k}}(\boldsymbol{r})\, \phi_{\alpha l}(z)\, \mathrm{e}^{-\mathrm{i}(E_{\alpha \boldsymbol{k}} + W_{\alpha l})t/\hbar}\, \mathrm{e}^{\,\Gamma_{\alpha l} t/\hbar} \tag{12.56}$$

$$J^z_{\alpha l \boldsymbol{k}}(\boldsymbol{r}, z, t) = -4e\frac{\hbar k_1}{m_{\mathrm{g}\alpha z}} |\chi_{\alpha \boldsymbol{k}}(\boldsymbol{r})|^2\, \Gamma^2_{\alpha l} |D_{\alpha l}|^2\, \mathrm{e}^{-2\,(k_2(z + t_{\mathrm{ox}}) - \Gamma_{\alpha l} t/\hbar)} \tag{12.57}$$

$$I_{\alpha l}(z, t) = -4e\frac{\hbar k_1}{m_{\mathrm{g}\alpha z}}\, \Gamma^2_{\alpha l} |D_{\alpha l}|^2\, \mathrm{e}^{-2\,(k_2(z + t_{\mathrm{ox}}) - \Gamma_{\alpha l} t/\hbar)} \tag{12.58}$$

$$\frac{dQ_{\alpha l}(t)}{dt} = -\int_0^\infty \mathrm{d}z\, \frac{\partial I_{\alpha l}(z, t)}{\partial z} = I_{\alpha l}(0, t)\,. \tag{12.59}$$

Assuming that the 'loading' states have zero amplitude at $t = -\infty$ and have contributed one electron at $t = 0$ – thereby adopting the normalization $\int_0^{+\infty} \mathrm{d}z |\phi_{\alpha l}(z)|^2 = 1$ once again – we now obtain

$$I_{\alpha l}(0, t) = \frac{\mathrm{d}Q_{\alpha l}(t)}{\mathrm{d}t} = -\frac{2e\Gamma_{\alpha l}}{\hbar}\, \mathrm{e}^{2\Gamma_{\alpha l} t/\hbar}\,. \tag{12.60}$$

Taking for granted also in this case the basic assumption $W_{\alpha l} << \Gamma_{\alpha l}$ we can similarly justify the identification

$$I_{\alpha l}(0, t) = I_{\alpha l}(-t_{\mathrm{ox}}, t) \qquad \forall\, t \leq 0, \tag{12.61}$$

leading to

$$|D_{\alpha l}|^2 = \frac{m_{\mathrm{g}\alpha z}}{2\hbar^2 k_1 \Gamma_{\alpha l}} \tag{12.62}$$

and, for $t \leq 0$,

$$J^z_{\alpha l \boldsymbol{k}}(\boldsymbol{r}, z, t) = J^z_{\alpha l}(z, t) = -\frac{1}{L_x L_y} \frac{e}{\tau_{\alpha l}}\, \mathrm{e}^{-2\,(k_2(z + t_{\mathrm{ox}}) - \Gamma_{\alpha l} t/\hbar)} \tag{12.63}$$

or, ignoring again any dependence on time and position,

$$J^z_{\alpha l} = J^z_{\alpha l}(z, t) = -\frac{1}{L_x L_y} \frac{e}{\tau_{\alpha l}}\,. \tag{12.64}$$

Within the scope of the present work, scattering by phonons, impurities or any other material dependent interactions is neglected. Moreover, electron–electron interaction is treated in the Hartree approximation that, in practice, amounts to a self-consistent solution of the one-particle Schrödinger equation and Poisson's equation. Therefore, bearing in mind that normal transport through the gate stack is limited by tunneling events and referring to the time reversal symmetry breaking between decaying and loading states, we propose the following boundary conditions for the statistical operator, satisfying the non-interacting[2]

$$
\begin{aligned}
\varrho &= \prod_{\alpha l \boldsymbol{k} \sigma} \varrho_{\alpha l \boldsymbol{k} \sigma} \\
\varrho_{\alpha l \boldsymbol{k} \sigma} &= \frac{1}{Z_{\alpha l \boldsymbol{k} \sigma}} \exp\left[-\beta\left(E_{\alpha \boldsymbol{k}} + W_{\alpha l} - \mu_{\alpha l}\right) c^{\dagger}_{\alpha l \boldsymbol{k} \sigma} c_{\alpha l \boldsymbol{k} \sigma}\right] \\
Z_{\alpha l \boldsymbol{k} \sigma} &= \operatorname{Tr} \exp\left[-\beta\left(E_{\alpha \boldsymbol{k}} + W_{\alpha l} - \mu_{\alpha l}\right) c^{\dagger}_{\alpha l \boldsymbol{k} \sigma} c_{\alpha l \boldsymbol{k} \sigma}\right] \\
\mu_{\alpha l} &= \begin{cases} E_{\mathrm{F}} - eV_{\mathrm{G}} & \text{if } k_1 > 0 \\ E_{\mathrm{F}} & \text{if } k_1 < 0, \end{cases}
\end{aligned}
\tag{12.65}
$$

where σ, $c^{\dagger}_{\alpha l \boldsymbol{k} \sigma}$ and $c_{\alpha l \boldsymbol{k} \sigma}$ respectively denote the spin index and electron creation- and annihilation operators. Imposing the boundary conditions we may conclude from (12.65) that the correlation functions of the type $\langle c^{\dagger} c \rangle$ are diagonal in all labels $\alpha, l, \boldsymbol{k}$ and σ. Consequently, the gate current density is given by

$$
\begin{aligned}
J_{\mathrm{G}} &= \sum_{\alpha l \boldsymbol{k} \sigma} J^{z}_{\alpha l} \left\langle c^{\dagger}_{\alpha l \boldsymbol{k} \sigma} c_{\alpha l \boldsymbol{k} \sigma} \right\rangle \\
&= -\frac{2e}{L_x L_y} \sum_{\alpha l} \frac{1}{\tau_{\alpha l}} \sum_{\boldsymbol{k}} \left[F(E_{\alpha \boldsymbol{k}} + W_{\alpha l}, E_{\mathrm{F}} - eV_{\mathrm{G}}) - F(E_{\alpha \boldsymbol{k}} + W_{\alpha l}, E_{\mathrm{F}})\right]
\end{aligned}
\tag{12.66}
$$

Evaluation of the two-dimensional integral over $\boldsymbol{k}$, provides us with the gate current as a sum of discrete contributions arising from the virtual resonances

$$
J_{\mathrm{G}} = -\frac{e}{\pi \hbar^2 \beta} \sum_{\alpha l} \frac{\sqrt{m_{\alpha x} m_{\alpha y}}}{\tau_{\alpha l}} \log \frac{1 + \exp\left(\beta\left(E_{\mathrm{F}} - W_{\alpha l} - eV_{\mathrm{G}}\right)\right)}{1 + \exp\left(\beta\left(E_{\mathrm{F}} - W_{\alpha l}\right)\right)} . \tag{12.67}
$$

12.3 Results of the Gate-Leakage Current Calculations

It is clear from (12.67) that the (inverse) lifetimes are the key quantities building up the new formula for the gate leakage current. These variables apparently replace the familiar transmission coefficients that would emerge from traveling states (running waves) contributing to the current in accumulation mode. As was discussed already in the previous section, this feature reflects the scope of nuclear decay theory which is a fair attempt to resolve

[2] The label 'non-interacting' refers to the absence of explicit inter-particle scattering.

the leakage current paradox. Although the latter theory produces a dynamical evolution of the one-particle wave functions, we have eventually inserted a time independent, yet non-equilibrium, statistical operator to calculate the averages and it would be desirable to verify the success of this procedure on the grounds of sound time-dependent non-equilibrium theory. The same recommendation can be made regarding a more systematic investigation of the agreement between the results of the present calculation and the simulations based on Bardeen's approach [65].

As far as the numerical evaluation is concerned, the above model has been implemented into SCALPEL [66], a computer program that was developed in the past to calculate subband energies and wave functions for inversion layers adjacent to thick oxide layers, i.e. without accounting for oxide penetration or tunneling currents. Although the inclusion of the insulating layers and the gate region required a considerable modification of the original code, the general framework of SCALPEL has remained unchanged and therefore we refer to [66, 67, 68] for further details.

In Fig. 12.5 we have compared the simulation results generated by SCALPEL with a gate current characteristic that was obtained from in-house measurements on a large MIS transistor with a NO insulator and grounded source and drain contacts. The latter serve as huge electron reservoirs capable of replacing the channel electrons (inversion) that participate in the gate tunneling current, such that the assumption on instantaneous injection or absorption compensating for migrating electrons is justified.

The following parameters are used: $T = 300K$, $t_{\rm ox} = 25$ Å, $m_{g\alpha x} = m_{g\alpha y} = m_{g\alpha z} = 0.32m_0$, $N_{\rm ox} = 3$, $m_{1,{\rm ox},\alpha} = \ldots m_{3,{\rm ox},\alpha} = 0.42m_0$. The barrier height and the dielectric constant of the NO layer are taken to be 3.15 eV

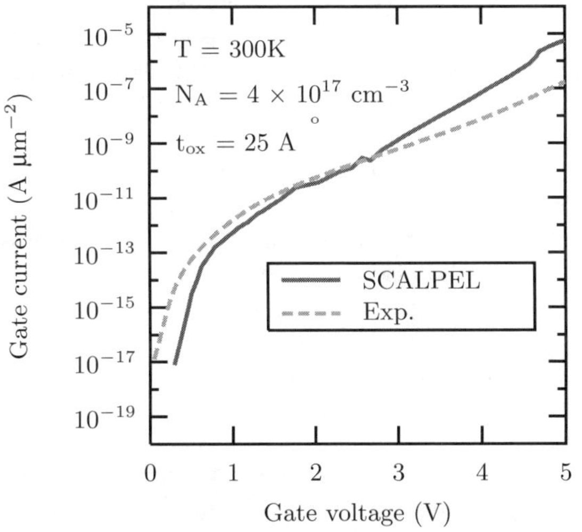

Fig. 12.5. Gate tunneling current versus gate voltage for an NO layer with thickness 25 Å. The substrate doping is 4×10^{17} cm^{-3} and T = 300 K

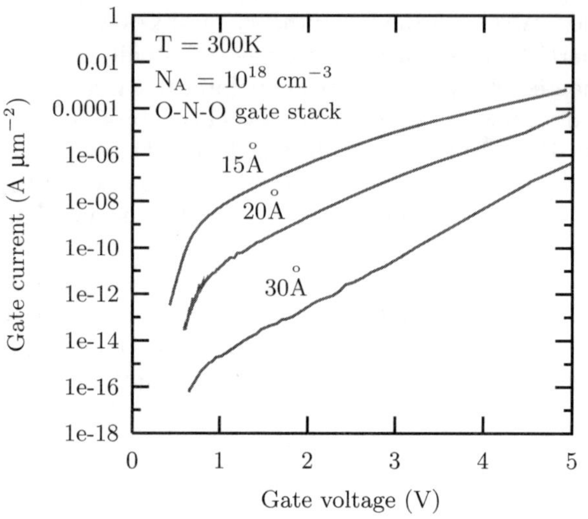

Fig. 12.6. Gate tunneling current versus gate voltage for an NO layer with thicknesses 15, 20 and 30 Å, the substrate doping being 10^{18} cm^{-3}. All other parameters are the same as in Fig. 12.5

and 3.9 respectively while the doping concentration N_A is 4×10^{17} cm^{-3}. Figure 12.6 shows current–voltage characteristics for oxide thicknesses of 15, 20 and 30 Å and $N_A = 10^{18}$ cm^{-3}.

The simulation results show a good agreement with the experimental data in the range 1–4 Volts. It should be noted that our present results are based on a set of 'default' material parameters [69, 70]. In particular for the effective electron mass in SiO_2, we used the results from Brar et al. [69]. The latter ones were obtained by measurements on *accumulation* layers. We suspect that the over-estimation of the gate leakage currents in the higher voltage regions is partly caused by poly depletion in the experiments such that a shift in the gate potential at the poly-insulator interface occurs. Another origin of the discrepancy may be found in the approximations that are used in the present method. The evaluation of the lifetimes of the states using the Breit–Wigner expansion becomes less accurate if the overlap increases.

It is also interesting to position the simulation results of the present method with respect to the outcome of other calculation schemes [71, 63, 64, 72]. We have compared the simulation data of the present method with data that were obtained using Bardeen's perturbation scheme and observed an excellent agreement [71]. Our results also agree very well with the data presented in [63, 64], which suggests that the transfer matrix method of the present method may be intimately connected to the transverse-resonant method that underlies the work of Lo, Buchanan and Taur.

In general, it can be concluded that all methods exploit the presence of specific states in the inversion layer that can be clearly identified. The identity is most visible in extreme quantization conditions such as a high substrate doping leading to a steep well and a thick insulator. If the insulator becomes too thin, say less than 5 Å or the well becomes wide, the expansion leading to equation (12.18) becomes either invalid or impractical. However, advanced processing (sub 100 nm) developments aim at shallow inversion layers and insulating layers having thicknesses in such a range that the present models will gain in importance.

Finally, although the quantitative results obtained by various methods may rather well agree for specific choices of the material parameters, it should be noted that the latter introduce a weakness in our approach, as well as in many others. In particular, all parameters and quantities related to the thin gate stack layers such as effective mass, barrier height, dielectric constant etc. are still a matter of debate not only when it comes to provide reliable numerical values but even when proper definitions are to be given. The concept of effective mass for instance is based on the band theory of bulk solids and its characteristic translation invariance. Nonetheless, it is taken for granted that we can still define a conduction band in ultra-thin layers containing less than 10 atoms in the perpendicular direction, thereby assigning meaningful effective mass tensors to the band minima and maxima. Also the values of permittivities and other layer dependent parameters may critically depend on the extremely small values of the layer thicknesses and a systematic investigation based on ab initio material calculations is needed.

Nevertheless, in spite of the lack of reliable material parameters characterizing the ultra-thin gate stack layers, the present theory provides a valuable alternative for existing gate current models. The conceptual simplicity of the theory allows for a straightforward numerical implementation into a dedicated, CPU-cost effective gate current subroutine.

13. Quantum Transport in Vertical Devices

The key to the growth of microelectronics during the last 30 years has been the drive to ever smaller dimensions of the devices using the principle of scaling introduced in the mid 1960's and early 1970's. The increase of the integration density and the related speed improvement due to smaller wiring and device dimensions and the reduction of the power dissipation per circuit are the most important advantages of scaling down the dimensions. In nowadays very-large-scale integrated (VLSI) circuits such as microprocessors and memory chips, the metal–semiconductor field-effect transistor (MOSFET) is the most important building block.

The enormous increase in semiconductor technologies has been quantified by Moore's law in 1965. The density of devices per integrated circuit has been doubled at regular intervals. This law sustained for more than three decades and led to the creation of the roadmaps for CMOS technology. Moore's law, by some called a self-fulfilling prophecy, is seen as a reliable method for predicting quantitative data sheets for the future generations of technologies in the semiconductor industry. At regular instances barriers have been encountered along the way, but innovations in technology have overcome these stumbling blocks such that down-scaling continues. At the time of the writing of this monograph, the research focuses on transistor channel lengths below 50 nanometer, while prototyping of products considers transistors with channel length around 0.1 micron. In particular, during the last decade significant progress has been achieved in the scaling of the metal–oxide–semiconductor field-effect transistor (MOSFET) down to semiconductor devices with sub-0.1 μm sizes. In [73, 74, 75], silicon-based MOSFETs with channel length shorter than 0.1 μm were fabricated and investigated for a wide range of temperatures.

It should be emphasized that Moore's law is phenomenological, i.e. its trueness is a bare observation. Ultimately, its applicability will end because technical barriers are unsurmountable, or economical laws may take over. In the end the real drive is obtaining market share and be profitable, however, if semiconductor production plants are more expensive than the income that they will generate, it is unlikely that these plants will ever be raised.

The success of CMOS technology lies also in its low cost per information processing unit. This low cost can be realized because the printing of masks for processing the integrated circuit are done by lithographic means. The alternative, e-beam writing is impractical for mass production. Future generations of mask printing equipment (steppers) will operate using deep UV beams or soft X rays, and a series of technological barriers have to be surpassed. In particular, it is not clear what material could serve as a suitable resist. Apart from the limitations that will be raised from technological feasibility, the physical operation mechanism by which the transistor works is reaching its limits. By decreasing the source–drain distance, the switching capabilities of the gate become limited, because the rounding of the source-channel and drain-channel voltages will touch. As a consequence, either a high stand-by or off-state current is present, or alternatively, a low drive current in the on-state is available. Both options usually fall outside the design rules for combining many transistors in a single circuit.

Due to above considerations, alternative device architectures are explored. A very revolutionary approach that has been advocated in the last years, is the vertical transistor. Although the original idea already dates back to the 1960's, new processing facilities have become available, that allow for a manufacturable process. The basic idea of the vertical transistor is to control the channel length and channel architecture by epitaxial growth techniques. With lithographical tools, the transistor channel and it geometrical dimension, and in particular the channel length are primarily determined by the smallest size that the printing tool or stepper can reproduce on the wafer (now of the order of 100 nm). Epitaxial techniques allow for arbitrary small channel lengths since the length (more precisely the 'height') of the channel is determined by the deposition time of the epitaxial layer. That the device architecture is really revolutionary follows from looking at the orientation of the gate with respect to the wafer. Conventional CMOS is essentially a planar technology, i.e. the current in the transition in on-state runs from source to drain parallel to the surface of the wafer. The gate engineering consists of building an appropriate stack of materials on top of the gate finger. On the other hand, in the vertical transistor, the current from source to drain flows in a direction orthogonal to the wafer surface. As a consequence, the design of the gate requires that a stack of material is grown on the side wall of the pillar that forms the source channel and drain.

13.1 Quantum Transport in a Cylindrical MOSFET

Following the description of the device architecture and processing steps that were discussed above, we will now consider a model that allows a detailed investigation of the current flowing through a cylindrical sub-0.1 μm MOSFET with a closed gate electrode [76].

The quantum mechanical features of the lateral charge transport are described by a Wigner distribution function that explicitly deals with electron scattering due to acoustic phonons and acceptor impurities. A numerical simulation is carried out to obtain a set of I–V characteristics for various channel lengths. It is demonstrated that inclusion of the collision term in the numerical simulation is important for low values of the source–drain voltage. The calculations have further shown that the scattering leads to an increase of the electron density in the channel thereby smoothing out the threshold kink in the I–V characteristics. An analysis of the electron phase-space distribution shows that scattering does not prevent electrons from flowing through the channel as a narrow stream, and that features of both ballistic and diffusive transport may be observed simultaneously.

Short channel effects together with random effects in the silicon substrate are the cause of a degradation of the threshold voltage and the appearance of uncontrollable charge and current in regions far from the gate electrode. Therefore in a sub-0.1 μm conventional MOSFET, the controlling ability of the gate electrode is substantially weakened. In order to achieve an improved device performance, a considerable attention has been paid to SOI (silicon-on-insulator) MOSFETs [77, 78, 79] and double-gate MOSFETs [80, 81]. As a result, a substantial reduction of the short-channel effect in the SOI-MOSFET was established as compared to the MOSFETs with the conventional geometry. In [82], the thermal equilibrium state of a cylindrical silicon-based MOSFET device with a closed gate electrode (CGE) is investigated. An advantage of using a CGE or all-around gate is the suppression of the floating body effect caused by external influences. Moreover, the short-channel effect in these devices can be even weaker than that in SOI and double-gate structures.

Here we present the investigation of quantum transport in a sub-0.1 μm CGE-MOSFET device. We describe a flexible 2D model which optimally combines analytical and numerical methods and describes the main features of the CGE-MOSFET. The theoretical modeling of the quantum transport features involves the use of the Wigner distribution function formalism [83, 84, 85]. Our description consists of the following steps: (1) the system is presented in terms of a one-electron Hamiltonian, (2) The quantum Liouville equation satisfied by the electron density matrix is transformed into a set of one dimensional equations for partial Wigner distribution functions, (3) a one-dimensional collision term is derived, (4) next we describe a numerical model to solve the equations which are derived for the partial Wigner distribution function.

13.2 The Hamiltonian of the System

We consider the cylindrical CGE-MOSFET structure [82] Fig. 13.1 described by cylindrical coordinates (r, ϕ, z), where the z-axis is chosen to be the symmetry axis. The structure contains a semiconductor pillar of Si with radius R and length L. The source and drain regions with the lengths L_s and L_d, respectively, are n-doped by phosphorus with concentration $N_D = 10^{20}$ cm^{-3},

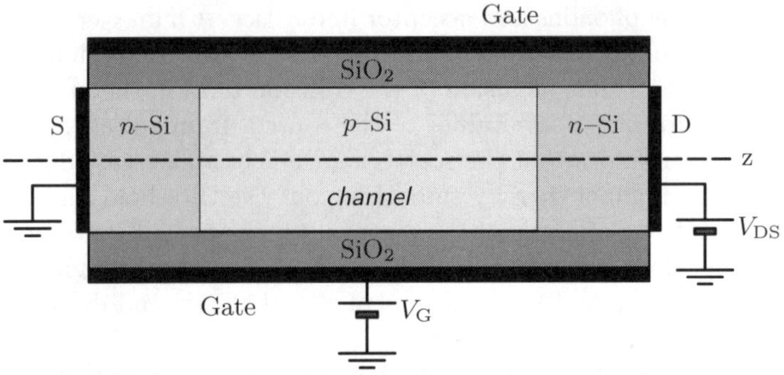

Fig. 13.1. Scheme of the cylindrical CGE-MOSFET

whereas the p-doped channel with the length L_{ch} has an acceptor concentration $N_{\mathrm{A}} = 10^{18}$ cm^{-3} (boron). The lateral surface of the semiconductor pillar is covered by the SiO_2 oxide layer, while the aluminum gate overlays the whole oxide layer. A cross-sectional view of the CGE-MOSFET is presented in Fig. 13.2. In the MOSFET structure under consideration, the translational symmetry of bulk silicon is broken due to the presence of the electrostatic potential. Electrons may occupy either one of the four 'heavy' valleys or one of the two 'light' valleys, in accordance with the masses related to the transverse or radial motion (perpendicular to the Si/SiO_2 interface). Since in the heavy valleys the x- and y-components of the effective mass tensor are different, the state density effective mass is $m^{\perp} = \sqrt{m_x m_y}$ is used to describe the transverse motion in the (x, y)-plane. The longitudinal (along the z-axis) effective mass is $m^{\|} = m_z$. Here, m_x, m_y and m_z are respectively the x-, y- and z-components of the effective mass tensor.

In the framework of the Hartree approximation, the many-electron problem is reduced to a description of one electron that moves in a potential that is created by all electrons. The electron radial motion, perpendicular

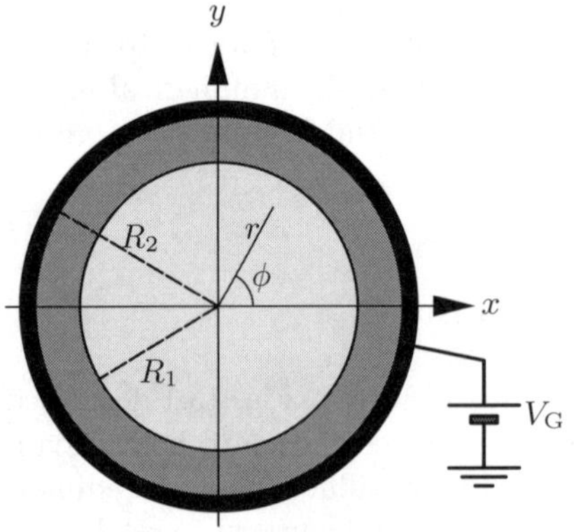

Fig. 13.2. Cross section of the cylindrical CGE-MOSFET

to the Si/SiO_2 interface in the inversion layer is expected to be faster than the longitudinal motion. Hence, the adiabatic approach can be used, i.e. it is assumed that along the z axis an electron moves in the potential averaged over the states of the radial motion.

In the semiconductor pillar, the electron motion is determined by the following Hamiltonian:

$$\hat{H}_\alpha = -\frac{\hbar^2}{2m_\alpha^\perp}\left[\frac{1}{r}\frac{\partial}{\partial r}\left(r\frac{\partial}{\partial r}\right) + \frac{1}{r^2}\frac{\partial^2}{\partial \phi^2}\right] - \frac{\hbar^2}{2m_\alpha^\parallel}\frac{\partial^2}{\partial z^2} + V(\boldsymbol{r}) , \tag{13.1}$$

where $V(\boldsymbol{r}) = V_\mathrm{b}(\boldsymbol{r}) + V_\mathrm{e}(\boldsymbol{r})$ is the potential energy associated with the energy barrier and the electrostatic field, respectively; $m_\alpha^\perp$ and $m_\alpha^\parallel$ are the effective masses of the transverse (in the (x, y)-plane) and longitudinal (along the z-axis) motion of an electron of the αth valley.

The electrostatic potential energy $V_\mathrm{e}(\boldsymbol{r})$ satisfies Poisson's equation

$$\Delta V_\mathrm{e}(\boldsymbol{r}) = \frac{e^2}{\varepsilon_0 \varepsilon_i}\left(-n(\boldsymbol{r}) + N_\mathrm{D}(\boldsymbol{r}) - N_\mathrm{A}(\boldsymbol{r})\right), \quad i = 1, 2 , \tag{13.2}$$

where ε_1 and ε_2 are the dielectric constants of the semiconductor and oxide layers, respectively; $n(\boldsymbol{r})$, $N_\mathrm{D}(\boldsymbol{r})$ and $N_\mathrm{A}(\boldsymbol{r})$ are, correspondingly, the concentrations of electrons, donors and acceptors as a function of $\boldsymbol{r} \equiv (r, \phi, z)$. The barrier potential $V_\mathrm{b}(\boldsymbol{r})$ is non-zero in the oxide layer only, wherein its value is constant V_b. In our calculations we assume that the source electrode is grounded whereas the potentials at the drain and gate electrodes equal V_D and V_G, respectively.

The study of the charge distribution in the cylindrical CGE-MOSFET structure in the state of the thermodynamical equilibrium [82] has shown that the concentration of holes is much lower than that of electrons so that electron transport is found to provide the main contribution to the current flowing through the MOSFET. For that reason, holes are neglected in the following transport calculations.

13.3 The Liouville Equation

We consider ballistic transport of electrons. Neglecting scattering processes and inter-valley transitions in the conduction band, we can write the one-electron density matrix as

$$\varrho(\boldsymbol{r}, \boldsymbol{r}') = \sum_\alpha \varrho_\alpha(\boldsymbol{r}, \boldsymbol{r}') , \tag{13.3}$$

where $\varrho_\alpha(\boldsymbol{r}, \boldsymbol{r}')$ is the density matrix of electrons residing in the αth valley satisfying Liouville's equation

$$\mathrm{i}\hbar\frac{\partial \varrho_\alpha}{\partial t} = [H_\alpha, \varrho_\alpha] . \tag{13.4}$$

In order to impose reasonable boundary conditions for the density matrix in the electrodes, it is convenient to describe the quantum transport along the z-axis in a phase-space representation. In particular, we rewrite (13.4) in terms of $\zeta = (z+z')/2$ and $\eta = z - z'$ coordinates and express the density matrix ϱ_α as

$$\varrho_\alpha(\boldsymbol{r},\boldsymbol{r}') = \sum_{lm,l'm'} \frac{1}{2\pi}\int_{-\infty}^{+\infty} \mathrm{d}k\, \mathrm{e}^{\mathrm{i}k\eta} f_{\alpha lml'm'}(\zeta,k) \times \Psi_{\alpha lm}(r,\phi,z)\,\Psi^*_{\alpha l'm'}(r',\phi',z') \qquad (13.5)$$

with a complete set of orthonormal functions $\Psi_{\alpha lm}(r,\phi,z)$. According to the cylindrical symmetry of the system, these functions take the following form:

$$\Psi_{\alpha lm}(r,\phi,z) = \frac{1}{\sqrt{2\pi}}\psi_{\alpha lm}(r,z)\,\mathrm{e}^{\mathrm{i}m\phi}\,. \qquad (13.6)$$

The functions $\psi_{\alpha lm}(r,z)$ are chosen to satisfy the equation

$$-\frac{\hbar^2}{2m_\alpha^\perp}\left[\frac{1}{r}\frac{\partial}{\partial r}\left(r\frac{\partial}{\partial r}\right) - \frac{m^2}{r^2}\right]\psi_{\alpha lm}(r,z) + V(r,z)\,\psi_{\alpha lm}(r,z) = \mathcal{E}_{\alpha lm}(z)\,\psi_{\alpha lm}(r,z)\,, \qquad (13.7)$$

which describes the radial motion of an electron. Here $\mathcal{E}_{\alpha lm}(z)$ are the eigenvalues of (13.7) for a given value of the z-coordinate which appears as a parameter. It will be shown, that $\mathcal{E}_{\alpha lm}(z)$ plays the role of an effective potential in the channel, and that $\Psi_{\alpha lm}(r,\phi,z)$ is the corresponding wave function of the transverse motion at a fixed z. Substituting the expansion (13.5) into (13.4), and using (13.7), we arrive at an equation for $f_{\alpha lml'm'}(\zeta,k)$

$$\frac{\partial f_{\alpha lml'm'}(\zeta,k)}{\partial t} = -\frac{\hbar k}{m_\alpha^\parallel}\frac{\partial}{\partial\zeta} f_{\alpha lml'm'}(\zeta,k) + \frac{1}{\hbar}\int_{-\infty}^{+\infty} W_{\alpha lml'm'}(\zeta,k-k') f_{\alpha lml'm'}(\zeta,k')\mathrm{d}k' - \sum_{l_1,l_1'}\int_{-\infty}^{+\infty} \hat{M}^{l_1 l_1'}_{\alpha lml'm'}(\zeta,k,k') f_{\alpha l_1 m l_1' m'}(\zeta,k')\mathrm{d}k'\,, \qquad (13.8)$$

where

$$W_{\alpha lml'm'}(\zeta,k-k') = \frac{1}{2\pi\mathrm{i}}\int_{-\infty}^{+\infty}\left(\mathcal{E}_{\alpha lm}(\zeta+\eta/2) - \mathcal{E}_{\alpha l'm'}(\zeta-\eta/2)\right)\mathrm{e}^{\mathrm{i}(k'-k)\eta}\mathrm{d}\eta \qquad (13.9)$$

$$\hat{M}^{l_1 l_1'}_{\alpha l m l' m'}(\zeta, k, k') = \frac{1}{2\pi} \int_{-\infty}^{+\infty} \left[\delta_{l' l_1'} \hat{\Gamma}_{\alpha l m l_1}(\zeta + \eta/2, k') + \delta_{l l_1} \hat{\Gamma}^*_{\alpha l' m' l_1'}(\zeta - \eta/2, k') \right] \times \mathrm{e}^{\mathrm{i}(k'-k)\eta} \mathrm{d}\eta \quad (13.10)$$

$$\hat{\Gamma}_{\alpha l m l_1}(z, k') = \frac{\hbar}{2m^{\|}_{\alpha}\mathrm{i}} b_{\alpha l m l_1}(z) + \frac{\hbar}{2m^{\|}_{\alpha}} c_{\alpha l m l_1}(z) \left(-\mathrm{i}\frac{\partial}{\partial \zeta} + 2k' \right) \quad (13.11)$$

and

$$b_{\alpha l m l_1}(z) = \int \psi^*_{\alpha l m}(r, z) \frac{\partial^2}{\partial z^2} \psi_{\alpha l_1 m}(r, z)\, r \mathrm{d}r \quad (13.12)$$

$$c_{\alpha l m l_1}(z) = \int \psi^*_{\alpha l m}(r, z) \frac{\partial}{\partial z} \psi_{\alpha l_1 m}(r, z)\, r \mathrm{d}r \,. \quad (13.13)$$

Note that (13.8) is similar to the Liouville equation for the Wigner distribution function, which is derived to model quantum transport in tunneling diodes (see [83]). The first drift term in the right-hand side of (13.8) is derived from the kinetic-energy operator of the longitudinal motion. It is exactly the same as the corresponding term of the Boltzmann equation. The second component plays the same role as the force term does in the Boltzmann equation. The last term in the right-hand side of (13.8) contains the operator $\hat{M}^{l_1 l_1'}_{\alpha l m l' m'}(\zeta, k, k')$, which mixes the functions $f_{\alpha l m l' m'}$ with different indexes l, l'. It appears because $\psi_{\alpha l m}(r, z)$ are not eigenfunctions of the Hamiltonian (13.1). The physical meaning of the operator $\hat{M}^{l_1 l_1'}_{\alpha l m l' m'}(\zeta, k, k')$ will be discussed below.

In order to solve (13.8), we need to specify boundary conditions for the functions $f_{\alpha l m l' m'}(\zeta, k)$. For a weak current, electrons incoming from both the source and the drain electrodes, are assumed to be maintained in thermal equilibrium. Comparing (13.5) with the corresponding expansion of the density matrix in the equilibrium state, one obtains the following boundary conditions

$$\begin{aligned} f_{\alpha l m l' m'}(0, k) &= 2\delta_{l l'}\delta_{m m'} \left[\exp\left(E_{\alpha l m k}\beta - E_{\mathrm{FS}}\beta \right) + 1 \right]^{-1}, \qquad k > 0 \\ f_{\alpha l m l' m'}(L, k) &= 2\delta_{l l'}\delta_{m m'} \left[\exp\left(E_{\alpha l m k}\beta - E_{\mathrm{FD}}\beta \right) + 1 \right]^{-1}, \qquad k < 0 \end{aligned} \quad (13.14)$$

where the total energy is $E_{\alpha l m k} = \hbar^2 k^2 / 2m^{\|}_{\alpha} + \mathcal{E}_{\alpha l m}(0)$ for an electron entering from the source electrode ($k > 0$) and $E_{\alpha l m k} = \hbar^2 k^2 / 2m^{\|}_{\alpha} + \mathcal{E}_{\alpha l m}(L)$ for an electron entering from the drain electrode ($k < 0$). In (13.14) $\beta = 1/k_{\mathrm{B}}T$ is the inverse thermal energy, while E_{FS} and E_{FD} are the Fermi energy levels in the source and in the drain, respectively. Note, that (13.14) meets the requirement of imposing only one boundary condition on the function $f_{\alpha l m l' m'}(\zeta, k)$ at a fixed value of k as (13.8) is a first order differential equation with respect to ζ. Generally speaking, the solution of (13.8) with the conditions (13.14) depends on the distance between the boundary position and the active device region. Let us estimate how far the boundary must be from the active device region in order to avoid this dependence. It is easy to show that the density matrix of the equilibrium state is a decaying function of $\eta = z - z'$. The decay length is of the order of the coherence length $\lambda_\mathrm{T} = \sqrt{\hbar^2/m_\alpha^{\|} k_\mathrm{B} T}$ at high temperature and of the inverse Fermi wave number $k_\mathrm{F}^{-1} = \sqrt{\hbar^2/2m_\alpha^{\|} E_\mathrm{F}}$ at low temperature. So, it is obvious, that the distance between the boundary and the channel must exceed the coherence length or the inverse Fermi wavenumber, i. e. $L_\mathrm{s}, L_\mathrm{d} \gg \lambda_\mathrm{T}$ or $L_\mathrm{s}, L_\mathrm{d} \gg k_\mathrm{F}^{-1}$. For example, at $T = 300$ K the coherence length $\lambda_\mathrm{T} \sim 3$ nm is much less than the source or drain lengths.

The functions $f_{\alpha l m l' m'}(\zeta, k)$, that are introduced in (13.5), are used in the calculations of the current and the electron density. The expression for the electron density follows directly from the density matrix as $n(\boldsymbol{r}) = \varrho(\boldsymbol{r}, \boldsymbol{r})$. In terms of the functions $f_{\alpha l m l' m'}(\zeta, k)$, the electron density can be written as follows:

$$n(\boldsymbol{r}) = \frac{1}{2\pi} \sum_{\alpha l m l' m'} \int_{-\infty}^{+\infty} f_{\alpha l m l' m'}(z, k) \mathrm{d}k \, \Psi_{\alpha l m}(r, \phi, z) \, \Psi^*_{\alpha l' m'}(r', \phi', z') \, . \tag{13.15}$$

The current density can be expressed in terms of the density matrix [86]

$$\boldsymbol{J}(\boldsymbol{r}) = \sum_\alpha \frac{e\hbar}{2m_\alpha \mathrm{i}} \left(\frac{\partial}{\partial \boldsymbol{r}} - \frac{\partial}{\partial \boldsymbol{r}'} \right) \varrho_\alpha(\boldsymbol{r}, \boldsymbol{r}') \bigg|_{\boldsymbol{r}=\boldsymbol{r}'} \, . \tag{13.16}$$

The total current, which flows through the cross-section of the structure at a point z, can be obtained by an integration over the transverse coordinates. Substituting the expansion (13.5) into (13.16) and integrating over r and φ, we find

$$\begin{aligned} I(z) = e \sum_{\alpha, l, m} \frac{1}{2\pi} \int_{-\infty}^{+\infty} \mathrm{d}k \frac{\hbar k}{m_\alpha^{\|}} f_{\alpha l m l m}(z, k) \\ - \frac{2e\hbar}{m_\alpha^{\|}} \sum_{\substack{\alpha l, m, l' \\ l' > l}} c_{\alpha m l l'}(z) \int_{-\infty}^{+\infty} \mathrm{d}k \, \Im \left[f_{\alpha l m l' m}(z, k) \right] \, , \end{aligned} \tag{13.17}$$

where $\Im[f]$ is the imaginary part of f. The first term in the right-hand side of (13.17) is similar to the expression for a current of the classical theory [86]. The second term, which depends on the non-diagonal functions $f_{\alpha l m l' m}$ only, takes into account the effects of intermixing between different states of the transverse motion.

The last term in the right-hand side of (13.8) takes into consideration the variation of the wavefunctions $\psi_{\alpha lm}(r,z)$ along the z-axis. In the source and drain regions, the electrostatic potential is essentially constant due to the high density of electrons. In these parts of the structure, the wavefunctions of the transverse motion are very weakly dependent on z, and consequently, the operator $\hat{M}^{l_1 l_1'}_{\alpha l m l' m'}$ has negligible effect. Inside the channel, electrons are strongly localized at the Si/SiO_2 interface as the positive gate voltage is applied. Earlier calculations, which we made for the case of equilibrium [82], have shown that in the channel the dependence of $\psi_{\alpha lm}(r,z)$ on z is weak, too. Therefore, in the channel the effect of the operator $\hat{M}^{l_1 l_1'}_{\alpha l m l' m'}$ is negligible. In the intermediate regions (the source-channel and the drain-channel), an increase of the contribution of the third term in the right-hand side of (13.8) is expected due to a sharp variation of $\psi_{\alpha lm}(r,z)$. Since $\hat{M}^{l_1 l_1'}_{\alpha l m l' m'}$ couples functions $f_{\alpha l m l' m'}(\zeta,k)$ with different quantum numbers $(\alpha l m)$, it can be interpreted as a collision operator, which describes transitions of electrons between different quantum states of the transverse motion. Thus, the third term in the right-hand side of (13.8) is significant only in the close vicinity of the p–n junctions. Therefore, this term is assumed to give a small contribution to the charge and current densities. Under the above assumption, we treat the last term in the right-hand side of (13.8) as a perturbation. Hereafter, we investigate the steady state of the system in a zeroth-order approximation with respect to the operator $\hat{M}^{l_1 l_1'}_{\alpha l m l' m'}$. Neglecting the latter, one finds that, due to the boundary conditions (13.14), all non-diagonal functions $f_{\alpha l m l' m'}(\zeta,k)$ ($l \neq l'$ or $m \neq m'$) need to be zero.

In the channel, the energy of the transverse motion can be approximately written in the form [82]

$$\mathcal{E}_{\alpha lm}(z) = \mathcal{E}_{\alpha l}(z) + \frac{\hbar^2 m^2}{2 m^{\perp}_{\alpha l} R^2_{\alpha l}} , \tag{13.18}$$

where $\mathcal{E}_{\alpha l}(z)$ is the energy associated with the radial size quantization and $\hbar^2 m^2 / 2 m^{\perp}_{\alpha l} R^2_{\alpha l}$ is the energy of the angular motion with averaged radius $R_{\alpha l}$. Hence, in (13.9) for the diagonal functions $f_{\alpha lmlm}(\zeta,k)$ the difference $\mathcal{E}_{\alpha lm}(\zeta+\eta/2) - \mathcal{E}_{\alpha lm}(\zeta-\eta/2)$ can be replaced by $\mathcal{E}_{\alpha l}(\zeta+\eta/2) - \mathcal{E}_{\alpha l}(\zeta-\eta/2)$. Furthermore, summation over m in (13.8) gives

$$\frac{\hbar k}{m^{\parallel}_{\alpha}} \frac{\partial}{\partial \zeta} f_{\alpha l}(\zeta,k) - \frac{1}{\hbar} \int\limits_{-\infty}^{+\infty} W_{\alpha l}(\zeta, k-k') f_{\alpha l}(\zeta,k') \mathrm{d}k' = 0 \tag{13.19}$$

with

$$f_{\alpha l}(\zeta,k) = \frac{1}{2\pi} \sum_{m} f_{\alpha lmlm}(\zeta,k) . \tag{13.20}$$

In (13.19) the following notation is used:

$$W_{\alpha l}(\zeta, k) = -\frac{1}{2\pi} \int_{-\infty}^{+\infty} (\mathcal{E}_{\alpha l}(\zeta + \eta/2) - \mathcal{E}_{\alpha l}(\zeta - \eta/2)) \sin(k\eta) \, d\eta \, . \quad (13.21)$$

The effective potential $\mathcal{E}_{\alpha l}(z)$ can be interpreted as the bottom of the subband (α, l) in the channel. The function $f_{\alpha l}(\zeta, k)$ is referred to as a partial Wigner distribution function describing electrons which are traveling through the channel in the inversion layer subband (α, l).

13.4 Electron Scattering

We consider the electron scattering from phonons and impurities. For this purpose we introduce a Boltzmann-like single collision term [86], which in the present case has the following form

$$St\, f_{\alpha lmk} = \sum_{\alpha' l' m' k'} (P_{\alpha lmk, \alpha' l' m' k'} f_{\alpha' l' m' k'} - P_{\alpha lmk, \alpha' l' m' k'} f_{\alpha lmk}) \, . \quad (13.22)$$

As was noted above, we neglect all transitions between quantum states with different sets of quantum numbers α and l. In the source and drain contacts the distribution of electrons over the quantum states of the angular motion corresponds to equilibrium. Consequently, due to the cylindrical symmetry of the system, we may fairly assume that across the whole structure the electron distribution is given by

$$f_{\alpha lmk}(z) = f_{\alpha l}(z, k)\, w_{\alpha lm} \, , \quad (13.23)$$

where

$$w_{\alpha lm} = \sqrt{\frac{\hbar^2 \beta}{2 m_\alpha^\perp R_{\alpha l}^2 \pi}} \exp\left(-\frac{\beta \hbar^2 m^2}{2 m_\alpha^\perp R_{\alpha l}^2} \right) \quad (13.24)$$

is the normalized Maxwellian distribution function with respect to the angular momentum m. The integration of the both sides of (13.22) over the angular momentum gives the one-dimensional collision term

$$St\; f_{\alpha l}(z, k) = \sum_{k'} (P_{\alpha l}(k, k') f_{\alpha l}(z, k') - P_{\alpha l}(k', k) f_{\alpha l}(z, k)) \, , \quad (13.25)$$

where

$$P_{\alpha l}(k, k') = \sum_{mm'} P_{\alpha lmk, \alpha lm'k'} w_{\alpha lm'} \, . \quad (13.26)$$

This collision term is directly incorporated into the one-dimensional Liouville equation (13.19) as

$$\tilde{W}_{\alpha l}(z, k, k') = W_{\alpha l}(z, k - k') + P_{\alpha l}(z, k, k') - \delta_{k,k'} \sum_{k'} P_{\alpha l}(z, k', k) \, , \quad (13.27)$$

where $\tilde{W}_{\alpha l}(z, k, k')$ is the modified force term in (13.19).

Here, we consider scattering by acceptor impurities and acoustic phonons described by a deformation potential. The scattering rates are evaluated according to Fermi's golden rule

$$P_{\alpha lmk,\alpha lm'k'} = \frac{2\pi}{\hbar} \left| \langle \alpha lm'k' | \hat{H}_{int} | \alpha lmk \rangle \right|^2 \delta \left(E_{\alpha lm'k'} - E_{\alpha lmk} \right) , \quad (13.28)$$

where $\hat{H}_{\rm int}$ is the Hamiltonian of the electron–phonon or the electron–impurity interaction. The potential of an ionized acceptor is modeled as $U(\boldsymbol{r}) = 4\pi e^2 R_{\rm s}^2/\varepsilon_1 \delta(\boldsymbol{r})$, where $R_{\rm s}$ determines a cross-section for scattering by an impurity. Consequently, the absolute value of the matrix element is

$$|\langle \alpha lm'k' | U(\boldsymbol{r} - \boldsymbol{r}_i) | \alpha lmk \rangle| = 4\pi e^2 R_{\rm s}^2/\varepsilon_1 \psi_{\alpha l}^2(r_i, z_i) . \quad (13.29)$$

Averaging this over a uniform distribution of acceptors results in the following scattering rate

$$P^{\rm im}_{\alpha lmk,\alpha lm'k'} = C_{\rm im} \int\limits_0^R \psi_{\alpha l}^4(r,z) \delta \left(E_{\alpha lm'k'} - E_{\alpha lmk} \right) r {\rm d}r , \quad (13.30)$$

where $C_{\rm im} = N_{\rm A} \left(4\pi e^2 R_{\rm s}^2/\varepsilon_1 \right)^2 /\hbar$ and $N_{\rm A}$ is the acceptor concentration.

At room temperature the rate of the scattering by acoustic phonons has the same form. Indeed, for $T = 300$ K the thermal energy $k_{\rm B}T \gg \hbar\omega_{\boldsymbol{q}}$, therefore the acoustic deformation potential scattering is approximately elastic, and the emission and absorption rates are equal to each other. For low energies we can approximate the phonon number as $N_{\boldsymbol{q}} \approx k_{\rm B}T/\hbar\omega_{\boldsymbol{q}} \gg 1$ and the phonon frequency $\omega_{\boldsymbol{q}} = v_{\rm S} q$, where $v_{\rm S}$ is the sound velocity. Assuming equipartition of energy in the acoustic modes, the scattering rate is

$$P^{\rm ph}_{\alpha lmk,\alpha lm'k'} = \frac{2\pi}{V} C_{\rm ph} \sum_{\boldsymbol{q}} \left| \langle \alpha lm'k' | {\rm e}^{{\rm i}\boldsymbol{q}\cdot\boldsymbol{r}} | \alpha lmk \rangle \right|^2 \delta \left(E_{\alpha lm'k'} - E_{\alpha lmk} \right) , \quad (13.31)$$

where the parameter $C_{\rm ph} = 4D_{\rm AC}^2 k_{\rm B}T/9\pi \varrho_{\rm Si} v_{\rm S}^2 \hbar$ and where D_{AC} and $\varrho_{\rm Si}$ are the deformation potential and the mass density, respectively. Integrating over $\boldsymbol{q}$ yields the scattering rate $P^{\rm ph}_{\alpha lmk,\alpha lm'k'}$ in the form (13.30) with $C_{\rm ph}$ instead of $C_{\rm im}$. The full scattering rate $P_{\alpha lmk,\alpha lm'k'} = P^{\rm im}_{\alpha lmk,\alpha lm'k'} + P^{\rm ph}_{\alpha lmk,\alpha lm'k'}$ is then inserted into (13.26) in order to obtain the one-dimensional scattering rate

$$P_{\alpha l}(z,k,k') = \left(C_{\rm im} + C_{\rm ph} \right) a_{\alpha l}(z) F \left(\frac{\hbar^2 k'^2}{2m_\alpha^{\parallel}} - \frac{\hbar^2 k^2}{2m_\alpha^{\parallel}} \right) , \quad (13.32)$$

where

$$a_{\alpha l}(z) = \sqrt{\frac{2m_\alpha^{\perp}}{\hbar^2 \pi \beta}} R_{\alpha l} \int\limits_0^R \psi_{\alpha l}^4(r,z)\, r\, {\rm d}r , \quad F(x) = {\rm e}^{-x/2} K_0(|x|/2) \quad (13.33)$$

and $K_0(x)$ is a McDonald function [87]. In the calculations of the scattering by acoustic phonons, the following values of parameters for Si are used: D_{AC} =9.2 eV, $\varrho_{\mathrm{Si}} = 2.3283 \cdot 10^3$ kg/m^3, $v_{\mathrm{S}} = 8.43 \cdot 10^5$ cm/s [88].

13.5 The Numerical Model

The system under consideration consists of regions with high (the source and drain) and low (the channel) concentrations of electrons. The corresponding electron distribution difference would produce a considerable inaccuracy if we would attempt to directly construct a finite-difference analog of (13.19). It is worth mentioning that, in the quasi-classical limit, i.e. $\mathcal{E}_{\alpha l}(\zeta+\eta/2) - \mathcal{E}_{\alpha l}(\zeta-\eta/2) \approx \frac{\partial \mathcal{E}_{\alpha l}(\zeta)}{\partial \zeta}\eta$, (13.19) leads to the Boltzmann equation with an effective potential which has the following exact solution in the equilibrium state ($E_{\mathrm{F}} = E_{\mathrm{FS}} = E_{\mathrm{FD}}$):

$$f_{\alpha l}^{\mathrm{eq}}(\zeta,k)=\frac{1}{\pi}\sum_m \left[\exp\left(\frac{\hbar^2 k^2}{2m_\alpha^{\|}}\beta + \mathcal{E}_{\alpha l}(\zeta)\beta + \frac{\hbar^2 m^2 \beta}{2m_{\alpha l}^{\perp} R_{\alpha l}^2(\zeta)} - E_{\mathrm{F}}\beta\right) + 1\right]^{-1} \tag{13.34}$$

For numerical calculations it is useful to write down the partial Wigner distribution function as $f_{\alpha l}(\zeta,k) = f_{\alpha l}^{\mathrm{eq}}(\zeta,k) + f_{\alpha l}^{\mathrm{d}}(\zeta,k)$. Inserting this into (13.19), one obtains the following equation for $f_{\alpha l}^{\mathrm{d}}(\zeta,k)$

$$\frac{\hbar k}{m_\alpha^{\|}}\frac{\partial}{\partial\zeta} f_{\alpha l}^{\mathrm{d}}(\zeta,k) - \frac{1}{\hbar}\int_{-\infty}^{+\infty} W_{\alpha l}(\zeta,k-k') f_{\alpha l}^{\mathrm{d}}(\zeta,k')\,\mathrm{d}k' = B_{\alpha l}(\zeta,k)\,, \tag{13.35}$$

where

$$B_{\alpha l}(\zeta,k)=\frac{1}{2\pi}\int_{-\infty}^{+\infty}\mathrm{d}k'\int_{-\infty}^{+\infty}\mathrm{d}\eta\left(\mathcal{E}_{\alpha l}(\zeta+\frac{\eta}{2}) - \mathcal{E}_{\alpha l}(\zeta-\frac{\eta}{2}) - \frac{\partial\mathcal{E}_{\alpha l}(\zeta)}{\partial\zeta}\eta\right) \times \sin\left[(k-k')\eta\right] f_{\alpha l}^{\mathrm{eq}}(\zeta,k')\,. \tag{13.36}$$

The unknown function $f_{\alpha l}^{\mathrm{d}}(\zeta,k)$ takes values of the same order throughout the whole system, and therefore is suitable for numerical computations. In the present work, we have used the finite-difference model, which is described in [83]. The position variable takes the set of discrete values $\zeta_i = \Delta\zeta i$ for $\{i = 0,\ldots,N_\zeta\}$. The values of k are also restricted to the discrete set $k_p = (2p - N_k - 1)\Delta k/2$ for $\{p = 1,\ldots,N_k\}$. On a discrete mesh, the first derivative $\frac{\partial f_{\alpha l}}{\partial\zeta}(\zeta_i,k_p)$ is approximated by the left-hand difference for $k_p > 0$ and the right-hand difference for $k_p < 0$. It was shown in [83], that such a choice of the finite-difference representation for the derivatives leads to a stable discrete model. Projecting the equation (13.35) onto the finite-difference basis gives a matrix equation $\mathbf{L} \cdot \mathbf{f} = \mathbf{b}$. In the matrix $\mathbf{L}$, only the diagonal blocks and one upper and one lower co-diagonal blocks are nonzero:

$$\mathbf{L} = \begin{pmatrix} A_1 & -E & 0 & \dots & 0 \\ -V & A_2 & -E & \dots & 0 \\ 0 & -V & A_3 & \dots & 0 \\ \vdots & \vdots & \vdots & \ddots & \vdots \\ 0 & 0 & 0 & \dots & A_{N_\zeta - 1} \end{pmatrix} . \tag{13.37}$$

Here, the $N_k \times N_k$ matrices A_i, E, and V are

$$\begin{aligned} [A_i]_{pp'} &= \delta_{pp'} - \frac{2m_\alpha^{\|}\Delta\zeta}{\hbar^2(2p - N_k - 1)\Delta k} W_{\alpha l}(\zeta_i, k_p - k_{p'}) \\ [E]_{pp'} &= \delta_{pp'}\theta\left\{\frac{N_k+1}{2} - p\right\} \\ [V]_{pp'} &= \delta_{pp'}\theta\left\{p - \frac{N_k+1}{2}\right\} \end{aligned} \tag{13.38}$$

and the vectors are

$$[f_i]_p = f_{\alpha l}(\zeta_i, k_p) \quad \text{and} \quad [b_i]_p = B_{\alpha l}(\zeta_i, k_p) \quad i = 1, N_{\zeta - 1} \quad i = 1, N_k \tag{13.39}$$

A recursive algorithm is used to solve the matrix equation $\mathbf{L} \cdot \mathbf{f} = \mathbf{b}$. Invoking downward elimination, we are dealing with $B_i = (A_i - VB_{i-1})^{-1} E$ and $N_i = (A_i - VB_{i-1})^{-1} (b_i + VN_{i-1})$ $(i = 1, \dots, N_\zeta)$ as relevant matrices and vectors. Then, upward elimination eventually yields the solution $f_i = B_i f_{i+1} + N_i$ $(i = N_\zeta - 1, \dots, 1)$. If an index of a matrix or a vector is smaller than 1 or larger than $N_\zeta - 1$, the corresponding term is supposed to vanish.

In the channel, the difference between effective potentials $\mathcal{E}_{\alpha l}(\zeta)$ with different (α, l) is of the order of or larger than the thermal energy $k_{\mathrm{B}}T$. Therefore, in the channel inversion layer only a few low-energy subbands must be taken into account. In the source and drain, however, many quantum states (α, l) of the radial motion are strongly populated by electrons. Therefore, we should account for all of them in order to calculate the charge distribution. Here, we can use the fact that, according to our approximation, the current flows only through the lowest subbands in the channel. Hence, only for these subbands the partial Wigner distribution function of electrons is non-equilibrium. In other subbands electrons are maintained in the state of equilibrium, even when a bias is applied. So, in (13.15) for the electron density, we can substitute functions $f_{\alpha lmlm}(z, k)$ of higher subbands by corresponding equilibrium functions. Formally, adding and subtracting the equilibrium functions for the lowest subbands in (13.15), we arrive at the following equation for the electron density

$$n(\boldsymbol{r})=n_{\mathrm{eq}}(\boldsymbol{r})+\frac{1}{2\pi}\sum_{\alpha l}\int_{-\infty}^{+\infty}\mathrm{d}k\left[f_{\alpha l}(z,k)\left|\psi_{\alpha l}(r,z)\right|^2-f_{\alpha l}^{\mathrm{eq}}(z,k)\left|\psi_{\alpha l}^{\mathrm{eq}}(r,z)\right|^2\right] \tag{13.40}$$

where $n_{\mathrm{eq}}(\boldsymbol{r})$ and $\psi_{\alpha l}^{\mathrm{eq}}(r,z)$ are the electron density and the wavefunction of the radial motion in the state of equilibrium, respectively. The summation on the right-hand side of (13.40) is performed only over the lowest subbands. Since the electrostatic potential does not penetrate into the source and drain, we suppose that the equilibrium electron density *in these regions* is well described by the Thomas–Fermi approximation:

$$n_{\mathrm{eq}}(\boldsymbol{r})=N_{\mathrm{C}}\frac{2}{\sqrt{\pi}}F_{1/2}\left(\beta\left(eV(r,z)+E_{\mathrm{F}}-E_{\mathrm{C}}\right)\right) \tag{13.41}$$

where the Fermi integral is

$$F_{1/2}(x)=\int_0^{\infty}\frac{\sqrt{t}\,\mathrm{d}t}{\exp(t-x)+1}\,. \tag{13.42}$$

Here N_{C} is the effective density of states in the conduction band and E_{F} is the Fermi level of the system in the state of equilibrium.

13.6 Numerical Results

During the device simulation three equations are solved self-consistently: (i) the Poisson equation (13.2), (ii) the equation for the wavefunction of the radial motion (13.7), and (iii) the equation for the partial Wigner distribution function (13.19). The methods of numerical solution of (13.2) and (13.7) are the same as for the equilibrium state [82]. The numerical model for (13.19) was described above. In the present calculations the four lowest subbands ($\alpha=1,2$ and $l=1,2$) are taken into account. The electron density in the channel is obtained from (13.15), whereas in the source and drain regions it is determined from (13.40). The calculations are performed for structures with a channel of radius $R=50$ nm and for various values of the length: $L_{\mathrm{ch}}=40$, 60, 70 and 80 nm. The lengths of the source and drain are $L_{\mathrm{s}}=L_{\mathrm{d}}=20$ nm. The width of the oxide layer is taken to be 4 nm and the barrier potential is $V_{\mathrm{b}}=3$ eV. All calculations are carried out with $N_\zeta=100$ and $N_k=100$. The partial Wigner distribution functions, which are obtained as a result of the self-consistent procedure, are then used to calculate the current according to (13.16).

We investigate the quantum transport with and without the electron scattering. The latter is so-called ballistic transport. The distribution of the electrostatic potential is represented in Fig. 13.3 for $V_{\mathrm{D}}=0.3$ V and $V_{\mathrm{G}}=1$ V. This picture is typical for the MOSFET structure, that is considered here. The cross-sections of the electrostatic potentials for $r=0,30,40,45,48,50$ nm

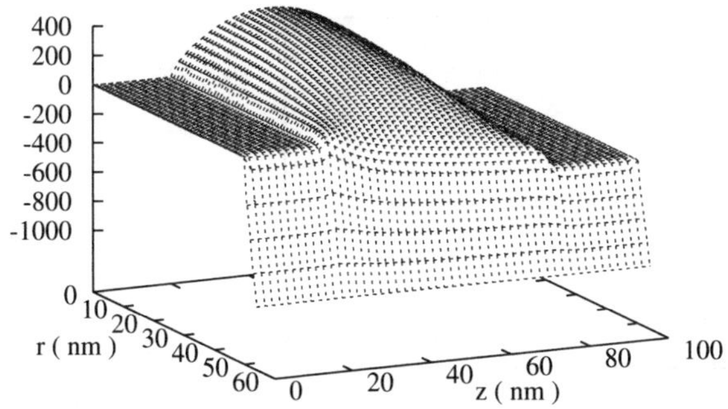

Fig. 13.3. Distribution of the electrostatic potential in the MOSFET with $L_{\rm ch}$ = 60 nm at $V_{\rm G} = 1$ V and $V_{\rm D} = 0.3$ V

are shown in Fig. 13.4. The main part of the applied gate voltage falls in the insulator (50 nm $< r <$ 54 nm). Along the cylinder axis in the channel, the electrostatic potential barrier for the electron increases up to about 0.4 eV. Since the potential along the cylinder axis is always high, the current mainly flows in a thin layer near the semiconductor–oxide interface. Varying $V_{\rm D}$ and $V_{\rm G}$ mainly changes the shape of this narrow path, and, as a consequence, influences the form of the effective potential $\mathcal{E}_{\alpha l}(z)$. As follows from Figs. 13.3 and 13.4, the radius of the pillar can be taken shorter without causing barrier degradation. At the p–n junctions (source-channel and drain-channel) the electrons meet barriers across the whole semiconductor. These barriers are found to persist even for high values of the applied source–drain voltage and prevent an electron flood from the side of the strongly doped source. The pattern of the electrostatic potentials differs mainly near the semiconductor–oxide interface, where the inversion layer is formed. In Fig. 13.5, the effective potential for the lowest inversion subband ($\alpha = 1, l = 1$) is plotted as a function of z for different applied bias $V_{\rm D} = 0,\ldots,0.5$ V, $V_{\rm G} = 1$ V, $L_{\rm ch} = 60$ nm. It is seen that the effective potential reproduces the distribution of the electrostatic potential near the semiconductor–oxide interface. In the case of ballistic transport (dashed curve), the applied drain–source voltage sharply drops near the drain-channel junction (Figs. 13.4 and 13.5).

The scattering of electrons (solid curve) smoothes out the applied voltage, which is now varying linearly along the whole channel. Note, *that the potential obtained by taking into account scattering is always higher than that of the ballistic case.* The explanation is that, due to scattering, the electron density in the channel rises and smoothes out [76]. Hence, the applied gate voltage is screened more effectively, and as a result, the potential exceeds that of the ballistic case. It should be noted that at equilibrium ($V_{\rm D} = 0$) the linear density and the effective potential for both cases (with and without scattering) are equal to each other. This result follows from the principle of detailed balance.

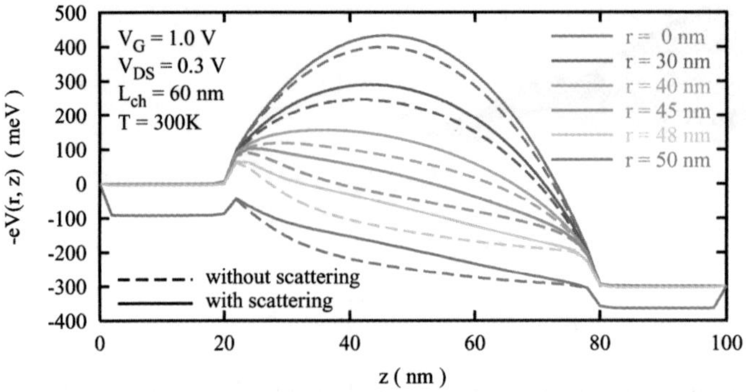

Fig. 13.4. Cross-sections of the electrostatic potentials calculated without scattering (*dashed curves*) and with scattering from acceptor impurities and from an acoustic deformation potential (*solid curves*)

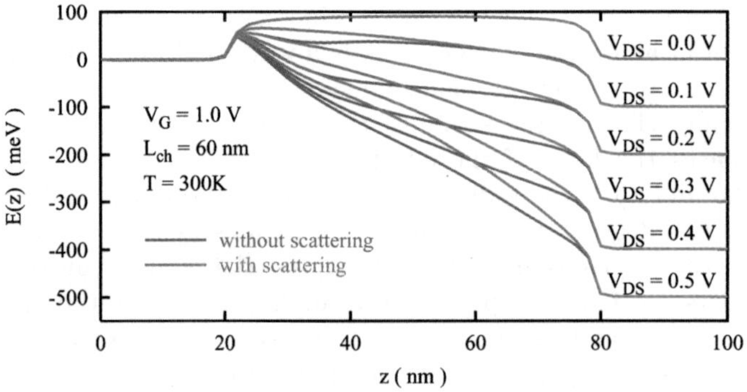

Fig. 13.5. Effective potential as a function of z for various V_{D}, $L_{\mathrm{ch}} = 60$ nm

The current–voltage characteristics (the current density $J_{\mathrm{DS}} = I/2\pi R$ vs. the source–drain voltage V_{D}) are shown in Figs. 13.6 and 13.7 for the structures with channel lengths $L_{\mathrm{ch}} = 40, \ldots, 80$ nm. At a threshold voltage $V_{\mathrm{D}} \approx 0.2$ V a kink in the I–V characteristics of the device is seen.
At subthreshold voltages $V_{\mathrm{D}} < 0.2$ V the derivative $\mathrm{d}V_{\mathrm{D}}/\mathrm{d}I$ gives the resistance of the structure. It is natural, that scattering enhances the resistance of the structure (solid curve) compared to the ballistic transport (dashed curve). Scattering is also found to smear the kink in the I–V characteristic. At a voltage $V_{\mathrm{D}} > 0.2$ V a saturation regime is reached. In this part of I–V characteristics, the current through the structure increases more slowly than it does at a subthreshold voltage. The slope of the I–V curve in the saturation regime rises when the length of the channel decreases. This effect is explained by a reduction of the p–n junction barrier potential as the length of the channel becomes shorter than the p–n junction width. In Fig. 13.7, one can see that, when the transistor is switched off ($V_{\mathrm{G}} < 0.5$ V), the influence of

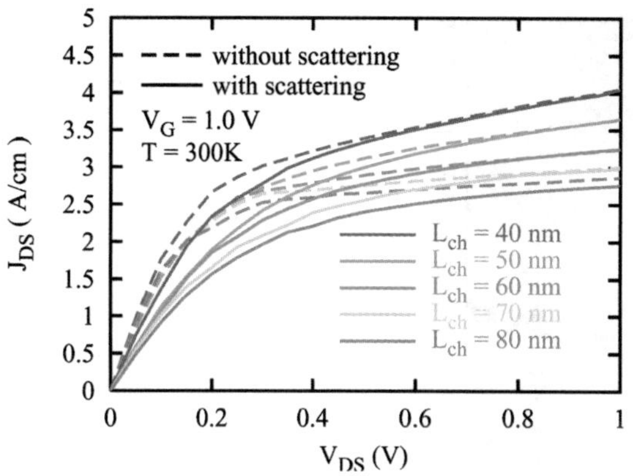

Fig. 13.6. Current–voltage characteristics at $V_\mathrm{G} = 1$ V for different channel lengths

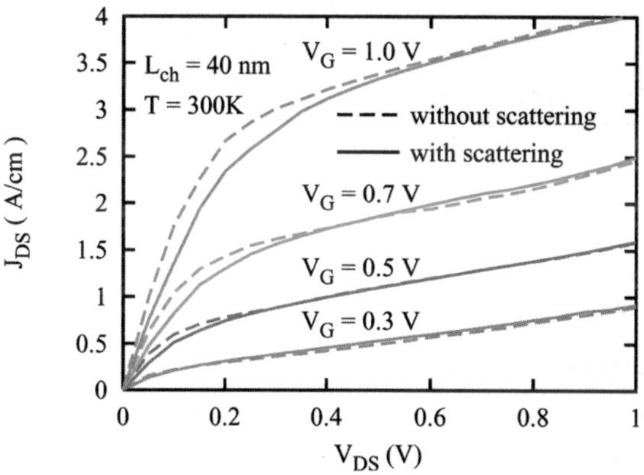

Fig. 13.7. Current–voltage characteristics for MOSFET with $L_\mathrm{ch} = 40$ nm

the scattering on the current is weak. This fact is due to a low concentration of electrons, resulting in a low amplitude of the scattering processes.

In Figs. 13.8 and 13.9 the contour plots of the partial Wigner distribution function ($\alpha = 1, l = 1$) are given for both cases (a – without and b – with scattering). The lighter regions in these plots indicate the higher density of electrons.

Far from the p–n-junction, where the effective potential varies almost linearly, the partial Wigner distribution function can be interpreted as a distribution of electrons in the phase space. When electrons travel in the inversion layer without scattering, their velocity increases monotonously along the whole channel. Therefore, in the phase-space representation the distribution of ballistic electrons has the shape of *a narrow stream in the channel* (Fig. 13.8). As it is expected, *scattering washes out the electron jet in the*

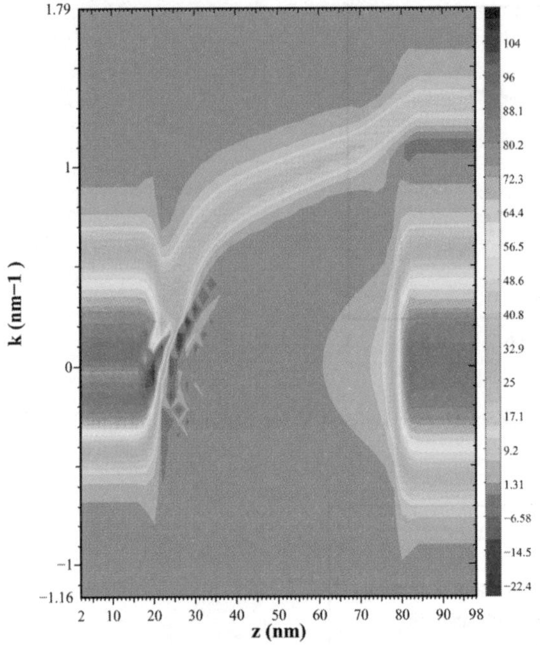

Fig. 13.8. Contour plots of the partial Wigner distribution function $f_{\alpha l}(z,k)$ for the lowest subband ($\alpha = 1, l = 1$) at $V_{\mathrm{G}} = 1$ V, $V_{\mathrm{D}} = 0.3$ V, $L_{\mathrm{ch}} = 60$ nm: without scattering

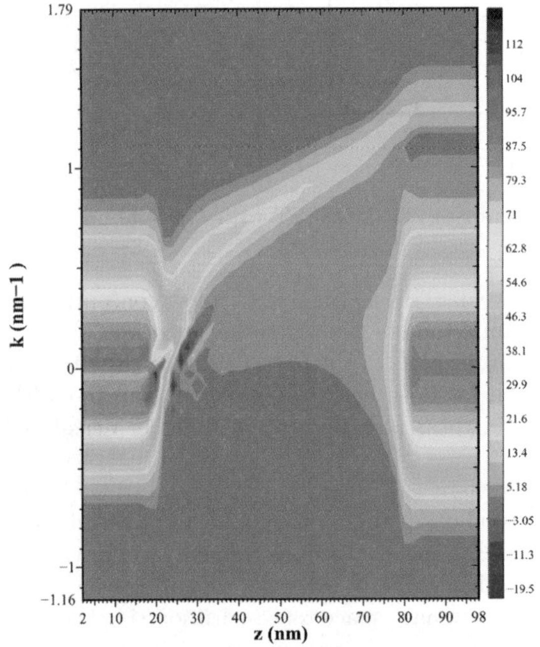

Fig. 13.9. Contour plots of the partial Wigner distribution function $f_{\alpha l}(z,k)$ for the lowest subband ($\alpha = 1, l = 1$) at $V_{\mathrm{G}} = 1$ V, $V_{\mathrm{D}} = 0.3$ V, $L_{\mathrm{ch}} = 60$ nm: with scattering

channel (see Fig. 13.9). It is worth mentioning that the electron stream in the channel does not disappear. It means that in this case the electron transport through the channel combines the features of both diffusive and ballistic motion.

Summary

A model for the detailed investigation of quantum transport in MOSFET devices was presented. The model employs the Wigner distribution function formalism allowing the account for electron scattering by impurities and phonons. Numerical simulations of a cylindrical sub-0.1 μm MOSFET structure were performed. I–V characteristics for different values of the channel length were obtained. It is seen that the slope of the I–V characteristic in the saturation regime rises as the channel length increases. This is due to the decrease of the p–n junction barrier potential.

Finally, the inclusion of a collision term in numerical simulation is important for low source–drain voltages. The calculations have shown that the scattering leads to an increase of the electron density in the channel and smoothes out the applied voltage along the entire channel. The analysis of the electron phase-space distribution in the channel has shown that, in spite of scattering, electrons are able to flow through the channel as a narrow stream although, to a certain extent, the scattering is seen to wash out this jet. Accordingly, features of both ballistic and diffusive transport are simultaneously encountered.

14. An Exactly Solvable Electron–Phonon System

Starting from Heisenberg's equations of motion, we calculate the exact time dependent drift velocity of a single electron, which is harmonically coupled to a phonon bath according to the Caldeira–Leggett (CL) Hamiltonian, and accelerated by a uniform electric field. Contrary to the conventional perturbational approach, the external field may be turned on abruptly at an arbitrary initial time, whence the drift velocity reflects the genuine physical response to the electric field. The electron motion is shown to be irreversible if, for appropriate phonon dispersion relations, the number of phonon modes approaches infinity. Although it has been devised for other purposes, the (CL) electron–phonon Hamiltonian, including the electric field term, provides an exactly solvable transport model, enlightening interesting features of non-equilibrium thermodynamics, such as entropy production.

Recent technological developments in the semiconductor processing led to devices that are operating far from thermodynamic equilibrium. Quantum mechanics and statistical mechanics provide unique tools for accurately modeling the motion of mobile carriers in modern semiconductor structures such as resonant tunneling transistors, NANOFETS, [89] etc.

The operation far from thermodynamic equilibrium is not fully understood and may prohibit a systematic calculation of important transport quantities appearing as time dependent expectation values, such as current–voltage characteristics and carrier drift velocities. In spite of the large computational facilities, it remains a formidable task to solve Liouville's equation [44] for the density matrix of an assembly of charged particles, interacting with phonons and other scattering agents. An alternative approach, bypassing the concept of the time dependent density matrix, is based on the solution of the Heisenberg equations of motion for the quantum-mechanical operators assigned to the physical quantities under consideration. It is well known that the explicit time dependence of such a solution completely determines the evolution of the corresponding statistical average, so that only the initial density matrix should be available [7]. Although this method is not widely used, it proves very convenient for studying the dynamical response of an electron under the influence of an electric field, which is counterbalanced by phonon scattering as described in the (CL) model [90]. This is demonstrated below. The transport equation which governs the dynamical evolution of the electron, is solved for two given functional forms of the electron–phonon coupling strengths. Moreover, the exact solution provides a natural way to encounter time irreversibility in the transport equation, which has yet been derived from the reversible Heisenberg equations. Finally, the entropy production in the electron–phonon system described by the (CL) model is discussed. The construction of the solution of this exactly-solvable model gives a clear insight into the cause of irreversibility [91].

14.1 The Time Dependent Drift Velocity for the CL Model

Any physical quantity A may be generally quantified by a time dependent statistical average of an operator $\hat{A}$, according to the prescription

$$A(t) = \langle \hat{A} \rangle_t = \mathrm{Tr}\left[\varrho(t)\hat{A}\right] , \quad \mathrm{Tr}[\varrho(t)] = 1 . \tag{14.1}$$

Instead of calculating the time dependent density matrix $\varrho(t)$, we may as well shift the explicit time dependence onto the level of the corresponding Heisenberg operator $\hat{A}(t)$ by solving the Heisenberg equation of motion

$$\mathrm{i}\hbar \frac{\mathrm{d}\hat{A}(t)}{\mathrm{d}t} = \left[\hat{A}(t), \hat{H}\right] . \tag{14.2}$$

$\hat{H}$ is the total Hamiltonian and is usually separated into two parts:

$$\hat{H} = \hat{H}_0 + \theta(t)\hat{H}' , \tag{14.3}$$

where $\hat{H}_0$ describes the physical system under consideration, including its internal interactions, $\hat{H}'$ represents the external interaction which drives the system out of its initial state, and $\theta(t)$ is the Heaviside step function. It now becomes clear from the well-known identity [7]

$$\mathrm{Tr}\left[\varrho(t)\hat{A}\right] = \mathrm{Tr}\left[\varrho(0)\hat{A}(t)\right] \tag{14.4}$$

that only the initial density matrix is needed to calculate the dynamical expectation value:

$$\langle \hat{A} \rangle_t = \mathrm{Tr}\left[\varrho(0)\hat{A}(t)\right] \equiv \langle \hat{A}(t) \rangle_0 . \tag{14.5}$$

In many cases, the physical system has attained a state of thermodynamic equilibrium or quasi-equilibrium before the external perturbation $\hat{H}'$ is turned on at $t = 0$, so that $\varrho(0)$ is simply given by

$$\varrho(0) = \frac{\mathrm{e}^{-\beta \hat{H}_0}}{\mathrm{Tr}\left[\mathrm{e}^{-\beta H_0}\right]} , \quad \beta = 1/k_\mathrm{B}T . \tag{14.6}$$

Unless stated otherwise, the full Hamiltonian $\hat{H}$ will refer to the (CL) model [90, 92, 93] throughout this chapter. The (CL) model has been thoroughly used during the past decade to investigate the influence of dissipative interactions on tunneling currents in Josephson junctions and similar devices. The Hamiltonian of the model may be considered a long-wavelength approximation of the 'realistic' Hamiltonian $\hat{\hat{H}}_0$, in which a deformation-potential type of electron–phonon interaction is proposed. The lattice dimension does not restrict the validity of the forthcoming calculations, but for the sake of simplicity we will consider only a one dimensional electron–phonon system, described by [94]:

$$\hat{\hat{H}}_0 = \frac{\hat{p}^2}{2m} + \sum_q \hbar\omega_q a_q^\dagger a_q + \mathrm{i} \sum_q v_q \left(\mathrm{e}^{\mathrm{i}q\hat{x}} a_q - \mathrm{e}^{-\mathrm{i}q\hat{x}} a_q^\dagger \right) , \tag{14.7}$$

where $\hat{x}$, $\hat{p}$ and m respectively denote the position and momentum operators and the mass of the electron, whereas $a_q^\dagger$ and a_q are creation and annihilation operators for the q-th phonon mode, having an energy $\hbar\omega_q$. The strength of the electron–phonon interaction is determined by the coupling constant v_q which is chosen to be real.

The (CL) Hamiltonian $\hat{H}_0$ may be obtained from $\hat{\hat{H}}_0$ by expanding the exponentials up to first order in $\hat{x}$ and by replacing the neglected higher order terms by an appropriate harmonic potential. Such a substitution may appear as an *ad hoc* manipulation, but it imposes the required lower bound on the interaction term. As a result, the (CL) Hamiltonian reads [90]

$$\begin{aligned} \hat{H}_0 = & \frac{\hat{p}^2}{2m} + \sum_q \hbar\omega_q a_q^\dagger a_q + \sum_q \left[\mathrm{i} v_q (a_q - a_q^\dagger) - F_q (a_q + a_q^\dagger)\, \hat{x} \right] \\ & + \hat{x}^2 \sum_q \frac{F_q^2}{\hbar\omega_q} \end{aligned} \tag{14.8}$$

with

$$F_q = q\, v_q \; . \tag{14.9}$$

Clearly, the crucial step leading to the (CL)-Hamiltonian (14.8) is the linearization of the exponential terms in (14.7). In principle, this manipulation can only be justified if the motion of the electron is confined to a space region which is small enough to guarantee that $|q\,\langle \hat{x} \rangle\,| \ll 1$ is valid for all relevant q-values. This condition may be reasonably satisfied in the case electron transport through short-channel devices with typical lengths of 100 nm or less. It should be noted however that the (CL)-Hamiltonian will be used throughout this chapter as a model Hamiltonian with a twofold purpose. Firstly, as it includes a microscopic mechanism which governs the electron's energy loss, it serves as basic tool for solving exactly the equations of motion, which do not contain any phenomenological damping terms. Secondly, it provides a test case for the investigation of time irreversibility and entropy production, without having to make any further approximation such as the truncation of the Lie-algebra of observables, or the a-priori introduction of a set of characteristic times according to Bogolyubov [95]. Therefore the formal extraction of the (CL)-electron–phonon Hamiltonian from its realistic counterpart (14.7) is merely mentioned to illustrate the origin of the former and does not claim any judgment concerning the replacement $\hat{\hat{H}}_0 \to \hat{H}_0$ whatsoever. The same remarks can be made for the choices of the interaction strength v_q: the proposed forms which also may look a bit unrealistic have been inserted in the equations in order to gain insight in the non-equilibrium features thereby avoiding any unnecessary computational complexity.

At $t = 0$, the external electric force F_0 starts accelerating the electron according to the perturbation term $\hat{H}'$:

$$\hat{H}' = -F_0 \,\hat{x} \; . \tag{14.10}$$

The equations of motion can be easily obtained by commuting each operator with $\hat{H} = \hat{H}_0 + \theta(t)\hat{H}'$

$$\begin{aligned} m\frac{\mathrm{d}^2\hat{x}(t)}{\mathrm{d}t^2} &= F_0\theta(t) + \sum_q F_q\left[a_q(t) + a_q^\dagger(t)\right] - 2\hat{x}(t)\sum_q \frac{F_q^2}{\hbar\omega_q} \\ \frac{\mathrm{d}a_q(t)}{\mathrm{d}t} &= -\mathrm{i}\omega_q a_q(t) - \left[v_q - \mathrm{i}F_q\,\hat{x}(t)\right]/\hbar \\ \frac{\mathrm{d}a_q^\dagger(t)}{\mathrm{d}t} &= \mathrm{i}\omega_q a_q^\dagger(t) - \left[v_q + \mathrm{i}F_q\,\hat{x}(t)\right]/\hbar \; . \end{aligned} \tag{14.11}$$

Bearing in mind the definition of the electron velocity operator

$$\hat{v}(t) \equiv \mathrm{d}\hat{x}(t)/\mathrm{d}t = -\mathrm{i}/\hbar\;[\hat{x}(t), H] \; , \tag{14.12}$$

we may combine and rearrange the equations (14.11) to yield an integro-differential equation for $\hat{v}(t)$

$$\begin{aligned} \frac{\mathrm{d}\hat{v}(t)}{\mathrm{d}t} = \frac{F_0}{m} + \frac{1}{m}\sum_q F_q &\left[\mathrm{e}^{-\mathrm{i}\omega_q t}a_q(0) + \mathrm{e}^{\mathrm{i}\omega_q t}a_q^\dagger(0) - \frac{2v_q}{\hbar\omega_q}\sin\omega_q t\right. \\ &\left. -\frac{2F_q}{\hbar\omega_q}\cos\omega_q t\,\hat{x}(0)\right] - \int_0^t \mathrm{d}t'\,\alpha(t-t')\,\hat{v}(t') \; . \end{aligned} \tag{14.13}$$

In the equation (14.13), $\alpha(t)$ plays the role of a memory function, which is determined completely by the phonon spectrum and the coupling strength:

$$\alpha(t) = \frac{2}{m}\sum_q \frac{F_q^2}{\hbar\omega_q}\cos\omega_q t \; . \tag{14.14}$$

As such, (14.13) is still an operator equation, which must be averaged with the help of the equilibrium density matrix $\varrho(0)$ in order to derive a transport equation for the drift velocity $v(t) \equiv \langle\hat{v}\rangle_t = \langle\hat{v}(t)\rangle_0$. The averages

$\langle\hat{x}(0)\rangle_0$, $\langle a_q(0)\rangle_0$ and $\langle a_q^\dagger(0)\rangle_0$ appearing as initial conditions, satisfy the following relations:

$$\langle a_q(0)\rangle_0 = \frac{F_q\,\langle\hat{x}(0)\rangle_0 + \mathrm{i}v_q}{\hbar\omega_q}$$
$$\langle a_q^\dagger(0)\rangle_0 = \frac{F_q\,\langle\hat{x}(0)\rangle_0 - \mathrm{i}v_q}{\hbar\omega_q} \tag{14.15}$$

which, in turn, can be derived by recalling that, for $t \le 0$, every single-time correlation function such as $\langle a_q(t)\rangle_0$ is an equilibrium averages and therefore does not depend on time. Finally, inserting the relations (14.15) into the averaged equation (14.13) we get to the following transport equation, which agrees with the classical equation, derived in [92]

$$\frac{\mathrm{d}v(t)}{\mathrm{d}t} = \frac{F_0}{m} - \int_0^t \mathrm{d}t'\,\alpha(t-t')\,v(t')\,. \tag{14.16}$$

The solution of the transport equation (14.16), can be obtained by applying Laplace transformation techniques, as will be briefly explained in the following section.

14.2 Solution of the Transport Equation

The transport equation (14.16) can be solved straightaway for $t > 0$ by taking Laplace transforms of both sides. In the frequency domain the solution reads

$$\tilde{v}(z) = \frac{F_0}{m}\frac{1}{z^2 + \mathrm{i}z\tilde{\alpha}(z)}$$
$$= \frac{F_0}{m}\frac{1}{z^2}\left(1 + \frac{2}{m}\sum_q \frac{F_q^2}{\hbar\omega_q}\frac{1}{z^2+\omega_q^2}\right)^{-1}, \tag{14.17}$$

where the tilde ˜ refers to a transformed quantity and z is the complex frequency variable. Obviously, the difficult part of the solution is to extract the drift velocity $v(t)$ from the inverse Laplace transform according to

$$v(t) = \frac{1}{2\pi\mathrm{i}}\int_\Gamma \mathrm{d}z\,\tilde{v}(z)\,\mathrm{e}^{zt}\,. \tag{14.18}$$

The integration path Γ is an arbitrary vertical line which must be chosen such that all the singularities of $\tilde{v}(z)$ are at its left-hand side. In order to determine the position of these singularities, we have to specify both the explicit form of the phonon dispersion relation and the q-dependence of the coupling strength. We may, for example, assume the presence of N acoustic phonon modes, described by a simple, linear dispersion relation

$$\omega_q = c\,|q|\,, \tag{14.19}$$

where c is the sound velocity and a discrete set of N allowed phonon wave vectors is dividing the first Brillouin zone $[-\frac{\pi}{a}, \frac{\pi}{a}]$ according to

$$q = \frac{2\pi}{Na} n \ , \ n = -N/2 + 1, \ldots, -1, 0, 1, \ldots, N/2 \ \ (N \text{ even}) \ . \tag{14.20}$$

Furthermore, we have adopted two functional forms for the interaction strength, the first one being

$$v_q^2 = \frac{D_{\mathrm{AC}}^2}{Na|q|} \ . \tag{14.21}$$

The constant D_{AC} may be interpreted as a deformation potential for acoustic phonons, whereas the imposed q-dependence favors the interaction with long-wavelength modes. Although one might rather expect a linear dependence on q [94] which could be invoked as well, the proposed form for v_q gives rise to a remarkable solution of the transport equation. But first it follows from (14.14),(14.19) and (14.21) that

$$\begin{aligned} \alpha(t) &= \frac{2D_{\mathrm{AC}}^2}{\hbar mcaN} \sum_q \cos cqt \\ &= \frac{2D_{\mathrm{AC}}^2}{\hbar mca} \frac{\sin \omega_{\mathrm{D}} t}{N \tan(\omega_{\mathrm{D}} t/N)} \ , \end{aligned} \tag{14.22}$$

where the free phonon system is assumed not to support higher frequencies than than $\omega_{\mathrm{D}} = \pi c/a$ which is of the order of the Debye frequency. Inspection of (14.22) reveals that the function $\alpha(t)$ is expressed as a sum of N cosine functions. As a consequence, the drift velocity $v(t)$ will also appear as a finite linear combination of periodic functions, thereby exhibiting reversible oscillating behavior. In general $v(t)$ may contain a 'runaway' component which corresponds to a uniform acceleration by the field. The pre-factor of this component equals the residue of $\tilde{v}(z)$ at $z = 0$ and is proportional to

$$\left(1 + \frac{2}{m} \sum_q \frac{F_q^2}{\hbar \omega_q^3}\right)^{-1} \ . \tag{14.23}$$

In the present case the sum over all modes diverges due to the occurrence of a soft mode ($\lim_{q \to 0} \omega_q = 0$) in the acoustic spectrum, so that the pre-factor vanishes. Physically such a soft mode reflects a macroscopic translation of the crystal with zero excitation energy, and gives rise to a restoring force, acting on the electron. On the other hand, if N, the number of phonon modes approaches infinity, the memory function $\alpha(t)$ decays with time according to

$$\alpha(t) = 2\lambda \frac{\sin \omega_{\mathrm{D}} t}{t} \ , \quad \lambda = \frac{D_{\mathrm{AC}}^2}{\pi \hbar mc^2} \ . \tag{14.24}$$

14.3 How to Invert Laplace Transforms

The inverse Laplace transform of $\tilde{v}(z)$ ((14.17) below) corresponds to a line integral to be carried out along the vertical path Γ shown in Fig. 14.1 and may be performed by considering the closed contour which consists of Γ, the large semi-circle, the two small circles and the two connecting line segments. In order to properly handle the present logarithmic function, it is convenient to define polar angles with respect to the imaginary axis according to

$$\begin{aligned} z - \mathrm{i}\omega_\mathrm{D} &= \mathrm{i} r_1 \mathrm{e}^{\mathrm{i}\theta_1} , \quad r_1 \equiv |z - \mathrm{i}\omega_\mathrm{D}| \\ z + \mathrm{i}\omega_\mathrm{D} &= \mathrm{i} r_2 \mathrm{e}^{\mathrm{i}\theta_2} , \quad r_2 \equiv |z + \mathrm{i}\omega_\mathrm{D}| , \end{aligned} \tag{14.25}$$

where $z = x + \mathrm{i}y$ is any point within the bounded region and $0 \le \theta_1, \theta_2 \le 2\pi$. The logarithmic function may then be expressed as

$$\log \frac{z + \mathrm{i}\omega_\mathrm{D}}{z - \mathrm{i}\omega_\mathrm{D}} = \log \frac{r_2}{r_1} + \mathrm{i}\,(\theta_2 - \theta_1 + 2\pi k) , \tag{14.26}$$

where the integer k is determined by requiring that

$$\lim_{z \to 0, x > 0} \tilde{\alpha}(z) = \tilde{\alpha}(0) = \pi\lambda , \tag{14.27}$$

whence it follows that $k = 0$. The abovementioned contour does not cross the branch line extending along the imaginary axis from $-\mathrm{i}\omega_\mathrm{D}$ to $\mathrm{i}\omega_\mathrm{D}$ whereas the bounded region does only contain two isolated singularities at $-\mathrm{i}\omega_0$ and $\mathrm{i}\omega_0$, as is illustrated in Fig. 14.1.

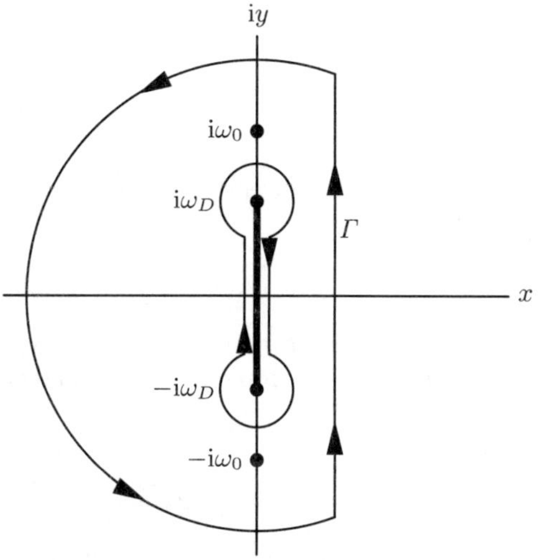

Fig. 14.1. Bromwich contour used to calculate the inverse Laplace transform of (14.17). The branch line connects $-\mathrm{i}\omega_\mathrm{D}$ with $\mathrm{i}\omega_\mathrm{D}$. Isolated singularities are found at $-\mathrm{i}\omega_0$ and $\mathrm{i}\omega_0$

14.4 Irreversibility and the Ohmic Case

We will demonstrate that in this case an irreversible solution of the transport equation may be obtained for appropriate values of the dimensionless parameter $\omega_{\mathrm{D}}/\lambda$. The Laplace transform of (14.24) is given by

$$\tilde{\alpha}(z) = -\mathrm{i}\lambda \log \frac{z+\mathrm{i}\omega_{\mathrm{D}}}{z-\mathrm{i}\omega_{\mathrm{D}}} . \tag{14.28}$$

In order to proceed, we need to investigate carefully the singularities of $\tilde{v}(z)$, or equivalently, the solutions of

$$z^2 - \mathrm{i}\lambda z \log \frac{z+\mathrm{i}\omega_{\mathrm{D}}}{z-\mathrm{i}\omega_{\mathrm{D}}} = 0 . \tag{14.29}$$

This analysis is elaborate but straightforward. The resulting drift velocity is given by

$$v(t) = 2\pi v_{\mathrm{S}} \int_0^{\omega_{\mathrm{D}}} \frac{\mathrm{d}\omega}{\omega} \frac{\sin \omega t}{\pi^2 + \left(\log \frac{\omega_{\mathrm{D}}+\omega}{\omega_{\mathrm{D}}-\omega} - \frac{\omega}{\lambda}\right)^2} + v_0 \sin \omega_0 t , \tag{14.30}$$

where

$$\begin{aligned} v_{\mathrm{S}} &= \frac{F_0}{\pi m \lambda} \\ v_0 &= \frac{2F_0}{m\omega_0\left(1 - \frac{2\lambda\omega_{\mathrm{D}}}{\omega_0^2 - \omega_D^2}\right)} \end{aligned} \tag{14.31}$$

and ω_0 denotes the (unique) solution of

$$\frac{\omega_0}{\lambda} = \log \frac{\omega_0+\omega_{\mathrm{D}}}{\omega_0-\omega_{\mathrm{D}}} , \quad \omega_0 > \omega_{\mathrm{D}} . \tag{14.32}$$

The long-time behavior of the drift velocity can be directly extracted from (14.30):

$$v(t) \overset{t\to\infty}{\longrightarrow} v_{\mathrm{S}} + v_0 \sin \omega_0 t . \tag{14.33}$$

Clearly, when the steady state is attained, the electron velocity may never take its initial value $v(0) = 0$ if the amplitude of the 'ripple' remains smaller than v_{S} or equivalently

$$2\pi\lambda < \omega_0 \left(1 - \frac{2\lambda\omega_{\mathrm{D}}}{\omega_0^2 - \omega_D^2}\right) . \tag{14.34}$$

In this light, the inequality (14.34) may be considered an irreversibility condition for the present functional forms of ω_q and v_q proposed in (14.19) and (14.21).

In Fig. 14.2 the drift velocity as a function of time is plotted for $\omega_{\mathrm{D}}/\lambda = 5$, 8 and 15 respectively, whereas λ is set to 0.1. The corresponding values for

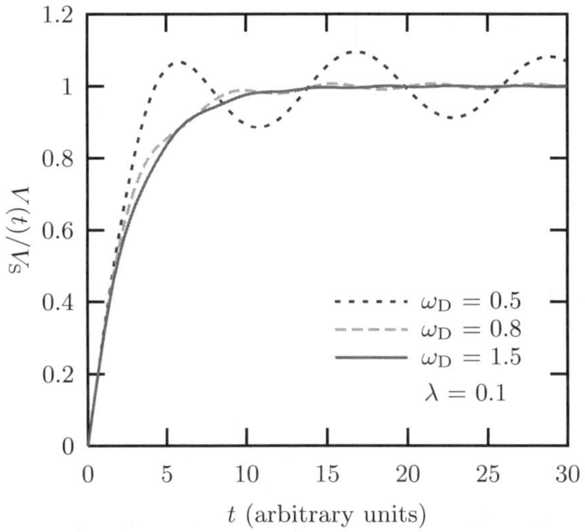

Fig. 14.2. Time dependent drift velocity versus time for three different values of ω_D and $\lambda = 0.1$

the difference $\omega_0 - \omega_D$ are 6.3×10^{-3}, 5.3×10^{-4} and 9.2×10^{-7}, so that the condition (14.34) is fulfilled in all three cases.

Finally, we have considered the so-called ohmic case, which is referred to in [90, 92, 93] when the transport equation (14.16) reduces to the classical equation of motion, supplied with a phenomenological frictional force

$$\frac{\mathrm{d}v(t)}{\mathrm{d}t} = -\frac{v(t)}{\tau} + \frac{F_0}{m} . \tag{14.35}$$

As was already pointed out by Caldeira and Leggett in another way, the two transport equations can only be made consistent if the memory function $\alpha(t)$ becomes Markovian, i.e. reduces to a delta function. Correspondingly for the abovementioned acoustic phonon model, the cut-off frequency ω_D should tend to infinity, which, in turn, requires the lattice constant to vanish. Hence the ohmic case is found to be consistent with a continuum description of the acoustic phonon field. When the limit $\omega_D \to \infty$ is taken in (14.30), the ripple amplitude tends to zero, while the integral can be evaluated analytically. The resulting drift velocity coincides with the solution of (14.35)

$$v(t) = v_S \left(1 - \mathrm{e}^{-t/\tau}\right) , \tag{14.36}$$

where the relaxation time τ is related to the deformation potential according to

$$\tau = \frac{1}{\pi\lambda} = \frac{\hbar m c^2}{D_{AC}^2} . \tag{14.37}$$

In a second example the cut-off frequency ω_D (or its corresponding wavenumber $q_D = \omega_D/c$) restricts the number of modes which are able to effectively interact with the electron, without imposing an upper bound on the phonon frequencies

$$v_q^2 = \frac{D_{\mathrm{AC}}^2}{Na|q|} \, \frac{1}{1+(q/q_{\mathrm{D}})^2} \, . \tag{14.38}$$

Converting the summation over discrete q-values to a an integral yields

$$\alpha(t) = \pi\lambda\omega_{\mathrm{D}} \, \mathrm{e}^{-\omega_{\mathrm{D}}|t|} \, . \tag{14.39}$$

The transport equation can now be rewritten as an ordinary second-order differential equation including a damping term

$$\frac{1}{\omega_{\mathrm{D}}} \frac{\mathrm{d}^2 v(t)}{\mathrm{d}t^2} + \frac{\mathrm{d}v(t)}{\mathrm{d}t} + \pi\lambda v(t) = \frac{F_0}{m} \, . \tag{14.40}$$

The trivial solutions of this equation will be omitted here. Two remarks however have to be made. First, the ohmic case again seems to correspond to the limit $\omega_{\mathrm{D}} \to \infty$. Next, it is clear that any oscillation, if present, is depressed as time increases unlimitedly. In other words, the presence of an infinite number of phonon modes is now a sufficient condition for time irreversibility.

14.5 Entropy Production

Since exact solutions of the transport equation based on the (CL) model are available, it is instructive to evaluate the so-called relative entropy. This quantity has been proposed earlier by many authors [96, 97], attempting to generalize the concept of entropy to the case of non-equilibrium states. In the Schrödinger picture it takes the form

$$S\left[\varrho(t), \varrho(0)\right] = -k_{\mathrm{B}} \operatorname{Tr}\left\{\varrho(t)\left[\log \varrho(0) - \log \varrho(t)\right]\right\} \, . \tag{14.41}$$

In our case the electron–phonon system is assumed to be in thermal equilibrium at time $t = 0$. Therefore the second term at the right-hand side of (14.41), which does not depend on time, can clearly be identified with the equilibrium entropy $S_0 = -k_{\mathrm{B}} \operatorname{Tr}[\varrho_0 \log \varrho_0]$, while the first term may be interpreted as the time dependent entropy:

$$\begin{aligned} S(t) &= -k_{\mathrm{B}} \operatorname{Tr}\left[\varrho(t) \log \varrho(0)\right] \\ &= \frac{1}{T} \left\{\operatorname{Tr}\left[\varrho(t) H_0\right] - \mathcal{F}_0\right\} \, . \end{aligned} \tag{14.42}$$

In the Heisenberg picture we may represent the relative entropy by

$$\begin{aligned} S\left[\varrho(t), \varrho(0)\right] &= S(t) - S_0 \\ &= -\frac{\mathcal{F}_0}{T} - S_0 + \frac{1}{T} \langle \hat{H}_0(t) \rangle_0 \, . \end{aligned} \tag{14.43}$$

$\mathcal{F}_0$ denotes the Helmholtz free energy of the system in equilibrium at temperature T and is given by

$$\mathcal{F}_0 = -\frac{1}{\beta} \log \operatorname{Tr} e^{-\beta \hat{H}_0} = \langle \hat{H}_0(0) \rangle_0 - T S_0 \, . \tag{14.44}$$

The entropy production, being the rate at which the entropy changes, is then simply proportional to the power, acquired from the electric field, as can be derived from

$$\begin{aligned} \frac{\mathrm{d}S(t)}{\mathrm{d}t} &= \frac{1}{T} \left\langle \frac{\mathrm{d}\hat{H}_0(t)}{\mathrm{d}t} \right\rangle_0 = -\frac{1}{T} \left\langle \frac{\mathrm{d}\hat{H}'(t)}{\mathrm{d}t} \right\rangle_0 \\ &= \frac{1}{T} F_0 \, v(t) \, . \end{aligned} \tag{14.45}$$

Obviously, the positivity of $\mathrm{d}S(t)/\mathrm{d}t$ is directly linked to the time irreversibility of the drift velocity and the related criteria, which have been formulated on rather intuitive grounds in the previous section. In particular, a constant entropy production is observed when the steady state is attained in the ohmic case. Equation (14.45) therefore seems to confirm the relevance of the relative entropy, as proposed in [96, 97] for the investigation of more general irreversible processes.

Summary

It was explained how the Heisenberg picture provides an alternative method to solve the transport problem for the (CL) model, and to obtain a solution, which is clarifying in a concrete manner some typical features of non-equilibrium statistical mechanics, such as the link between the entropy production and power dissipation. It is however tempting to look at the previous solution as the zero-th order solution of a perturbational scheme, in which the perturbation $\Delta\hat{H}$ would cover to the difference between the 'real' electron–phonon interaction and that of the (CL) Hamiltonian H_{CL}, i.e.

$$\Delta\hat{H} = \hat{H} - \hat{H}_{\mathrm{CL}} = \hat{\hat{H}}_0 - \hat{H}_0 \, . \tag{14.46}$$

In such an approach, the external field is treated rigorously as it has been included in both $\hat{H}_{\mathrm{CL}}$ and the exact Hamiltonian $\hat{H}$. As a consequence, all the 'unperturbed' expectation values refer to non-equilibrium states, whereas the perturbation term merely introduces the non-linear contributions of the electron–phonon coupling. Contrary to conventional perturbation theory, there are no systematic computational recipes, such as Wick's theorem which could greatly simplify the evaluation of the zeroth order non-equilibrium averages. Nevertheless, the suggested perturbation scheme may provide an alternative tool for studying important time dependent non-equilibrium phenomena such as transient transport characteristics in advanced semiconductors.

15. Open Versus Closed Systems

Quantum transport in open systems is obscured by the fact that the number of carriers is not well defined. We cannot even define an expectation values for the number of particles, since particles permanently leaving the device, whereas others enter. Crudely speaking, we imagine a device connected to two or more leads, and that carriers enter the device by one lead and leave the device through another lead. This picture suggests that we can discriminate or distinguish particle by their location in space. However, quantum-mechanical treatment should avoid any attempt to distinguish identical particles and therefore it is not appropriate to associate some carriers with restricted parts (the leads) of the system and others with the device. Another drawback of open systems is that the Hilbert space (or classically, the phase space) cannot be constructed. Again, this shortcoming is not due to the fact that we do not know the number of particles, but to the fact that particles are leaving and entering the system.

These problems can be circumvented by converting the full system to a closed one by bending the leads to each other such that at all times we are dealing with the same set of particles. Physically, such a modeling also makes sense because we may assume that the device is driven by a battery and therefore the carriers are running around in a closed loop. The first step in the program is:

Adapt the geometry of the leads by deforming the geometry or by inserting periodic boundary conditions for the carriers. A few examples will illustrate this further.

Quantum Point Contact. A point contact consists a two dimensional electron gas that is pushed through a very narrow slit. In Fig. 15.1 the situation is sketched. On the left there is a high potential, whereas on the right there is a low potential. One of the most remarkable properties is that the conductance in quantized. A geometrical modification is shown in Fig. 15.2. We have bended the lead into a circular shape without introducing curvature into the leads.

Quantum MOSFET. A quantum MOSFET is scaled to a size where quantum interferences become important. The channel length is much less than 0.1 μ. The device is sketched in Fig. 15.3.

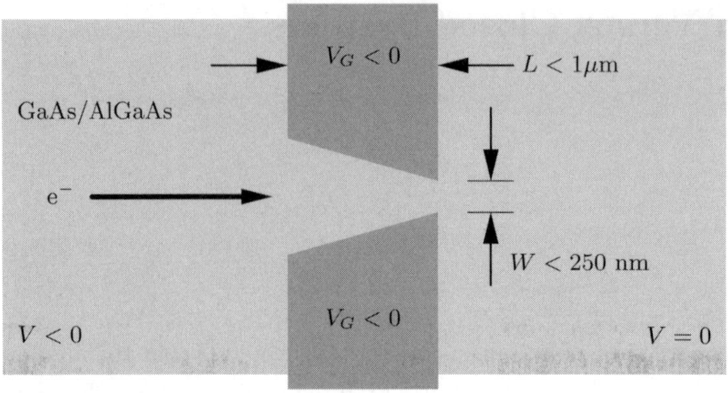

Fig. 15.1. Sketch of the quantum point contact

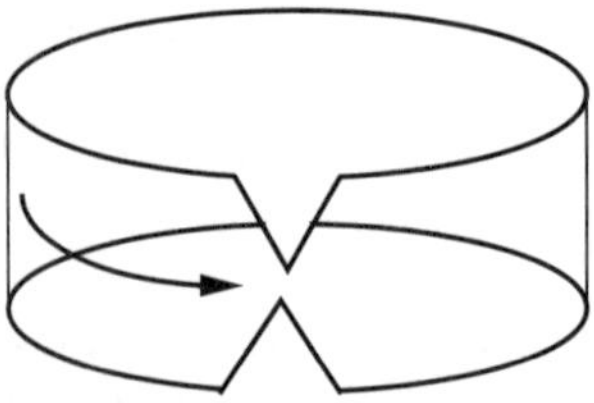

Fig. 15.2. Sketch of the quantum point contact after closing the system

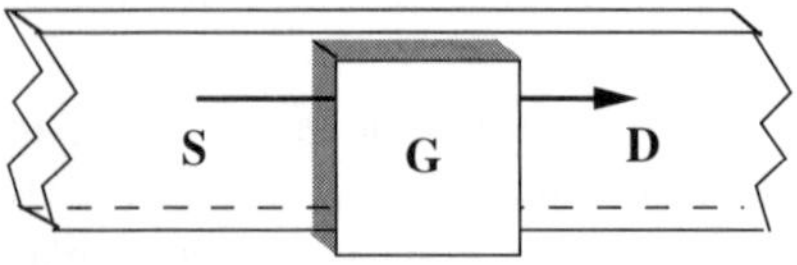

Fig. 15.3. Sketch of the quantum MOSFET

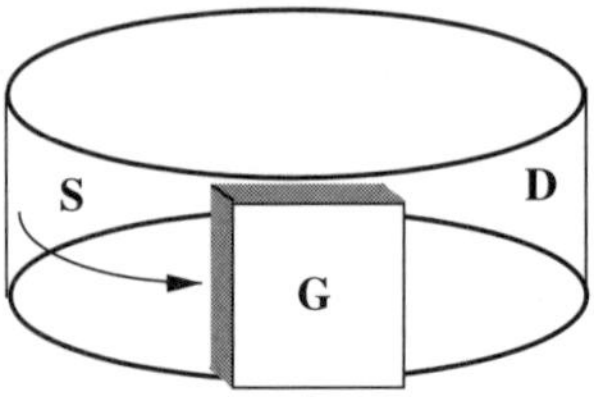

Fig. 15.4. Sketch of the quantum MOSFET after connecting source and drain

The source and drain regions are equally doped, such that there is no additional jump generated by folding them towards each other as is demonstrated in Fig. 15.4. The Quantum MOSFET can be further abstracted by ignoring the source and drain parts all together and fully concentrate on the channel modeling assuming a non-conservative circular electric field which drives the channel current and a fully circular gate which modulates the current in the channel by the gate bias. The model is illustrated in Fig. 15.5.

Resonant Tunnel Diode. The resonant tunnel device has two or more sharp doping layers such the a resonating valley is in between. The doping

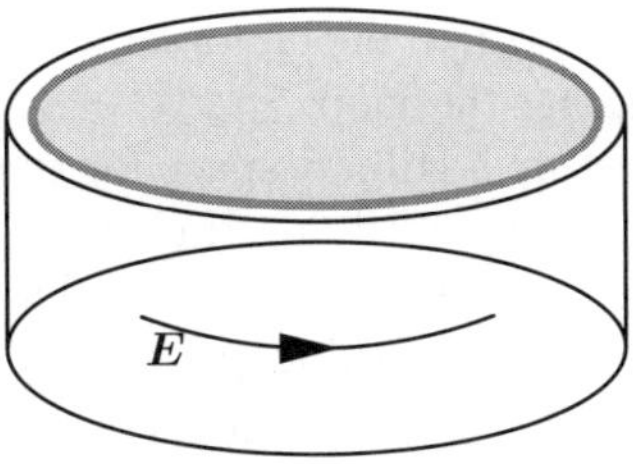

Fig. 15.5. Sketch of the quantum MOSFET after removal of the source and drain

profile along the device is shown in Fig. 15.6. After the modification, the most convenient geometry is again a loop but we may neglect all intrinsic curvature of the loop. Such a modeling corresponds to a donut in a space of dimension $d > 3$ (Fig. 15.7).

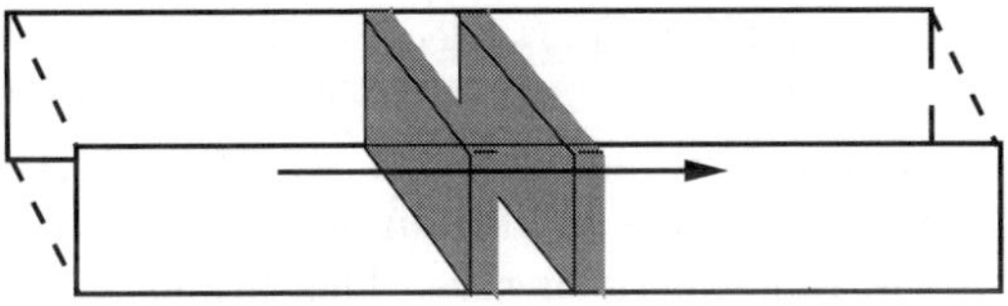

Fig. 15.6. Sketch of the resonant tunnel device

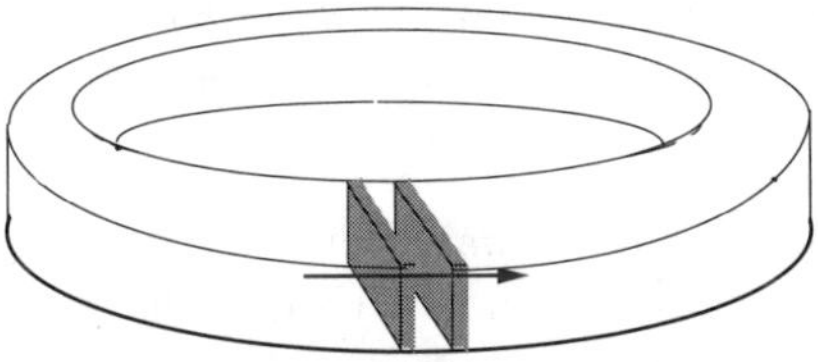

Fig. 15.7. Sketch of the resonant tunnel device after modification

Of course, in a closed system we must adapt the force, such that a current is sustained, otherwise we would end up in a situation that all net currents vanish and the system is in equilibrium. For that purpose it is needed to consider forces which are generated by a *non-conservative* potential. In particular, we want

$$\oint_{\text{circuit}} \boldsymbol{F}.\boldsymbol{dl} \neq 0 \tag{15.1}$$

The second step in the program is:

The driving force is provided by a non-conservative field along the closed but not simply connected geometry.

This picture mimics the battery modeling very accurately. The potential which is picked up after one rotation just corresponds to the electromotive force of the battery. Of course, the detailed physical origin of the force is not relevant. All that is needed is a sustained force along the loop.

Since the system is closed we are now able to set up a Hilbert space for the devices and perform several calculations, varying from identifying Hamiltonians, single-particle wave functions, and equilibrium statistical ensembles. The introduction of the non-equilibrium transport description is done by switching on the driving force at time $t = 0$. This assumption brings us to the third step in the program:

For $t < 0$, the system is in statistical equilibrium and there is no driving force.

At $t = 0$ the force is immediately switched on at full strength. So after $t = 0$ we have the Hamiltonian

$$\hat{H} = \hat{H}_{\rm in} + \hat{H}_{\rm ext} \tag{15.2}$$

where $\hat{H}_{\rm ext}$ is the external driving force part. This part needs not to be treated as a quantum field but may be considered as an external classical background field. The total Hamiltonian determines the time evolution after $t = 0$. All expectation values are now calculated from the equilibrium density operator $\hat{\varrho}_0$, which is time-independent in the Heisenberg picture. This brings us to the fourth step in the program:

The calculation of the expectation values at time $t > 0$ are done by solving Heisenberg's equation of motion using the equilibrium density operator at $t = 0$.

The driving force can be chosen in an intelligent way. Being an external field, it may be taken constant along a loop of the geometry. Using a flat strip with periodic boundary conditions for the carrier wave functions, we can not use periodic boundary conditions for the potential of the external field, but

$$\begin{aligned} V(L) &= V(0) + \Delta V \\ \Delta V &= \int_0^L \boldsymbol{dl}.\boldsymbol{F} \end{aligned} \tag{15.3}$$

This is another way of implementing a non-conservative force. For the carriers this problem does not pose itself, because the boundary conditions are only used at $t \leq 0$. Before the external field is switched on a self-consistent determination of the Poisson field and the carrier fields is required, due to the built-in potentials of the devices under consideration.

16. Conductance Quantization

Exploiting the possibility of down scaling the sizes of modern semiconductor devices, their insulating layers and the connecting wires, one has gradually entered the range where quantum physics dominates the descriptive base for understanding experimental results. In particular, coherent transport through so-called quantum devices such as quantum wires, quantum dots and quantum point contacts has been regarded as a substantial support for the Landauer picture of transport in disordered solids [98, 99, 100, 101, 102]. Moreover, for more than two decades numerous authors have investigated charge transport through large normal or superconducting metallic contacts which are separated by narrow semiconducting or insulating layers the size of which are below the mean free paths related to typical scattering mechanisms. In spite of the agreement between the electrical characterization of mesoscopic conductors and the Landauer–Büttiker theory predicting the conductance in those devices, conceptual doubt has persisted regarding the derivation of the famous Landauer–Büttiker formula from first principles of quantum mechanics and statistical physics [105, 106, 107].

We will argue that a simple, closed electric circuit, considered as a multiply connected manifold with a single hole and acted upon by an externally applied DC electromotive force, which is kept constant in time can exhibit Landauer–Büttiker conductance if the following conditions are met: (*1*) quantization of the magnetic flux generated by the electric current flowing through the circuit, (*2*) spatial localization of the driving electric field in a simply connected subregion of the circuit [108].

Clearly, the underlying model requires a totally different approach compared with conventional studies of coherent transport treating the electric circuit as an open system of charge carriers: the latter are injected from a reservoir and, after some ballistic propagation through the mesoscopic area they are seen to disappear in another reservoir. On the contrary, our model is describing the electric circuit as a closed, torus-shaped region confining the charge carriers to its interior – as is essentially realized in a real circuit – but allowing energy to be exchanged with the environment. It is interesting to note that naive arguments based on topology, may already give rise to conductance quantization. Starting from the definition of the current as the amount of charge ΔQ, e.g. one electron, which crosses a surface per unit time, we may consider a time lap corresponding to one revolution of the charge. The gain in energy in one revolution is $\Delta W = eV_\varepsilon$. It is tempting to consider ΔW and Δt as conjugated variables and invoke quantum mechanics through a Heisenberg uncertainty relation, i.e.

$$I = \frac{\Delta Q}{\Delta t}, \qquad \Delta W = eV_\varepsilon , \qquad \Delta W \, \Delta t = h , \tag{16.1}$$

which after elimination of Δt results into a linear current–voltage relation, $V_\varepsilon = R_\text{K} I$, where $R_\text{K} = h/e^2$ is von Klitzing's resistance. Such arguments are encouraging for pursuing the inclusion of topological aspects into the description of quantum conductance, but much more elaborate details need to be provided.

In order to uncover the subtle details which are encountered in a quantum-mechanical description of circuits, we will first present the analysis of a simple conducting ring being a prototype for a more general 'quantum circuit'. Although there exists already an extensive literature about quantum devices and nanostructures, the full circuit aspects are usually addressed by transfer matrix methods and reservoir-type boundary conditions. Unfortunately, the reservoirs can obscure the underlying physics, since only a qualitative separation is possible between the devices under consideration and the reservoirs. It should also be realized that the reservoirs represent equilibrium systems, weakly coupled to devices under consideration [56]. The weak coupling is realized by evaluation of the wave-function overlap which is generated by collapse of an infinite high potential barrier wall to a finite or zero value. From the point of view of potential perturbation theory, such modifications are difficult to handle. An illustration of the entangling of device – and reservoir physics is provided by the present modeling of the quantum-point contact. The conductance is regarded as a property of the full device. However, the entropy increase associated with transport through the point contact is assigned to the different state functions of the reservoirs. As a consequence, resistance and entropy increase are *physically* decoupled. We will address the question whether this conclusion can be sustained. For that purpose we will abandon in our modeling the reservoirs and include non-conservative force fields along closed loops in the circuit which should provide the electromotive forces (emf).

16.1 Circuit Topology, Non-Conservative Fields and Dissipationless Transport

We consider a closed electric circuit that comprises a three dimensional multiply connected region Ω encircling exactly one hole, i.e. a torus-shaped region confining an ensemble of electrons. Consequently, all one-electron wave-functions and the electron field operator $\psi(\boldsymbol{r}, t)$ are assumed to vanish at the boundary surface $\partial\Omega$:

$$\psi(\boldsymbol{r}, t) = 0 \qquad \forall \boldsymbol{r} \in \Omega \,. \tag{16.2}$$

As shown in Fig. 16.1, the circuit consists of four regions:

- a so-called active region Ω_A that is restricted by the cross-sectioned surfaces $\Sigma_{1\mathrm{A}}$ and $\Sigma_{2\mathrm{A}}$, representing any arbitrary mesoscopic area such as a quantum point contact, a quantum dot or a narrow energy barrier,
- a 'battery' region Ω_B representing the seat of an externally applied DC electromotive force V_ε,
- two ideally conducting leads $\Omega_{1\mathrm{L}}$ and $\Omega_{2\mathrm{L}}$ connecting the battery to the active region.

The motion of the electron ensemble is driven by an electric field $\boldsymbol{E}(\boldsymbol{r})$ which is invoked by the external emf V_ε

$$V_\varepsilon = \oint_\Gamma \boldsymbol{E}(\boldsymbol{r}) \cdot \mathrm{d}\boldsymbol{r} \,, \tag{16.3}$$

where Γ is an arbitrary closed curve in the interior of Ω, encircling the 'hole' of the circuit once and only once. Under the assumptions that no magnetic field lines are penetrating in the circuit region Ω, we may conclude from the third Maxwell equation that in the interior of Ω the total electric field must be irrotational yet non-conservative, since it exists in a multiply connected region. Moreover, due to Stokes' theorem, the emf which is nothing but the circulation of the electric field along the closed curve Γ must be independent on any particular choice of Γ. As even the leads are assumed to have no resistance, the electric field is identically vanishing in both lead regions. In general, the electric field in the circuit region may be decomposed into a conservative and non-conservative part:

$$\boldsymbol{E}(\boldsymbol{r}) \;=\; \boldsymbol{E}_\mathrm{C}(\boldsymbol{r}) \,+\, \boldsymbol{E}_\mathrm{NC}(\boldsymbol{r}) \tag{16.4}$$

with

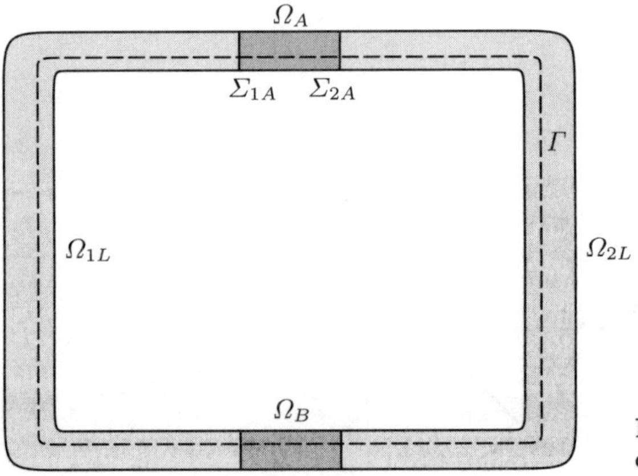

Fig. 16.1. Torus-shaped electric circuit

$$\boldsymbol{E}_{\mathrm{C}}(\boldsymbol{r}) = -\nabla V(\boldsymbol{r})$$

$$\oint_{\Gamma} \boldsymbol{E}_{\mathrm{NC}}(\boldsymbol{r}) \cdot \mathrm{d}\boldsymbol{r} = V_{\varepsilon} \,. \tag{16.5}$$

Here, the conservative component $\boldsymbol{E}_{\mathrm{C}}$ is derived from an electrostatic potential V taking fixed values V_1 and V_2 within the equipotential volumes $\Omega_{1\mathrm{L}}$ and $\Omega_{2\mathrm{L}}$ and exhibiting relatively large drops in the active region and the battery region (see Fig. 16.2). However, as the entire circuit is assumed to be scattering free, we may as well neglect the internal resistance of the battery and choose the non-conservative component $\boldsymbol{E}_{\mathrm{NC}}$ to counteract the conservative field in the battery region:

$$\begin{aligned} \boldsymbol{E}_{\mathrm{NC}}(\boldsymbol{r}) &= -\boldsymbol{E}_{\mathrm{C}}(\boldsymbol{r}) \qquad &&\text{for} \quad \boldsymbol{r} \in \Omega_A \\ &= 0 &&\text{elsewhere.} \end{aligned} \tag{16.6}$$

In other words, the total transport field $\boldsymbol{E}$ vanishes everywhere in the circuit except for the active region where the electric field strength may take rather huge values. Furthermore, the potential difference $V_1 - V_2$ is maintained by the emf as can be seen from

$$V_{\varepsilon} = \oint_{\Gamma} \boldsymbol{E}(\boldsymbol{r}) \cdot \mathrm{d}\boldsymbol{r} = \int_{\Sigma_{1\mathrm{A}}}^{\Sigma_{2\mathrm{A}}} \boldsymbol{E}_{\mathrm{C}}(\boldsymbol{r}) \cdot \mathrm{d}\boldsymbol{r} = V_1 - V_2 \,. \tag{16.7}$$

Clearly, the present transport model requires a totally different approach compared with conventional studies of coherent transport that treat the electric circuit as an open system of charge carriers.
The latter ones are injected from a reservoir and, after some ballistic propagation through the mesoscopic area they are seen to disappear in another reservoir. On the contrary, our model describes the electric circuit as a closed, torus-shaped region confining the charge carriers to its interior – as is essentially realized in a real circuit – and allowing them to extract energy from an external electric field which is forcing the carriers back to the active region.

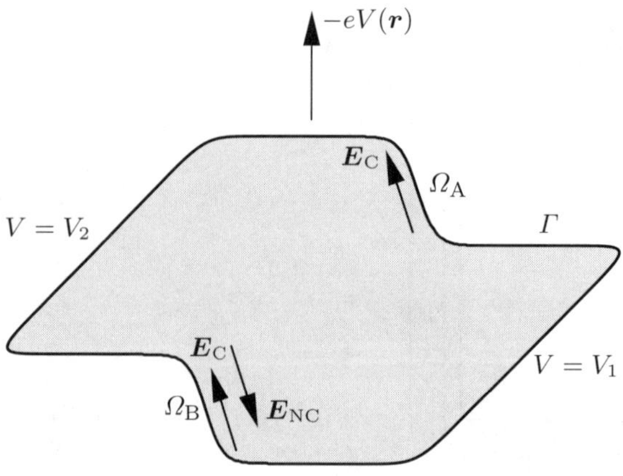

Fig. 16.2. Potential energy profile along Γ

The above vector potential enters the gauge-invariant Hamiltonian that is governing the dynamics of the confined electron system. Its most general second quantized formulation, though in the effective mass approximation, reads:

$$\begin{aligned}\hat{H}_t = & \int_{\Omega} \mathrm{d}\tau \, \psi^{\dagger}(\boldsymbol{r}) \left[\frac{1}{2m} (\boldsymbol{p} + e\boldsymbol{A}(\boldsymbol{r},t))^2 + U(\boldsymbol{r}) \right] \psi(\boldsymbol{r}) \\ & - e \int_{\Omega} \mathrm{d}\tau \, \psi^{\dagger}(\boldsymbol{r}) \psi(\boldsymbol{r}) \, V(\boldsymbol{r},t) \, + \, H' \, , \end{aligned} \tag{16.8}$$

where $U(\boldsymbol{r})$ denotes the internal potential energy due to confinement effects, potential barriers etc., and H' represents the relevant scattering agents such as phonons, impurities, lattice imperfections etc. and their interaction with the electron ensemble. However, since we address the question whether the DC current flowing through a closed circuit is restricted by the Landauer–Büttiker conductance or can grow unlimitedly without the presence of scattering, the interaction term H' is set equal to zero. Consequently, the Heisenberg equation governing the time evolution of the electron field operator is given by

$$\begin{aligned} \mathrm{i}\hbar \, \frac{\partial \psi(\boldsymbol{r},t)}{\partial t} = & \left[\frac{1}{2m} (\boldsymbol{p} + e\boldsymbol{A}(\boldsymbol{r},t))^2 + U(\boldsymbol{r}) \right] \psi(\boldsymbol{r},t) \\ & - eV(\boldsymbol{r},t) \, \psi(\boldsymbol{r},t) \, . \end{aligned} \tag{16.9}$$

The gauge-invariant current-density operator is

$$\begin{aligned} \hat{\boldsymbol{J}}(\boldsymbol{r},t) = & \frac{\mathrm{i}e\hbar}{2m} \left[\psi^{\dagger}(\boldsymbol{r}) \nabla \psi(\boldsymbol{r}) - (\nabla \psi^{\dagger}(\boldsymbol{r})) \psi(\boldsymbol{r}) \right] \\ & + \frac{e}{m} \, \psi^{\dagger}(\boldsymbol{r}) \psi(\boldsymbol{r}) \boldsymbol{A}(\boldsymbol{r},t) \, . \end{aligned} \tag{16.10}$$

16.2 Quantum Rings

We consider a simple one-dimensional ring with radius R. We firstly assume that a uniform electric field $\boldsymbol{E}(\phi) = E(\phi)\boldsymbol{e}_{\phi}$ and a uniform charge distribution $\varrho(\phi) = -e(n - \bar{n}) = 0$, exists on the ring. The emf of the electric field, V_{ε} completely specifies the uniform magnitude of the electric field:

$$E(\phi) = \frac{V_{\varepsilon}}{2\pi R} \, . \tag{16.11}$$

We may imagine such an electric field to be generated by a time-dependent magnetic-field 'needle' $\boldsymbol{B}(t) = -V_{\varepsilon} t \delta(\boldsymbol{r}) \boldsymbol{e}_{\mathrm{z}}$. Then we obtain

$$\boldsymbol{A} = -\boldsymbol{E}t = -\frac{V_{\varepsilon} t}{2\pi R} \, \boldsymbol{e}_{\phi}, \qquad V(\boldsymbol{r},t) = 0 \, . \tag{16.12}$$

The one-electron Hamiltonian becomes

$$H_t = \frac{(\boldsymbol{p}+e\boldsymbol{A})^2}{2m} = \frac{(\boldsymbol{p}-e\boldsymbol{E}t)^2}{2m}$$
$$= \frac{\hbar^2}{2mR^2}\left(\mathrm{i}\frac{\partial}{\partial\phi} + \frac{eV_\varepsilon t}{2\pi\hbar}\right)^2 . \tag{16.13}$$

This Hamiltonian suggests to define a unit of time according to

$$\tau = \frac{2\pi\hbar}{eV_\varepsilon} . \tag{16.14}$$

The equilibrium situation corresponds to $V_\varepsilon = 0$ and the complete set of orthogonal wave functions and energy eigenvalues are

$$u_n(\phi) = \frac{1}{\sqrt{2\pi R}}\,\mathrm{e}^{in\phi} \tag{16.15}$$

$$\varepsilon_n = \frac{\hbar^2 n^2}{2mR^2} , \qquad n = 0,\ \pm 1,\ \pm 2 \ldots \tag{16.16}$$

with normalization

$$\langle u_n | u_n \rangle = R\int_0^{2\pi} |u_n(\phi)|^2 \mathrm{d}\phi = 1 \tag{16.17}$$

The field operators may be expanded as follows:

$$\psi(\phi,t) = \sum_{n\sigma} c_{n\sigma}(t)\; u_n(\phi) ,$$
$$c_{n\sigma}(t) = c_{n\sigma}(0)\; e^{-\mathrm{i}\varepsilon_n t/\hbar} . \tag{16.18}$$

Let is now consider transport, i.e. $V_\varepsilon \neq 0$. The electron field can still be expanded in the functions $\{u_n(\phi)\}$, but with a modified evolution of the coefficients, $c_{n\sigma}(t)$. Since

$$\mathrm{i}\hbar\,\frac{\partial\psi(\phi,t)}{\partial t} = H_t\;\psi(\phi,t) \tag{16.19}$$

and

$$\left(\mathrm{i}\frac{\partial}{\partial\phi} + \frac{t}{\tau}\right) u_n(\phi) \;=\; \left(\frac{t}{\tau} - n\right) u_n(\phi) \tag{16.20}$$

we obtain

$$\mathrm{i}\hbar\sum_n \frac{\mathrm{d}c_{n\sigma}(t)}{dt}\; u_n(\phi) \;=\; \frac{\hbar^2}{2mR^2}\sum_n \left(n - \frac{t}{\tau}\right)^2 c_{n\sigma}(t) u_n(\phi) . \tag{16.21}$$

For all n, we find the following time evolution

$$\mathrm{i}\hbar\,\frac{\mathrm{d}c_{n\sigma}(t)}{dt} = \varepsilon_n\left(1 - \frac{t}{n\tau}\right)^2 c_{n\sigma}(t) . \tag{16.22}$$

The solution of (16.22) is

$$c_{n\sigma}(t) = c_{n\sigma}(0)\exp\left[-\frac{\mathrm{i}\varepsilon_n t}{\hbar}\left(1-\frac{t}{n\tau}+\frac{t^2}{3n^2\tau^2}\right)\right] . \tag{16.23}$$

Note that all modes are decoupled and that this result may be interpreted as that the energy in each mode is time dependent. The operator for the current density is given by

$$\begin{aligned}\hat{\boldsymbol{J}}(\phi,t)=&\sum_{n,n',\sigma}\left\{\frac{\mathrm{i}e\hbar}{2mR}\left[u_n^*(\phi)\frac{du_n(\phi)}{\mathrm{d}\phi}-\frac{\mathrm{d}u_n^*(\phi)}{\mathrm{d}\phi}u_n(\phi)\right]\boldsymbol{e}_\phi\right.\\ &\left.-\frac{e^2}{m}\,u_n^*(\phi)u_n(\phi)\,\boldsymbol{A}(\phi,t)\right\}\times\,c^\dagger_{n'\sigma}(t)c_{n\sigma}(t)\\ \equiv&\, J(\phi,t)\,\boldsymbol{e}_\phi\end{aligned}$$

$$\hat{J}(\phi,t)=\frac{e\hbar}{4\pi mR^2}\sum_{n,n',\sigma}\left(\frac{2t}{\tau}-n-n'\right)\,\mathrm{e}^{\mathrm{i}(n-n')\phi}\,c^\dagger_{n'\sigma}(t)c_{n\sigma}(t) . \tag{16.24}$$

In the one-dimensional model we find the current as the expectation value of $\hat{J}(\phi,t)$:

$$\begin{aligned}I(\phi,t)=&\left\langle\hat{J}(\phi,t)\right\rangle_0\\ =&\frac{e\hbar}{4\pi mR^2}\sum_{n,n',\sigma}\left(\frac{2t}{\tau}-n-n'\right)\,\mathrm{e}^{\mathrm{i}(n-n')\phi}\,\langle c^\dagger_{n'\sigma}(t)c_{n\sigma}(t)\rangle_0 .\end{aligned} \tag{16.25}$$

Finally, by using (16.23), (16.25) and

$$\langle c^\dagger_{n'\sigma}(0)c_{n\sigma}(0)\rangle_0 = F(\varepsilon_n)\delta_{nn'} , \tag{16.26}$$

we arrive at the following result for the current:

$$\begin{aligned}I(\phi,t) =& I(t)=2\times\frac{e\hbar}{4\pi mR^2}\sum_{n,n'}\left(\frac{2t}{\tau}-n-n'\right)\,\mathrm{e}^{\mathrm{i}(n-n')\phi}\,F(\varepsilon_n)\delta_{nn'}\\ =&\frac{e\hbar}{\pi mR^2}\sum_n\left(\frac{t}{\tau}-n\right)\,F(\varepsilon_n) .\end{aligned} \tag{16.27}$$

Since $F(\varepsilon_n)$ is an even function of n we conclude that $\sum_n nF(\varepsilon_n)=0$. The number of electrons on the ring is $\sum_n F(\varepsilon_n)=N$ and therefore we arrive at the conclusion that for a uniformly charged ring with uniform electric field we have classical or ballistic transport.

$$I(t) = \frac{e^2NV_\varepsilon}{2\pi^2 mR^2}\,t . \tag{16.28}$$

The motion on uniform rings is not affected by quantum conductance effects.

Next, we discuss the case of a one-dimensional ring that is affected by a non-uniform electric field. Although in general the decomposition of the electric field in a conservative and a non-conservative component is not unique, it should be noted that the non-uniform component can always be absorbed in the conservative part while the non-conservative component has a constant magnitude along the ring. This can easily be seen by exploiting the periodicity of the total electric field and expanding the latter in a Fourier series:

$$E(\phi,t) = \sum_{n=-\infty}^{+\infty} E_n(t)\, \mathrm{e}^{\mathrm{i}n\phi} . \tag{16.29}$$

The uniform Fourier component directly relates to the emf, which is time-independent:

$$\begin{aligned} E_0 = \frac{1}{2\pi}\int_{-\pi}^{\pi} E(\phi)\, \mathrm{d}\phi &= \frac{1}{2\pi R}\oint_{\Gamma} \boldsymbol{E}\cdot \mathrm{d}\boldsymbol{r} \\ &= \frac{V_\varepsilon}{2\pi R} . \end{aligned} \tag{16.30}$$

In other words, the total electrical field can be written as the sum of a static non-conservative component and a non-uniform time-dependent conservative part:

$$E(\phi,t) = E_{\mathrm{NC}} + E_{\mathrm{C}}(\phi,t)$$

$$E_{\mathrm{NC}} = E_0 = \frac{V_\varepsilon}{2\pi R} \tag{16.31}$$

$$\begin{aligned} E_{\mathrm{C}}(\phi,t) &= \sum_{n\neq 0} E_n(t)\, \mathrm{e}^{\mathrm{i}n\phi} \\ &= -\frac{1}{R}\frac{\partial V(\phi,t)}{\partial \phi} , \end{aligned} \tag{16.32}$$

where the expansion of the scalar potential $V(\phi,t)$ directly follows from (16.31):

$$V(\phi,t) = -\mathrm{i}\sum_{n\neq 0} \frac{E_n(t)}{n} \mathrm{e}^{\mathrm{i}n\phi} . \tag{16.33}$$

As expected, the scalar potential exhibits full periodicity in ϕ.

On the other hand, a similar potential cannot be found to reproduce the non-conservative field, since such a potential would be linear in ϕ and therefore multi-valued.

While the non-conservative field enters the one-electron Hamiltonian through the time unit 'τ', the scalar potential $V(\phi,t)$ should be included as an additional potential energy term:

$$H_t = \frac{\hbar^2}{2mR^2}\left(\mathrm{i}\frac{\partial}{\partial\phi} + \frac{t}{\tau}\right)^2 + U(\phi) - eV(\phi,t)\ , \tag{16.34}$$

where $U(\phi)$ denotes any 'built-in' potential energy that is already present before the external electric field is turned, thereby reflecting the broken rotational symmetry of the unperturbed ring states. If the electron–electron interaction is treated in the Hartree approximation – which is the case throughout this book – the scalar potential is determined by the average local charge density governing Poisson's equation

$$\frac{1}{R^2}\frac{\partial^2 V(\phi,t)}{\partial\phi^2} = \frac{e}{\varepsilon_\mathrm{S}}\left(n(\phi,t) - \bar{n}\right)\ , \tag{16.35}$$

where $e\bar{n}$ represents a uniform, positive background charge density ensuring electrical neutrality of the ring.

We will now consider a ring with a rotational symmetry breaking potential profile $U(\phi)$ and address the question whether a steady current will emerge after switching on the non-conservative electric field. From the angular momentum balance equation

$$\frac{\mathrm{d}\boldsymbol{L}(t)}{\mathrm{d}t} = -e\int_\Omega \mathrm{d}\tau\, n(\boldsymbol{r},t)\,\boldsymbol{r}\times\left[E_\mathrm{NC}(\boldsymbol{r}) + E_\mathrm{C}(\boldsymbol{r},t)\right]\ , \tag{16.36}$$

it follows for a one-dimensional ring

$$\begin{aligned}\frac{\mathrm{d}L_z(t)}{\mathrm{d}t} &= -eR^2\int_{-\pi}^{\pi}\mathrm{d}\phi\, n(\phi,t)\left[E_\mathrm{NC}(\phi) + E_\mathrm{C}(\phi,t)\right]\\ &= -eR^2\int_{-\pi}^{\pi}\mathrm{d}\phi\, n(\phi,t)\left[\frac{V_\varepsilon}{2\pi R} + E_\mathrm{C}(\phi,t)\right]\ .\end{aligned} \tag{16.37}$$

The occurrence of a steady state requires

$$\lim_{t\to\infty}\frac{\mathrm{d}L_z(t)}{\mathrm{d}t} = 0\ , \tag{16.38}$$

which, in turn, implies that

$$\lim_{t\to\infty}\langle E_\mathrm{C}(t)\rangle = -\frac{V_\varepsilon}{2\pi R} \tag{16.39}$$

and $\langle E_\mathrm{C}(t)\rangle$ is the average electric field with respect to the electron concentration

$$\langle E_\mathrm{C}(t)\rangle \equiv \frac{\int_{-\pi}^{\pi}\mathrm{d}\phi\, n(\phi,t)E_\mathrm{C}(\phi,t)}{\int_{-\pi}^{\pi}\mathrm{d}\phi\, n(\phi,t)}\ . \tag{16.40}$$

This situation can never be realized for a uniform ring since at all times we have that $n(\phi,t) = 1/2\pi R$ and therefore $\langle E_\mathrm{C}(t)\rangle = 0$. However, the lack of rotational symmetry due to presence of $U(\phi)$ and $V(\phi,t)$ induces a gradient of the electron concentration which is related to the average field $\langle E_\mathrm{C}(t)\rangle$:

$$\langle E_C(t)\rangle = \frac{\int_{-\pi}^{\pi} \mathrm{d}\phi\, V(\phi,t)\dfrac{\mathrm{d}n(\phi,t)}{\mathrm{d}\phi}}{\int_{-\pi}^{\pi} \mathrm{d}\phi\, n(\phi,t)} . \tag{16.41}$$

Nonetheless, the fact that a non-uniform electric field drives the current in a ring is not a sufficient condition for realizing a saturation of the current if time progresses. In general, the Heisenberg equation is

$$\mathrm{i}\hbar\frac{\partial\psi}{\partial t} = \frac{\hbar^2}{2mR^2}\left(\mathrm{i}\frac{\partial}{\partial\phi} + \frac{t}{\tau}\right)^2 + U(\phi)\psi . \tag{16.42}$$

The (unitary) transformation function

$$\Omega(\phi,t) = \exp\left(-\mathrm{i}\frac{t}{\tau}\phi\right) \tag{16.43}$$

maps the Heisenberg equation onto the following equivalent problem:

$$\zeta = \Omega\psi \tag{16.44}$$

$$\mathrm{i}\hbar\frac{\partial\zeta}{\partial t} = -\frac{\hbar^2}{2mR^2}\frac{\partial^2\zeta}{\partial\phi^2} + U(\phi)\,\zeta + \frac{\hbar}{\tau}\phi\,\zeta . \tag{16.45}$$

Furthermore, the uniqueness of ψ implies that

$$\zeta(-\pi,t) = \exp\left(2\pi\mathrm{i}\frac{t}{\tau}\right)\zeta(\pi,t) . \tag{16.46}$$

The current-density operator in terms of the ζ-field is

$$\hat{J}(\phi,t) = \frac{\mathrm{i}e\hbar}{2mR}\left[\zeta^\dagger\frac{\mathrm{d}\zeta}{\mathrm{d}\phi} - \frac{\mathrm{d}\zeta^\dagger}{\mathrm{d}\phi}\zeta\right] . \tag{16.47}$$

while the expansion of ψ translates into

$$\begin{aligned}\psi(\phi,t) &= \sum_n u_n(\phi)\, c_n(t)\\ \zeta(\phi,t) &= \sum_n \mathrm{e}^{-\mathrm{i}\phi t/\tau}\, u_n(\phi)\, c_n(t) .\end{aligned} \tag{16.48}$$

The derivatives $\mathrm{d}\zeta/\mathrm{d}\phi$ and $\mathrm{d}\zeta^\dagger/\mathrm{d}\phi$ in J induce terms which are linearly increasing in t, and therefore correspond to currents that grow unlimited in time.

In order to realize a saturating current, we need to refine the physical description. In the next section, we include the induced magnetic flux trapped by the closed circuit. It is argued that flux quantization provides a current-saturating mechanism.

16.3 Hamiltonian and Current Response

Under the effective mass approximation the most general second-quantized Hamiltonian describing an ensemble of free electrons acted upon by an electromagnetic field reads

$$\hat{H}_{\mathrm{E}} = \int_{\Omega} \mathrm{d}\tau\, \psi^{\dagger}(\boldsymbol{r}) \left[\frac{1}{2m} (\boldsymbol{p} + e\boldsymbol{A}(\boldsymbol{r},t))^2 + U(\boldsymbol{r}) \; - \; eV(\boldsymbol{r}) \right] \psi(\boldsymbol{r}) \, , \quad (16.49)$$

where m is the electron mass and $U(\boldsymbol{r})$ represents any internal potential profile (such as a built-in potential or an energy barrier). The vector potential $\boldsymbol{A}$ consists of an irrotational part $\boldsymbol{A}_{\mathrm{ex}}$, related to the non-conservative, external DC field $\boldsymbol{E}_{\mathrm{NC}}$ and an induced component $\boldsymbol{A}_{\mathrm{in}}$ corresponding to the magnetic field which is generated by the moving electrons and the induced electromotive forces. Under the assumption that the external emf is switched on at some initial time $t = 0$, we may write according to elementary electrodynamics:

$$\boldsymbol{A}(\boldsymbol{r},t) = \boldsymbol{A}_{\mathrm{ex}}(\boldsymbol{r},t) + \boldsymbol{A}_{\mathrm{in}}(\boldsymbol{r},t) \quad (16.50)$$

$$\boldsymbol{A}_{\mathrm{ex}}(\boldsymbol{r},t) = -\boldsymbol{E}_{\mathrm{NC}}(\boldsymbol{r})\, t \quad (16.51)$$

$$\boldsymbol{B}(\boldsymbol{r},t) = \nabla \times \boldsymbol{A}_{\mathrm{in}}(\boldsymbol{r},t) \, . \quad (16.52)$$

The current response can be obtained from a self consistent solution of Maxwell's equations and the quantum dynamical equation yielding the time dependent ensemble average of the gauge invariant current density operator given in (16.10). Defining the total current $I(t)$ in the usual way as the electron charge passing the cross section $\Sigma_{1\mathrm{A}}$ per unit time, we may express $I(t)$ in the Heisenberg picture as follows:

$$I(t) = \int_{\Sigma_{1\mathrm{A}}} \boldsymbol{J}(\boldsymbol{r},t) \cdot \mathrm{d}\boldsymbol{S} \, , \quad (16.53)$$

where $\boldsymbol{J}(\boldsymbol{r},t) = \left\langle \hat{\boldsymbol{J}}(\boldsymbol{r},t) \right\rangle_0$ is the time-dependent ensemble average of the Heisenberg operator $\hat{\boldsymbol{J}}(\boldsymbol{r},t)$ and $\langle \ldots \rangle_0$ represents the Gibbs ensemble average describing the equilibrium state of the circuit for a given temperature $T = 1/k_{\mathrm{B}}\beta$ and a given chemical potential μ at all times $t \leq 0$

$$\left\langle \hat{\boldsymbol{J}}(\boldsymbol{r},t) \right\rangle_0 = \frac{\mathrm{Tr} \left[\hat{\boldsymbol{J}}(\boldsymbol{r},t) \exp(-\beta(\hat{H}_{\mathrm{E}} - \mu \hat{N})) \right]}{\mathrm{Tr} \exp(-\beta(\hat{H}_{\mathrm{E}} - \mu \hat{N}))}$$

$$\hat{N} = \int_{\Omega} \mathrm{d}\tau\, \psi^{\dagger}(\boldsymbol{r}) \psi(\boldsymbol{r})$$

$$N = \langle \hat{N} \rangle_0 \, . \quad (16.54)$$

For the present investigation however, it proves convenient to focus on the equation of motion for $\langle \hat{H}_{\mathrm{E}}(t) \rangle_0$ describing the rate at which the electron ensemble is extracting energy from the power supply. A lengthy and cumbersome but straightforward calculation based on the the Heisenberg equation governing the time evolution of the electron field operator

$$i\hbar \frac{\partial \psi(\boldsymbol{r}, t)}{\partial t} = \left[\frac{1}{2m} \left(\boldsymbol{p} + e\boldsymbol{A}(\boldsymbol{r}, t) \right)^2 + U(\boldsymbol{r}) - eV(\boldsymbol{r}, t) \right] \psi(\boldsymbol{r}, t) \tag{16.55}$$

and its Hermitian conjugate yields the quantum mechanical representation of the classical energy-rate equation ($t > 0$)

$$\frac{\mathrm{d}\,\langle \hat{H}_{\mathrm{E}}(t) \rangle_0}{\mathrm{d}t} = \int_{\Omega} \mathrm{d}\tau\, \boldsymbol{J}(\boldsymbol{r}, t) \cdot \boldsymbol{E}(\boldsymbol{r}, t) - \frac{\mathrm{i}}{\hbar} \left\langle \left[\hat{H}_{\mathrm{E}}(t), H'(t) \right] \right\rangle_0 , \tag{16.56}$$

where $\boldsymbol{E}(\boldsymbol{r}, \mathrm{t})$ generally denotes the total electric field, including the externally applied transport field, the self consistent field in the active region and the induced electric field due to possible changes of the magnetic flux trapped by a curve in the circuit region. As was announced in the introduction, the interaction term H' generally representing all elastic and inelastic scattering mechanisms has been switched off in the present work in order to investigate the existence of another current limiting mechanism.

After a first glance at equation (16.56), in which we have put $H' = 0$, it might appear that the absence of any dissipative scattering mechanism would inevitably lead to both an unlimited increase of the electron energy and an unbounded acceleration of the electron ensemble due to steady absorption of energy supplied by the non-conservative electric field. In a classical circuit, this is the expected scenario. Indeed, if H' could be switched off in some Gedankenexperiment, for instance, if one considered a simple LR circuit governed by the well-known formula

$$I(t) = \frac{V_\varepsilon}{R} (1 - \mathrm{e}^{-Rt/L}) , \tag{16.57}$$

the current would trivially diverge as $V_\varepsilon t/L$ if the series resistance R were put equal to zero.

For a quantum circuit however, we propose the existence of a closed curve Γ in the interior of Ω such that neither the vector potential $\boldsymbol{A}$ nor electron field operator are identically vanishing along Γ and the induced magnetic flux Φ trapped by Γ is quantized. In particular, the magnetic flux trapped by Γ is supposed to be an integer multiple of the elementary flux quantum $\Phi_0 = h/e$. Clearly, the above proposed flux quantization is very well known to occur in superconducting rings and plays an important role also when it comes to trace quantum interference phenomena such as the Aharonov–Bohm effect [109]. Flux quantization for a superconducting ring is a direct consequence of the requirement that paired electrons be described by single-valued wave functions together with the observation that the presence of an irrotational vector potential (Meissner effect) can be fully absorbed in a phase shift $\Delta\phi = (2e/\hbar) \oint_\Gamma \boldsymbol{A}(\boldsymbol{r}) \cdot \mathrm{d}\boldsymbol{r} = 2e\Phi/\hbar$ acquired by an electron after a virtual revolution along Γ [110].

In this work we have omitted all interactions that may destroy the coherence of the electron transport in the circuit and in this respect one might be tempted to consider Ω as an artificial superconducting circuit for which flux quantization needs not be imposed as an external constraint as it is already enforced through the superconducting features – upon a simple replacement of the Cooper pair charge $2e$ by a single electron charge e.

On the other hand, we wish also to exploit the results of this work to understand the conductance mechanisms of mesoscopic structures that are embedded in more realistic circuits, containing non perfectly conducting leads and areas which are exposed to the magnetic field caused by the current flow. Since the total vector potential will no longer be irrotational due to the presence of an induced magnetic field in the interior of those circuits, flux quantization cannot simply emerge from the phase shift argument in the very same way as for the superconducting circuit, and therefore it needs to be postulated explicitly. Moreover, the magnetic flux trapped by a superconducting ring equals both the flux induced by the supercurrent and the external magnetic flux, whereas the flux quantization proposed in this section is restricted to the induced magnetic field.

As a major consequence of the conjectured universality of flux quantization, the induced magnetic flux trapped by the closed loop Γ can only increase or decrease with steps of magnitude Φ_0. Hence the electric current carried by the free electron gas of our model can change only after a minimal time τ_0 required to realize exactly one creation or absorption of an elementary flux quantum, since if such a current change took place at an earlier instant, it would produce a proportional magnetic flux change that would be smaller than Φ_0. (A similar argument is proposed by 't Hooft in [111].)

In other words, the transient electric current, after the power is switched on, will be characterized by a discrete time series $\{t_1, t_2, \ldots, t_n, \ldots\}$ or $t_n = n\tau_0$, such that all dynamical quantities remain constant between two subsequent time instants and energy extraction from the power supply takes place at these discrete time instants only.

The characteristic time τ_0 can be easily calculated by comparing the energy $\Delta H_{\mathrm{E}n}$ extracted from the external field during a time interval $[t_n - \frac{1}{2}\tau_0, t_n + \frac{1}{2}\tau_0]$ with the corresponding magnetic energy increase ΔU_{M} of the circuit. Integrating the energy rate equation (16.56) from $t_n - \frac{1}{2}\tau_0$ to $t_n + \frac{1}{2}\tau_0$, we may express $\Delta H_{\mathrm{E}n}$ as follows:

$$\Delta H_{\mathrm{E}n} = \int_{t_n - \frac{1}{2}\tau_0}^{t_n + \frac{1}{2}\tau_0} \mathrm{d}t \int_{\Omega} \mathrm{d}\tau \; \boldsymbol{J}(\boldsymbol{r}, t) \cdot \boldsymbol{E}(\boldsymbol{r}, t) \; . \tag{16.58}$$

During $[t_n - \frac{1}{2}\tau_0, t_n + \frac{1}{2}\tau_0]$, the charge density remains unchanged before and after the jump at $t = t_n$ and consequently, the current density is solenoidal, while the external electric field is irrotational. Hence, according to a recent integral theorem for multiply connected regions [52], we may disentangle the right-hand side of (16.58):

$$\int_{t_n-\frac{1}{2}\tau_0}^{t_n+\frac{1}{2}\tau_0} \mathrm{d}t \int_{\Omega} \mathrm{d}\tau \, \boldsymbol{J}(\boldsymbol{r},t)\cdot\boldsymbol{E}(\boldsymbol{r},t) = \frac{1}{2}\left[I_{n-1}+I_n\right]V_\varepsilon\tau_0 \tag{16.59}$$

where $I_n = \int_{\Sigma_{1A}} \boldsymbol{J}(\boldsymbol{r},t_n)\cdot\mathrm{d}\boldsymbol{S}$ is the net current entering the cross section Σ_{1A} at a time t_n. On the other hand the flux change $\Delta\Phi_n$ associated with the jump $\Delta I_n \equiv I_n - I_{n-1}$, reads

$$\Delta\Phi_n = L\Delta I_n \, , \tag{16.60}$$

where L is the inductance of the circuit. Since $\Delta\Phi_n$ has to equal Φ_0, we obtain the increased magnetic energy of the circuit:

$$\begin{aligned}\Delta U_{\mathrm{M}} &= \frac{1}{2}LI_n^2 - \frac{1}{2}LI_{n-1}^2 \\ &= \frac{1}{2}(I_{n-1}+I_n)\Phi_0 \, .\end{aligned} \tag{16.61}$$

Combining (16.58), (16.59) and (16.61) and putting $\Delta U_{\mathrm{M}} = \Delta H_{E,n}$, we derive the following result:

$$\tau_0 = \frac{\Phi_0}{V_\varepsilon} \, . \tag{16.62}$$

We are now in a position to show that the interplay between flux quantization and the topology of the transport electric field explains the limitation of the electric current even in the absence of scattering. Since the total transport field is non-zero only in the active region Ω_A, only the charge Q_n consisting of electrons residing in Ω_A at $t = t_n$ will feel the action of the electric field during the interval $[t_n - \frac{1}{2}\tau_0, t_n + \frac{1}{2}\tau_0]$. For the sake of simplicity, we have considered a circuit in which the electric current is carried by M occupied transverse modes at low temperatures [114]:

$$I_n = \sum_{k=1}^{M} I_{nk} \qquad n = 1, 2, \ldots \tag{16.63}$$

For instance, in a mesoscopic one dimensional ring, these modes would be simply the energy eigenstates of the ring, whereas they would correspond to the discrete resonances emerging in the continuous spectrum of a circuit with large leads connected by a quantum point contact. Apart from the current sequence $I_{1k}, I_{2k}, I_{3k}, \ldots$ that is increasing due to the steady energy supply, we may also define for each transmission mode k a sequence of charge packets $Q_{1k}, Q_{2k}, Q_{3k}, \ldots$ that are brought into the active region by the current carried by the kth mode as well as a sequence of 'dwell times' $\{\Delta t_{1k}, \Delta t_{2k}, \Delta t_{3k}, \ldots\}$ representing the time spent by the charge packets in the active region after having entered the latter at $t = t_n$. The superposition of all charge packets is nothing but the total charge residing in the active region between $t = t_n$ and $t = t_{n+1}$:

$$Q_n \equiv -e \int_{\Omega_A} \mathrm{d}\tau \left\langle \psi^\dagger(\boldsymbol{r}, t_n)\psi(\boldsymbol{r}, t_n)\right\rangle_0 = \sum_{k=1}^{M} Q_{nk} \,. \tag{16.64}$$

Obviously, the three sequences are linked through

$$\Delta t_{nk} = \frac{Q_{nk}}{I_{nk}} \,; \quad n = 1, 2, \dots \,; \quad k = 1, 2, \dots, M \,. \tag{16.65}$$

Considering for simplicity a ballistic conductor for which each transmission mode provides full transmission, we have $Q_{nk} = -2e$ where the factor 2 accounts for spin degeneracy. Hence, the increasing current sequence translates to a decreasing sequence of dwell times:

$$\Delta t_{1k} \geq \Delta t_{2k} \geq \Delta t_{3k} \geq \dots \tag{16.66}$$

Since τ_0 is non-negative, for each k there must exist a positive integer n_k such that

$$\Delta t_{n_k k} \geq \tau_0 \geq \Delta t_{n_k+1 k} \,. \tag{16.67}$$

The intuitive interpretation is clear: at $t = t_{n_k}$, the electrons of the kth mode are still residing sufficiently long in the active region to generate a final elementary flux change; as of $t = t_{n_k+1}$, electrons are traveling too fast through the finite active region and are no longer able to collect energy from the external field because that would require them to spend at least a time τ_0 in Ω_A. As a consequence, the current is no longer increasing and $I_{n_k k} = I_{n_k+1 k} = I_{n_k+2 k} = \dots$. Together with the inequalities (16.67) or, equivalently, with

$$\frac{2e}{|I_{n_k k}|} \geq \tau_0 \geq \frac{2e}{|I_{n_k+1 k}|} \tag{16.68}$$

we are left with

$$|I_{n_k k}| = |I_{n_k+1 k}| = \frac{2e}{\tau_0} = 2e\frac{eV_\varepsilon}{h} \tag{16.69}$$

from which we may finally infer a quantized conductance G:

$$G \equiv \frac{1}{V_\varepsilon}\sum_k |I_{n_k k}| = \frac{2e^2}{h}M \,, \tag{16.70}$$

which is the well-known Landauer–Büttiker formula.

16.4 Open Versus Closed Circuits

We will now discuss both the approach itself as it was adopted to derive (16.70) and its relation to other, more conventional theories leading to Landauer–Büttiker type conductance formulas.

To the best of our knowledge, most – if not all – theoretical investigations of quantum transport in mesoscopic structures are based on the reservoir concept [115, 112, 113, 116] in which a mesoscopic structure, representing the active area, is squeezed between two huge, half-open particle reservoirs. This remark does not apply to superconducting devices and mesoscopic rings carrying persistent currents [117], for which the torus-like topology is a natural feature. Playing the role of leads, the reservoirs contain two distinct, thermalized electron gases characterized by two different chemical potentials, the difference of which is assumed to equal the applied voltage [106]. We believe that there are at least three good reasons to abandon the reservoir concept when quantum transport is addressed, even if the latter seems to take place only in a small region of the electric circuit. First, the artificial subdivision of the circuit in three distinct parts amounts to the assignment of three sets of quantum states to the three regions: two continuous spectra which are supposed to mimic the semi-infinite reservoirs and a discrete spectrum for the mesocopic structure. The latter is generally confining a relatively small number of electrons to reside in a nanometer scale area. If the three regions were perfectly separated by infinite walls, they would constitute three distinguished quantum systems requiring also separate quantum descriptions and the definition of distinct Hilbert spaces. However, if a communication channel is established, no matter how narrow it is, the three regions should be regarded as one single quantum mechanical entity the dynamics of which is to be described in a unique Hilbert space of states where discrete transmission modes are appearing as sharp, enumerable resonances of a continuous spectrum as is explained in any decent textbook on quantum mechanics [20, 19, 21]. From the many-particle point of view, the study of the system dynamics is a formidable task since one has to deal with an open-ended system which is loosing and gaining particles in a rather uncontrollable way. Nevertheless, ever since the introduction of the transfer Hamiltonian formalism by Bardeen [56], numerous transport calculations have been relying on the possibility of treating carrier transport as a set of transitions between 'reservoir states' and 'mesoscopic area states'. Here, we do not want to contribute to the on-going discussion as to whether the transfer Hamiltonian formalism is appropriate for studying quantum transport or not, and we definitely do not criticize results which are corroborated by experiment, but we strongly believe that the formalism is descriptive rather than explanatory.

A second, even more striking observation reveals that the open-circuit topology is not accounting for the non-conservative nature of the driving electric field. Being transmitted from reservoir 1 to reservoir 2 by the local electrostatic field of the active region, the charged particles are never returning to reservoir 1. Consequently, the pumping action of the battery mimicing both the energy supply and the maintenance of the electrostatic potential difference is not incorporated at all. On the contrary, the explicit requirement that the reservoirs be thermalized and have fixed chemical potentials is supposed to maintain the *chemical* potential difference and hence the applied voltage. To our feeling, such an approach is hardly appropriate to probe energy limiting mechanisms as any conservative field is already limiting the energy increase itself. This point can be illustrated by the example of a billiard ball moving without friction in a gravitational field. If the ball is leaving a horizontal platform (reservoir 1) with some velocity v_1 to roll down from a frictionless hill of finite height and width, thereafter arriving at another horizontal platform (reservoir 2) with velocity v_2, the velocity increase is trivially finite as it is acquired at the expense of a finite potential energy decrease. This observation however does not learn us anything about the time evolution of the velocity of the ball in a non-conservative field, if the ball is bound to continue forever its rectilinear motion on platform 2 of the open system.

Finally, the approach adopted in this work completely avoids the necessity of introducing the concept of 'contact resistance'. The latter refers to the interface between the mesoscopic area and the huge contact reservoirs which should be responsible for the dissipation of energy that cannot be relaxed to the environment when the particles are still residing in the scattering free mesoscopic area. Although it is at least unclear how to identify such interfaces, it is nevertheless generally proposed [115] that the dissipation results from a mismatch between the continuous spectra of the reservoirs and the discrete spectrum of the mesoscopic area, comparable to traffic jam due to a local reduction of the number of available lanes. In other words, the whole explanation of conductance quantization would have to rely on the questionable division of the (open) circuit into spatial subregions. Furthermore, even if the dissipation is related to inelastic scattering events taking place in the reservoirs in the close vicinity of such an interface, it remains an open question how the interaction between the charge carriers and the scattering agents (phonons, impurities, alloys, etc.) and the corresponding coupling strengths which are typical material parameters, can give rise to a resistance that can be expressed solely in terms of fundamental constants (e, $\hbar$) and a set of transmission matrix elements. As a matter of fact, we have found the last problem a strong incentive to look for alternative mechanisms to explain the phenomenon of quantized conductance.

16.5 Energy Dissipation Versus Current Limitation

Quite remarkably, the establishment of a stationary state in which a finite current is flowing through an electric circuit in response to a given electromotive force V_ε is relying on current limitation, or equivalently on the phenomenon that the electrons can extract energy from the external V_ε only for a limited number of cycles. As was explained in the previous section, this limitation in turn relies on the existence of a characteristic time τ_0 an electron should spend in the field region to induce a flux jump and to extract from the external field the corresponding energy packet. This is probably the most striking difference with other treatments which still allow for energy dissipation in the conductance process whereas, in our model, unlimited gain of energy is prohibited by the selection rule for energy extraction.

16.6 Flux Quantization

The flux quantization postulate which is clearly the price we had to pay in this work is inspired by recent work in the field of the fractional quantum Hall effect in two-dimensional gases acted upon by a perpendicular magnetic field where each electron is viewed as composite of a charged boson and a flux tube containing a odd number of flux quanta [104]. Also the argument of Laughlin's Gedankenexperiment [118] invokes quantized flux changes to recover the von Klitzing resistance in a metallic ribbon bent into a circular loop. While Laughlin is considering flux changes associated with the external magnetic field and being related to a flow of electrons from one edge to the other (in the direction of the Hall voltage), the present model is addressing the magnetic field produced by the current flowing through the loop.

Moreover, recent successful attempts [119] to discover striking connections between the quantum Hall effect and superconductivity have suggested that magnetic fields impinging on (2D) electron systems may be characterized by a number of flux quanta and therefore seem to line up with the flux quantization picture.

On the other hand, we do realize that the assumption of flux quantization along a characteristic circuit trajectory may lead to far-reaching consequences on both the theoretical and experimental level. On the theoretical side, we may justify the basic assumption on the existence of the curve Γ by observing that the magnetic field generated by a toroidally confined, stationary current distribution circulates around any cross-section of the circuit, as predicted by the fourth Maxwell equation $\nabla \times \boldsymbol{B} = \mu_0 \boldsymbol{J}$. Hence, all transverse components of $\boldsymbol{B}$ are bound to change sign within the circuit region and therefore they must vanish along a closed curve Γ inside the circuit. The latter therefore defines a region where $\boldsymbol{A}$ is irrotational whereas neither $\boldsymbol{A}$ nor the field operators are identically vanishing, so that the phase argument leading to a quantized flux threaded by Γ can be repeated.

From the experimental point of view, thorough investigations should be conducted to trace directly or indirectly the presence of flux quantization in closed circuits subjected to localized driving electric fields. Obviously, macroscopic circuits in which mesoscopic active areas are embedded cannot be experimentally accessed as a whole without the disturbing presence of scattering events in the conducting leads. On the other hand, mesoscopic metallic rings interrupted by one or two tunnel barriers – such as the Aharonov–Bohm interferometers discussed in [120] and [121] – may provide an appropriate experimental setup for studying changes of the magnetic flux. In such devices one may generate a constant V_ε along a closed trajectory by applying a linearly growing magnetic field piercing the hole of the ring.

16.7 Localization of the Electric Field

In addition to flux quantization, the topology of the electric field plays a crucial role when it comes to obtain current limitation. In particular, it is required that the driving electric field governing the electron motion in the circuit be localized in a finite, simply connected region of the circuit. This observation has been made already a few years ago by Fenton [106] who pointed out that for an arbitrary open circuit, the Landauer–Büttiker conductance regime can be realized only for strictly localized transport fields, whereas a uniform field would inevitably yield the Drude–Lorenz conductivity which, in the absence of scattering, would lead to zero resistance. The same conclusions can be drawn from our model. The connection between the Landauer–Büttiker conductance regime and the requirement of having localized fields is already demonstrated in the previous section, where the finiteness of the active region Ω guarantees that the dwell times become lower than τ_0 after a finite number of cycles. Clearly, this situation cannot occur if the electric field is uniform along the circuit or at least non vanishing in the whole circuit region, since then the dwell times would increase without limit and the quantized flux changes would be unable to prevent the electrons from unlimited energy extraction.

16.8 A Quantum Lenz' Law?

The postulate of flux quantization proposed in this work, should be properly embedded in a suitable quantum field theory, the dynamical solution of which should encompass the Landauer–Büttiker conductance regime in a natural way. Nevertheless, the quantization of the corresponding fields will follow another path than that of familiar quantum electrodynamics (QED), the main reason being that, apart from the local field operators associated with the electrons, we have now also to quantize global canonical variables associated with the electromagnetic field, whereas other components of the latter may or may not remain classical. In the light of this work, the induced magnetic flux is an obvious example of such a global quantity to be quantized. We believe that an appropriate choice of global canonical variables will eventually lead to a useful quantum circuit theory (QCT) unifying all well-known features of classical circuits as well as the characteristics of quantum devices which are to be included in real circuits. Finally, it should be noticed that the above mentioned quantization procedure explicitly affects the Maxwell equations expressed in integral form. In particular, Faraday's induction law relates the total emf in a circuit (external and induced) to the change of a quantized flux, the discrete time evolution of which is given by:

$$\Phi(t) = \Phi_0 \sum_{n=0}^{\infty} \alpha_n \theta(t - n\tau_0) , \tag{16.71}$$

where the coefficients α_n can take only the values 0, ± 1 and have to be determined by a full time dependent solution of the dynamical equations. As an illustration, we may obtain a quantum mechanical version of Lenz' law by taking the time derivative of (16.71):

$$L\frac{dI(t)}{dt} = \Phi_0 \sum_{n=0}^{\infty} \alpha_n \delta(t - n\tau_0) . \tag{16.72}$$

17. Transport in Quantum Wires

17.1 Balance Equations for an Imperfect Quantum Wire

In various circumstances, the electric field in an active device region may not only be inhomogeneous, but also too large for applying linear response theory. In such cases the difference between the initial state of the circuit, that is often an equilibrium state, and the final steady state cannot be regarded as being caused by a small perturbation. However, as was explained in Chap. 10, if only the steady state is addressed, one may construct a heuristic initial density matrix ϱ_0 corresponding to a non-equilibrium state that is 'close' to the real steady state. By 'close' we mean that, by construction, both the heuristic density matrix and the true density matrix are sharing a finite set of global observables. The underlying idea is that in such a framework the time evolution leading from the heuristic state to the true steady state can be treated perturbatively as far as the finite set of common observables is concerned. In the case of homogeneous and static fields [48, 49, 50, 123, 124, 68] the most natural candidate for ϱ_0 turned out to be based on a boosted Fermi–Dirac distribution ϱ_0^B describing a non-interacting electron gas at an elevated electron temperature T_e and carrying a uniform drift velocity $\boldsymbol{v}_D$, i.e.

$$F(E(\boldsymbol{k}) - \hbar\boldsymbol{k}\cdot\boldsymbol{v}_D - \mu, T_e) = \left[1 + \exp\left(\frac{E(\boldsymbol{k}) - \hbar\boldsymbol{k}\cdot v_D - \mu}{k_B T_e}\right)\right]^{-1}, \tag{17.1}$$

where $E(\boldsymbol{k})$ and μ respectively denote the one-electron dispersion relation and the equilibrium chemical potential or Fermi level. Here, the 'set of common observables' just consists of $\boldsymbol{v}_D$ and T_e. he aim of the present chapter however is to illustrate how the balance equation approach can be exploited to describe transport in system where the electric field has a significant localized component. As an example we will treat a circuit consisting of an long quantum wire which is perturbed by a single, localized impurity and a phonon bath. In particular, we will investigate the local properties of energy dissipation in the presence of the impurity. For simplicity, we will assume that the impurity can be modeled by a delta function

$$U_{imp}(x) = \Lambda\delta(x), \tag{17.2}$$

where x is the (cyclic) coordinate along the transport directions, whereas the electrons of the wire are confined in the transverse y- and z-directions by a parabolic confinement potential:

$$U_{\text{conf}}(y,z) = \frac{1}{2} m\omega^2 (y^2 + z^2) \,. \tag{17.3}$$

The one-electron energy eigenstates including the effect of the impurity are solutions to the Schrödinger equation

$$\left[-\frac{\hbar^2}{2m} \nabla^2 + U(x,y,z) \right] \phi(x,y,z) = \varepsilon \phi(x,y,z) \tag{17.4}$$

with

$$U(x,y,z) = U_{\text{imp}}(x) + U_{\text{conf}}(y,z) \tag{17.5}$$

Due to the structure of $U(x)$, separation of variables directly leads to factorized wave functions:

$$\phi^{(\lambda)}_{k,n_y,n_z}(x,y,z) = \varphi^{(\lambda)}_k(x) F_{n_y,n_z}(y,z) \,, \tag{17.6}$$

where the energy spectrum is given by

$$\varepsilon_{k,n_y,n_z} = \frac{\hbar^2 k^2}{2m} + \hbar\omega(n_y + n_z + 1) \,. \tag{17.7}$$

The variable k is a positive wave number associated with the longitudinal motion, while the positive integers n_y and n_z are labeling the harmonic oscillator modes related to the lateral confinement. Correspondingly, the envelope functions $F_{n_y,n_z}(y,z)$ are taken to be products of normalized Hermite functions. Finally, λ indicates the propagation direction of the current carried by the eigenfunction $\varphi^{(\lambda)}_k(x,y,z)$. The right (+) and left (-) traveling solutions are respectively represented by

$$\varphi^{(+)}_k(x) = \begin{cases} \mathrm{e}^{\mathrm{i}kx}\,\mathrm{e}^{\mathrm{i}\delta(k)} - \mathrm{i}R^{1/2}(k)\,\mathrm{e}^{-\mathrm{i}kx} & x \le 0 \\ T^{1/2}(k)\,\mathrm{e}^{\mathrm{i}kx} & x \ge 0 \end{cases} \tag{17.8}$$

and

$$\varphi^{(-)}_k(x) = \begin{cases} T^{1/2}(k)\,\mathrm{e}^{-\mathrm{i}kx} & x \le 0 \\ \mathrm{e}^{\mathrm{i}\delta(k)}\,\mathrm{e}^{-\mathrm{i}kx} - \mathrm{i}R^{1/2}(k)\,\mathrm{e}^{\mathrm{i}kx} & x \ge 0 \,. \end{cases} \tag{17.9}$$

$T(k)$ and $R(k)$ are the reflection and transmission coefficients for a delta potential which are well-known from elementary quantum mechanics.

In order to keep the calculations as simple as possible, we have further assumed that the lattice temperature is low enough, say 10 K, and also that the confinement potential is sufficiently narrow such that the electrons of the wire are occupying only the ground state of the subband ladders, corresponding to $n_y = n_z = 0$. The corresponding subband wave function reads

$$F_{00}(y,z) = N_0^2 \mathrm{e}^{-\alpha^2 (y^2+z^2)} \,, \tag{17.10}$$

where $\alpha = \sqrt{m\omega/\hbar}$ and N_0 is a normalization coefficient. Using the above one-electron wave functions and adopting the interaction between electrons and acoustic phonons in the same way as in chapter (11), we may now equally calculate the dissipated power P and the frictional force F_x appearing in the balance equations. However, contrary to the case of the long-channel MOSFET where a uniform electric field was assumed, the presence of a scattering barrier in principle causes the formation of an electric dipole layer in the barrier region. Since the external field is only specified by its electromotive force V_ε, we should in principle solve Poisson's equation in order to determine the spatial distribution of the scalar potential $V(x)$ that is compatible with the charge density emerging from the balance equations. However, since the scattering barrier is ideally represented by a delta function, we may consider the dipole charge approximately being described by the derivative of a delta function. Equivalently, the total electric field $E_x(x)$ could be regarded as being decomposed in a uniform component E_{H} and a delta function with strength V_δ:

$$E_x(x) = E_{\mathrm{H}} + V_\delta \delta(x) , \tag{17.11}$$

where V_δ can be related to the dipole moment of the total charge density $e(n_0 - n(x))$ where n_0 is the average electron concentration in the wire. Multiplying both sides of the first Maxwell equation

$$\frac{\mathrm{d}E_x(x)}{\mathrm{d}x} = \frac{e}{\varepsilon_{\mathrm{S}}}\,(n_0 - n(x)) \tag{17.12}$$

with x and integrating over an interval $[-a/2, a/2]$ surrounding the dipole charge, we find

$$V_\delta = \frac{d}{\varepsilon_{\mathrm{S}}} , \tag{17.13}$$

where d denotes the dipole moment

$$d = e\int_{-a/2}^{a/2} \mathrm{d}x\, x\,(n_0 - n(x)) . \tag{17.14}$$

Having determined V_δ through its relation with the charge density, we may fix E_{H} by the boundary condition

$$\oint_{\mathrm{wire}} E_x(x)\mathrm{d}x = V_\varepsilon . \tag{17.15}$$

Hence,

$$E_{\mathrm{H}} = \frac{V_\varepsilon - V_\delta}{L_x} . \tag{17.16}$$

The resulting energy and momentum balance equations have the following form:

$$IV_\varepsilon + P(\gamma_0, T_\mathrm{e}) = 0 \tag{17.17}$$

$$\frac{1}{2\pi}\int_0^\infty \mathrm{d}k\,\{eV_\delta T(k) + e(V_\varepsilon - V_\delta)\}\sum_\lambda F_\mathrm{e}(\varepsilon_k,\lambda)$$
$$-\frac{\Lambda}{\pi}\int_0^\infty \mathrm{d}k\,k\,R^{1/2}(k)\,T^{1/2}(k)\sum_\lambda F_\mathrm{e}(\varepsilon_k,\lambda) - F_x(\gamma_0,T_\mathrm{e}) = 0 \tag{17.18}$$

where $P(\gamma_0, T_\mathrm{e}) = -(\mathrm{i}/\hbar)\,\langle\,[H_\mathrm{e}, H']\,\rangle_\mathrm{S}$ denotes the power absorbed by the phonons and $F_x(\gamma_0, T_\mathrm{e}) = -(\mathrm{i}/\hbar)\,\langle[P_x, H']\rangle_\mathrm{S}$ is the electron–phonon frictional force, while $F_\mathrm{e}(\varepsilon_k, \lambda)$ denotes the boosted Fermi–Dirac distribution function, i.e.

$$F_\mathrm{e}(\varepsilon_k,\lambda) = \left[1 + \exp\left(\frac{\varepsilon_k - \gamma_0 I_k^{(\lambda)} - \mu}{k_\mathrm{B}T_\mathrm{e}}\right)\right]^{-1}. \tag{17.19}$$

17.2 Current–Voltage Characteristics and Local Energy Dissipation

Below some typical numerical results are presented for a silicon quantum wire of length $L_x = 1$ mm and lateral widths $L_y = L_z = 10$ nm, while the lattice temperature is taken to be 10 K. Figure 17.1 shows the current I as a function of the applied voltage V_ε ranging between 10^{-4} and 10 V. Increasing the voltage up to a threshold value (about 1 V for a pure wire), the response of the quantum wire is in the ohmic regime where the current is proportional to the applied voltage, while the response becomes non-linear beyond the threshold. There however, the obtained results may become questionable because more subbands may start getting occupied while also other inelastic scattering mechanisms may contribute to the relaxation of the electron energy.

In order to get insight in the spatial dependence of energy dissipation, we have calculated the local energy density profile $w(x)$ of the electrons flowing through the quantum wire as a byproduct of the solutions to the balance equations. The result is presented in Fig. 17.2 for different values of the barrier height Λ at $V_\varepsilon = 10$ mV. For sufficiently small values of the barrier height Λ one observes that the energy density exhibits almost no fluctuations. Moreover, one can readily see that an energy drop occurs at $x = 0$ for values of Λ that are not too large.

In the region $x < 0$ the electrons acquire on the average a larger energy density than for $x > 0$ in this case. This energy density drop first increases for increasing values of the barrier height Λ, but when Λ becomes sufficiently large the energy drop disappears and only the oscillations in the energy density remain. This behaviour for large values of Λ is due to the fact that the elastic scatterer effectively isolates the left from the right part of the wire. As a result an electron gas at equilibrium is realized, where, due to the presence

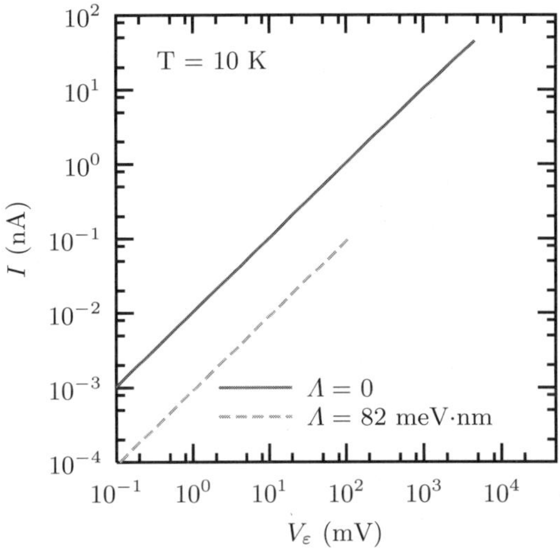

Fig. 17.1. Current–voltage characteristics for a quantum wire with and without elastic scattering

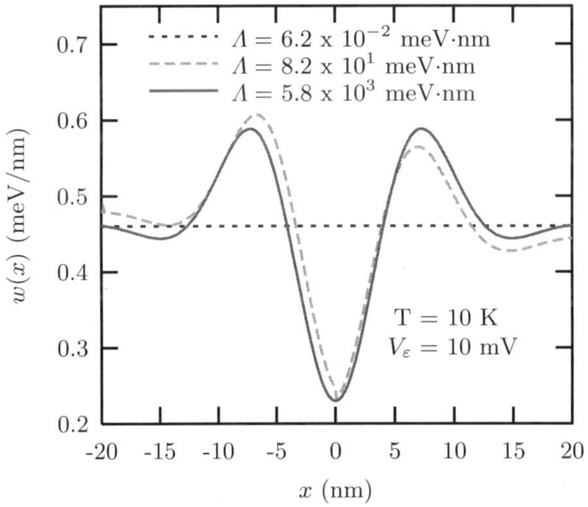

Fig. 17.2. Local electron energy density $w(x)$ for different values of the barrier height Λ

of an elastic scatterer, depletion and Friedel oscillations occur in the vicinity of $x = 0$ at low values of the lattice temperature.

Although the interaction model adopted in the present calculation was highly simplified, we have been able to investigate qualitatively how the presence of an elastic scattering barrier changes the electron wave functions which, in turn, influence the electron–phonon coupling, and therefore affect the local dissipation of energy governed by inelastic scattering mechanisms.

18. Future Work

This last chapter is devoted to some speculative suggestions for further theoretical research. It also gives the reader a number of suggestions that may trigger his drive to contribute to the deeper understanding of nanoscale devices. The inclusion of a chapter with above title prints a 'time stamp' on this monograph. As a consequence, the work will become out-of-date in the near or distant future, either because the work is realized or because the work turns out to be irrelevant. Yet we take the risk to include this chapter because we believe that finalizing the reading of this book should be rather the beginning of scientific exploration than an end.

18.1 Constructing Non-equilibrium Ensembles

In the eighth chapter, non-equilibrium statistical physics, has been presented, starting from the density operator. It turned out that irreversibility is plugged in by a projection mechanism, such that at each projection information is lost and therefore entropy increases. From a pragmatic point of view such a method may be legitimate, however a flavor of arbitrariness is still present and therefore we may attempt to construct non-equilibrium density operators in a less ad hoc manner.

In the introduction we have suggested that d'Alembert ingenious trick to replace dynamics by statics might be mimiced in order to set up a formalism for non-equilibrium thermodynamics. Here we will describe such an attempt. In fact, since the non-equilibrium issues concern both variations in space and time, it is beneficial to consider first a related problem: the covariant generalization of the path-integral quantization procedure. Several quantum-mechanical formulas are placed in a new context. As a side result, a theoretical frame is given for the treatment of non-conservative systems and non-equilibrium thermodynamic systems.

Relativity has merged space and time into a unified geometrical structure: the (possibly curved) Minkowski space. Such a level of unification has never been achieved in quantum theory. In particular, the *postulates* of quantum theory make a clear distinction in the treatment of the time coordinate and the spatial coordinates. To be more explicit, the time coordinate lies at the basis of the description of the dynamical evolution of the quantum system, whereas, the spatial coordinates are linked to degrees of freedom of the system, either as observables or as pointers which count the degrees of freedom as is the case in a local field theory. There are reported several approaches in the literature to set up quantum theories in which the Minkowski space acts as the underlying geometry. The most common approach for introducing relativistic covariance into quantum theory is by performing the operator substitutions

$$H \to \mathrm{i}\hbar\frac{\partial}{\partial t}\,, \qquad \boldsymbol{p} \to -\mathrm{i}\hbar\nabla \tag{18.1}$$

and the construction of covariant wave equations. It is demonstrated that measurable results obey relativistic covariance [125].

In the path-integral approach, the Feynman postulates for *non-relativistic* quantum mechanics are applied to continuous systems (fields) and the action integral becomes a Lorentz scalar, such that covariance is obtained. One may suspect that Feynman considers in his famous Review of Modern Physics paper [23], non-relativistic systems for tutorial purposes, but the generalization of the Feynman postulates to non-relativistic systems is very subtle. The reason is that in non-relativistic mechanics there is no limitation to the velocity, whereas in relativity the velocity of light is an absolute limit. Remarkably enough, applying the path-integral method to relativistic systems requires that all space–time paths needs to be considered, i.e. also paths which have segments with velocity greater than the velocity of light. This fact was noted by Papadopoulos and Devreese [126].

An attempt to set up a quantum theory for space–time paths in Minkowski space was presented by Feynman [127]. In this approach, the paths are parameterized using an invariant variable, τ, and the trajectories are mappings of τ to a four-vector function, $x^\mu = x^\mu(\tau)$. The action integral and the Klein–Gordon equation are ($\hbar = c = 1$)

$$S = \int d\tau\, \frac{1}{2}\dot{x}_\mu \dot{x}^\mu \tag{18.2}$$

$$\mathrm{i}\frac{\partial\phi}{\partial\tau} = -\partial^2\phi\,. \tag{18.3}$$

The mass term is recovered by identifying the usual Klein–Gordon solution $\phi(x)$ as

$$\phi(x) = \int d\tau\, \mathrm{e}^{\mathrm{i}m^2\tau}\phi(x,\tau)\,. \tag{18.4}$$

Later considerations in higher-dimensional field theories [128] have led to similar equations, where τ is treated on equal footing with x^μ, i.e.

$$\frac{\partial^2\phi}{\partial\tau^2} = \partial^2\phi\,, \qquad \phi(x) = \int d\tau\, \mathrm{e}^{\mathrm{i}m\tau}\phi(x,\tau)\,. \tag{18.5}$$

Another approach was initiated by Tomonaga[129] and Schwinger [130]. In this formulation, the Minkowski space is subdivided into three-dimensional space-like hyper planes, $\sigma(x)$, and the dynamical evolution is described by locally varying these space-like surfaces, i.e. with $\partial/\partial t \to \delta/\delta\sigma(x)$ one finds

$$\mathrm{i}\frac{\delta}{\delta\sigma(x)}\Psi\left[\sigma\right] = -\mathcal{L}_{\mathrm{int}}\Psi\left[\sigma\right] \; . \tag{18.6}$$

The interaction representation is essential to arrive at a covariant equation, since at least in quantum electrodynamics, $\mathcal{L}_{\mathrm{int}} = -\mathcal{H}_{\mathrm{int}}$ is invariant. The covariant generalization of the constant time planes to space-like surfaces implies that the quantum theory is independent of the choice of the basis in the Minkowski space.

Here we will present yet another approach to a covariant formulation of the postulates of quantum theory. The method starts from Feynman's non-relativistic postulates and considers constrained paths in extended phase space. Using the quantization methods of Faddeev and Popov [132], we arrive at a covariant extension of Feynman's postulates.

18.1.1 Covariance in Classical Physics

We will consider a system with N generalized coordinates $q_1, q_2, \ldots, q_N$. The potential in the action integral may be explicitly time-dependent.

$$\begin{aligned} S &= \int L(q,\dot{q},t)\mathrm{d}t \\ &= \int_{t_0}^{t_1}\left(\frac{1}{2}\sum_{n=1}^{N} m_n\,\dot{q}_n^2\right)\mathrm{d}t - \int_{t_0}^{t_1} V(q_1,\ldots,q_N,t)\,\mathrm{d}t \; . \end{aligned} \tag{18.7}$$

This integral may be reparameterized by defining a new coordinate $q_0(s) = t$, i.e.

$$\begin{aligned} S &= \int_{s_0}^{s_1}\left(\frac{1}{2}\sum_{n=1}^{N} m_n\,q_n'^2\frac{1}{q_0'}\right)\mathrm{d}s - \int_{s_0}^{s_1} V(q_0,q_1,\ldots,q_N)\,q_0'\,\mathrm{d}s \\ &= \int \mathrm{d}s\,\tilde{L}(q_0,q_1,\ldots,q_N,q_0',q_1',\ldots,q_N') \end{aligned} \tag{18.8}$$

and

$$q_i' = \frac{\mathrm{d}q_i}{\mathrm{d}s} \; . \tag{18.9}$$

Defining the conjugate momentum

$$p_0 = \frac{\partial\tilde{L}}{\partial q_0'} \tag{18.10}$$

one obtains that $p_0 = -H$. The remaining canonical momenta are

$$p_i = \frac{\partial\tilde{L}}{\partial q_i'} = m_i\frac{q_i'}{q_0'} = m_i\dot{q}_i \; . \tag{18.11}$$

Moreover, the generalized Hamiltonian becomes

$$\begin{aligned}\tilde{H}=&\sum_{n=o}^{N} p_n q_n' - \tilde{L} \\ =&q_0' \left(-H + \frac{1}{2}\sum_{n=1}^{N} m_n \dot{q}_n^2 + V(q_0, q_1, \ldots, q_N) \right) = 0 \,, \end{aligned} \tag{18.12}$$

which implies that

$$\tilde{L} = \sum_{n=0}^{N} p_n q_n' \,. \tag{18.13}$$

The fact that $K = p_0 + H = 0$, implies that not all conjugate momenta are independent and therefore the equations of motion must be obtained using the method of Lagrange multipliers. Variation of the action $S = \int \mathrm{d}s\,(\tilde{L} - \lambda K)$ gives

$$q_n' = \lambda \frac{\partial K}{\partial p_n}, \qquad p_n' = -\lambda \frac{\partial K}{\partial q_n}, \qquad K = 0 \,. \tag{18.14}$$

An observable may in general be a function of all the generalized coordinates, i.e.

$$F = F(q_0, \ldots, q_N, p_0, \ldots, p_N) \tag{18.15}$$

or in conventional notation: $F = F(\boldsymbol{q}, \boldsymbol{p}, t, -H)$, i.e. an explicit time-dependence may be included. The change of F as a function of the parameter s is

$$\begin{aligned}F' =& \sum_{n=0}^{N} \left(\frac{\partial F}{\partial q_n} q_n' + \frac{\partial F}{\partial p_n} p_n' \right) \\ =& \lambda \sum_{n=0}^{N} \left(\frac{\partial F}{\partial q_n}\frac{\partial K}{\partial p_n} - \frac{\partial F}{\partial p_n}\frac{\partial K}{\partial q_n} \right) \\ =& \lambda\,[F, K] \,, \end{aligned} \tag{18.16}$$

where the last equality defines the Poisson bracket in the extended phase space. In particular, the following fundamental Poisson brackets will be used in the next section.

$$[q_k, q_l] = 0 \,, \qquad [p_k, p_l] = 0 \,, \qquad [q_k, p_l] = \delta_{kl} \,. \tag{18.17}$$

Relativistic covariance can be realized by elimination of the preferred choice of p_0 in the constraint $K = 0$. By selecting a manifest covariant constraint, relativistic covariant dynamics can be obtained. For a relativistic particle, the constraint can be taken as [2]

$$K = \frac{1}{2}\left(p^2 - m^2 c^2\right) = 0 \tag{18.18}$$

where $p = (E/c, \boldsymbol{p})$ is the relativistic four momentum and $p^2 = (E/c)^2 - |\boldsymbol{p}|^2$.

18.1.2 Canonical Quantization

In the canonical quantization procedure, variables will be replaced by Hermitian operators. The application of this prescription to the variable set $\{q_0, q_1, \ldots, q_N, p_0, p_1, \ldots, p_N\}$ implies that the time variable also becomes an *operator*.

$$q_k \to \hat{q}_k \,, \qquad p_k \to \hat{p}_k \,. \tag{18.19}$$

Furthermore, by using the substitution prescription for the Poisson brackets [18], i.e.

$$[A, B]_{\mathrm{pb}} \to -\frac{\mathrm{i}}{\hbar}\,[A, B] \tag{18.20}$$

we obtain

$$[\hat{q}_k, \hat{p}_l] = \mathrm{i}\hbar\delta_{kl} \,. \tag{18.21}$$

Using the constraint $K = p_0 + H = 0$, we derive that the pair consisting of the time operator $\hat{t} = \hat{q}_0$ and the Hamiltonian $\hat{p}_0 = -\hat{H}$, obey the following commutation relation

$$\left[\hat{t}, \hat{H}\right] = -\mathrm{i}\hbar \,. \tag{18.22}$$

In the representation in which we use the eigenstates of the time operator, i.e. the 'diagonal-time' representation, the Hamiltonian becomes $\hat{H} = \mathrm{i}\hbar\frac{\partial}{\partial t}$, and the classical constraint $K = 0$ translates to

$$\mathrm{i}\hbar\frac{\partial}{\partial t} - \hat{H} = \hat{0} \,. \tag{18.23}$$

Therefore we may interpret the Schrödinger equation as a *constraint*. The specific role that the time coordinate plays in formulating the postulates of quantum mechanics may be traced to the specific choice of the constraint $K = 0$. This situation may be compared with the quantization scheme of gauge theories. In the latter ones, the axial gauge is very specific in the sense that of the four gauge potentials, there is selected one which is constrained to be identically zero. Such a choice is highly non-symmetrical but reduces the remaining degrees of freedom. The axial gauges are also special because the ghost couplings disappear. This analogy can be pursued further, as will be demonstrated in the next section. It is also noted that the commutation rules in (18.21) are the starting point for a covariant derivation of the Heisenberg uncertainty relations.

18.1.3 Path-Integral Quantization

By formulating the quantization procedure in extended phase space, well established results should be reproduced. In particular, the transition amplitude for a system going from a state (q_a, t_a) to a state (q_b, t_b) is [131, 133]

$$\langle q_b, t_b | q_a, t_a \rangle = \int \prod_{i=1}^{N} [\mathrm{d}q\, \mathrm{d}p] \exp\left(\frac{\mathrm{i}}{\hbar} \int_{q_a(t_a)}^{q_b(t_b)} \mathrm{d}t \left[\sum_{i=1}^{N} p_i \dot{q}_i - H \right] \right) . \quad (18.24)$$

A mere *renaming*, using $p_0 = -H$ and $q_0 = t(s) = s$, allows us to write

$$\langle q_b | q_a \rangle = \int \prod_{i=1}^{N} [\mathrm{d}q\, \mathrm{d}p] \exp\left(\frac{\mathrm{i}}{\hbar} \int_{q_a(s_a)}^{q_b(s_b)} \mathrm{d}s \sum_{i=0}^{N} p_i q_i' \right) . \quad (18.25)$$

Furthermore, we can insert two more functional delta functions for enlarging the path-integral measure.

$$\langle q_b | q_a \rangle = \int \prod_{i=0}^{N} [\mathrm{d}q\, \mathrm{d}p]\, \delta\left(q_0 - t(s)\right) \delta\left(p_0 + H\right) \times \exp\left(\frac{\mathrm{i}}{\hbar} \int_{q_a(s_a)}^{q_b(s_b)} \mathrm{d}s \sum_{i=0}^{N} p_i q_i' \right) . \quad (18.26)$$

We may now replace the specific constraint $K = p_0 + H$ and apply an arbitrary constraint $K = 0$. Then a corresponding additional condition [133], analogous to $q_0 = t(s)$, must be selected, i.e. $C = 0$, such that the Poisson bracket $[K, C] \neq 0$, and the transition amplitude finally becomes

$$\langle q_b | q_a \rangle = \int \prod_{i=0}^{N} [\mathrm{d}q\, \mathrm{d}p]\, \delta\left(K\right) \delta\left(C\right) \mid \det[K, C] \mid \times \exp\left(\frac{\mathrm{i}}{\hbar} \int_{q_a(s_a)}^{q_b(s_b)} \mathrm{d}s \sum_{i=0}^{N} p_i q_i' \right) , \quad (18.27)$$

where the determinant is the Jacobian of the mapping to the new constraint. For the specific constraint this Jacobian is a constant and therefore it decouples from the quantization procedure, c.f. the axial gauge in Yang–Mills theories, but in general this term may not be omitted. Equation (18.27) is the starting point for the quantization in extended phase space in which the time coordinate and space coordinates are equivalent.

To illustrate these ideas, we consider a relativistic scalar particle. A covariant Lagrangian is given by the following expression

$$L = \frac{1}{2} m \frac{\mathrm{d}q_\mu}{\mathrm{d}s} \frac{\mathrm{d}q^\mu}{\mathrm{d}s} . \quad (18.28)$$

The generalized Hamiltonian is $K = 1/2(p^2 - m_0^2 c^2) = 0$ as was noted in (18.18). In particular, we may select a covariant condition for C, i.e.

$$C = p\dot{q} . \quad (18.29)$$

The constraint corresponds to a constant-phase constraint and leads to the following Poisson brackets:

$$[K, C] = -p^2 . \tag{18.30}$$

Using a Faddeev–Popov ghost variable $c(s)$ with complex conjugate $c^*(s)$, we can rewrite the determinant for the last constraint as

$$|\det [K, C]| = \int [\mathrm{d}c\, \mathrm{d}c^*] \exp \int \mathrm{d}s\, c^*(s)\, p^2(s)\, c(s) . \tag{18.31}$$

Therefore, the relativistic covariant path integral for a scalar particle becomes after using Fourier representations for the constraints

$$\langle q_b | q_a \rangle = \int \prod_{i=0}^{4} [\mathrm{d}q\, \mathrm{d}p][\mathrm{d}\lambda\, \mathrm{d}\omega][\mathrm{d}c\, \mathrm{d}c^*] \exp \left(\frac{\mathrm{i}}{\hbar} \int_{q_a(s_a)}^{q_b(s_b)} \mathrm{d}s \right.$$
$$\left. \times \left[\sum_{i=0}^{N} p_i q_i' - \hbar\lambda \left(p^2 - m^2\right) - \hbar\omega\, p \cdot q - \mathrm{i}\hbar\, c^*(s) p^2(s) c(s) \right] \right) . \tag{18.32}$$

The ghost variables c and c^* are independent Grassmann variables and obey anti-commutation rules

$$\{c^*(s), c^*(s')\} = \{c^*(s), c(s')\} = \{c(s), c(s')\} = 0 . \tag{18.33}$$

It is interesting to consider the classical limit of this quantization prescription. In particular, we consider the constraints $p_0 + H = 0$ and $q_0 - t(s) = 0$ with $t(s) = s$. The delta-functionals in (18.26) can be represented by Fourier integrals:

$$\langle q_b | q_a \rangle = \int \prod_{i=0}^{N} [\mathrm{d}q\, \mathrm{d}p][\mathrm{d}\lambda\, \mathrm{d}\omega] \exp \left(\frac{\mathrm{i}}{\hbar} \int_{q_a(s_a)}^{q_b(s_b)} \mathrm{d}s \right.$$
$$\left. \times \left[\sum_{i=0}^{N} p_i q_i' - \hbar\lambda \left(p_0 + H\right) - \hbar\omega \left(q_0 - t(s)\right) \right] \right) . \tag{18.34}$$

A stationary action generates the following equations of motion

$$p_k' = \frac{\partial}{\partial q_k} (\lambda K + \omega C) \qquad q_k' = -\frac{\partial}{\partial p_k} (\lambda K + \omega C) \tag{18.35}$$

$$K = p_0 + H = 0 \qquad C = q_0 - t(s) = 0 . \tag{18.36}$$

Evaluation of p_0' and q_0' gives $\omega = 0$. Therefore, the part ωC of the action is usual ignored in classical mechanics [2]. Nevertheless, in general this term should be included.

Summarizing, we have presented a method for the quantization of dynamical systems in which the role of the time coordinate and the space coordinates are treated on equivalent footing. The key observation is that the dynamics in extended space is constrained, i.e. not all degrees of freedom are independent, which implies that Faddeev–Popov quantization procedure is required. The constrained dynamics suggests a formal analogy with the Yang–Mills theories. In fact, for the free relativistic particle there is a continuous symmetry of the action corresponding to τ-reparameterizing invariance in (18.2). Siegel has used this symmetry to pursue the similarity with the Yang–Mills theory and also introduced a Faddeev–Popov term as well as a ghost coupling in the action in order to define a quantum theory for the relativistic particle [134]. Siegel furthermore exploits BRST invariance to construct the complete Lagrangian. His theory differs from ours in the specific choice of the auxiliary constraint and corresponding ghost coupling. By formulating the quantum theory in extended phase space we have also obtained a new method for quantizing non-conservative systems, since an explicit time dependence of the Hamiltonian is permitted in this formalism.

18.1.4 Guessing a Density Function

The N-particle system follows a trajectory in the $(2N+2)$-dimensional extended phase space. In an analogous line of reasoning we may now include statistical considerations and imagine many copies of the physical system that are distributed with a density function $\varrho = \varrho(q_0, q_1, \ldots, q_N; p_0, p_1, \ldots, p_N)$. Each copy corresponds to a trajectory in phase space. From a microscopic point of view, all trajectories fall on the hyper plane determined by the constraints $K = 0$ and $C = 0$. We now take the (bold) step that such detailed knowledge concerning the constraints, is not required: only the expectation values of these constraints are needed. In other words, the constraints are relaxed and every point in extended phase space may be visited by a ensemble trajectory. Of course, the probability that this happens depends on the density function. The entropy principle allows us to find an explicit expression for the density function. By introducing two Lagrange multipliers for the two expectation values, we obtain that

$$\varrho(q_0, q_1, \ldots, q_N; p_0, p_1, \ldots, p_N) = C \, \exp\left(-\lambda K - \omega C\right) . \tag{18.37}$$

The constant C may be taken

$$Z = C^{-1} = \int [\mathrm{d}p \, \mathrm{d}q] \, \varrho(p, q) . \tag{18.38}$$

The 'micro-canonical' conditions $K \equiv 0$ and $C \equiv 0$ are reobtained if $\lambda, \omega \to \infty$. Defining the entropy as the $\langle \log \varrho \rangle$, we find

$$S = \left(\frac{\partial}{\partial \lambda} + \frac{\partial}{\partial \omega} \right) Z . \tag{18.39}$$

The above considerations conclude our suggestions for further progress in the field of non-equilibrium statistical mechanics. It is clear that only a preliminary scheme is given and much further work is needed.

18.2 Quantum Circuit Theory

Coming back to the area of electric circuit theory, it goes without saying that a sound theory is required not only to support and to refine the concept of flux quantization for non-superconducting circuits, but also to bridge the gap between a rigorous, microscopic transport description and the global circuit model that is to reflect the quantum mechanical features of coherent transport through an electric circuit or part of it. Such a theory which could be called 'quantum circuit theory' (QCT) might emerge as an extension of the good old theory of QED that would generalize the quantization of the electromagnetic field on two levels: not only should one address non-trivial topologies such as toroidal regions in which finite currents may flow and finite charges may be induced, but also an appropriate set of conjugate observables describing the global circuit properties should be defined. In view of the previous considerations regarding the magnetic flux trapped by the circuit, a natural pair of variables could be the flux of the electric displacement field $\boldsymbol{D}$ through a cross section Σ_0 crossing the circuit in the interior of the active region and the magnetic flux threaded by the loop Γ_0:

$$\Phi_{\mathrm{D}} = \int_{\Sigma_0} \boldsymbol{D} \cdot \mathrm{d}\boldsymbol{S} \tag{18.40}$$

$$\Phi_{\mathrm{M}} = \oint_{\Gamma_0} \boldsymbol{A} \cdot \mathrm{d}\boldsymbol{r} \,. \tag{18.41}$$

Taking the electric displacement field instead of the electric field itself to construct a 'partner' for Φ_{M} has mainly to do with the requirement that the the product of two conjugate variables have the dimension of an action ($\propto \hbar$). Assuming that $\boldsymbol{D}$ vanishes outside the active region Ω_{A}, one may consider the latter as a leaky capacitor the plates of which are separated by Σ_0 such that, according to Gauss' law, Φ_{D} would equal the charge accumulated on one plate, say Q_{A} (see Fig. 18.1).

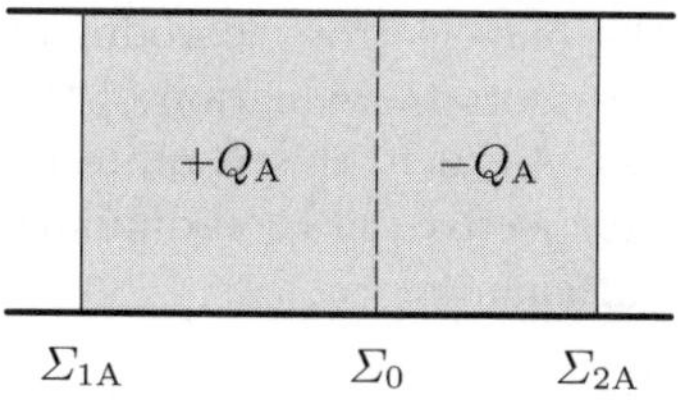

Fig. 18.1. Cross section Σ_0 separating positive and negative charges in the active region Ω_{A}

Canonical quantization would then imply the replacement $\Phi_{\mathrm{D}} \to \hat{\Phi}_{\mathrm{D}}$ and $\Phi_{\mathrm{M}} \to \hat{\Phi}_{\mathrm{M}}$, imposing

$$\left[\hat{\Phi}_{\mathrm{D}}, \hat{\Phi}_{\mathrm{M}}\right] = \mathrm{i}\hbar$$

$$\left[\hat{\Phi}_{\mathrm{D}}, \hat{\Phi}_{\mathrm{D}}\right] = \left[\hat{\Phi}_{\mathrm{M}}, \hat{\Phi}_{\mathrm{M}}\right] = 0 \,. \tag{18.42}$$

It is now tempting to propose a phenomenological expression like

$$\hat{H} = \frac{\hat{\Phi}_{\mathrm{D}}^2}{2C} + \frac{\hat{\Phi}_{\mathrm{M}}^2}{2L} + \hat{\Phi}_{\mathrm{D}} V_\varepsilon - \hat{\Phi}_{\mathrm{M}} \hat{I} \tag{18.43}$$

for a circuit Hamiltonian describing the interaction between the electromagnetic field variables $(\hat{\Phi}_{\mathrm{D}}, \hat{\Phi}_{\mathrm{M}})$ and the electron current operator $\hat{I} = \int_{\Sigma_0} \hat{\boldsymbol{J}} \cdot \mathrm{d}\boldsymbol{S}$ under the constraint $Q_{\mathrm{A}} = \langle \Phi_{\mathrm{D}} \rangle$, and to derive the corresponding Heisenberg equations of motion with the help of the commutation relations (18.42):

$$\frac{\mathrm{d}\hat{\Phi}_{\mathrm{D}}(t)}{\mathrm{d}t} = -\frac{\mathrm{i}}{\hbar}\left[\hat{\Phi}_{\mathrm{D}}(t), \hat{H}\right] = \frac{\hat{\Phi}_{\mathrm{M}}(t)}{L} - \hat{I}(t) \tag{18.44}$$

$$\frac{\mathrm{d}\hat{\Phi}_{\mathrm{M}}(t)}{\mathrm{d}t} = -\frac{\mathrm{i}}{\hbar}\left[\hat{\Phi}_{\mathrm{M}}(t), \hat{H}\right] = -\frac{\hat{\Phi}_{\mathrm{D}}(t)}{C} - V_\varepsilon \, . \tag{18.45}$$

At first sight, the above equations are satisfied by meaningful steady-state solutions that may be obtained by setting the long-time averages $\langle \ldots \rangle_S = \lim_{t\to\infty} \langle \ldots \rangle_t$ of $\mathrm{d}\hat{\Phi}_{\mathrm{D}}(t)/\mathrm{d}t$ and $\mathrm{d}\hat{\Phi}_{\mathrm{M}}(t)/\mathrm{d}t$ equal to zero. Indeed, the resulting equations

$$\langle \hat{I} \rangle_S = \frac{\langle \hat{\Phi}_{\mathrm{M}} \rangle_S}{L} \tag{18.46}$$

$$\frac{Q_A}{C} = \frac{\langle \Phi_{\mathrm{D}} \rangle_S}{C} = -V_\varepsilon \tag{18.47}$$

are restating the familiar result that the steady-state of the circuit is determined by a current that is proportional to the magnetic flux, while the capacitor voltage tends to the externally applied electromotive force.

However, in order to investigate whether the quantum dynamics generated by the proposed Hamiltonian eventually leads to the Landauer–Büttiker formula or not, would require us to give a meaningful definition of the inductance and capacitance coefficients L and C as well as a recipe to calculate the statistical averages in a straightforward manner. Clearly, this can only be accomplished if a full microscopic investigation of the circuit is performed including both the self-consistent solution of the one-electron Schrödinger equation and the fourth Maxwell equation, and a rigorous evaluation of the dynamical, quantum-statistical ensemble averages. As such, this is quite an elaborate task which, however, may open new perspectives in the boundary region between electromagnetism and quantum mechanics.

References

1. H. Goldstein: *Classical Mechanics* (John Wiley & and Sons, Inc., New York 1963)
2. C. Lanczos: *The variational Principles of Mechanics*, 4th edn. (University of Toronto Press, Toronto 1977)
3. R.P. Feyman, R. Leighton, M. Sands: *The Feyman Lectures on Physics, Vol. I* (Addison-Wesley Publ. Comp. Inc., Reading MA 1963)
4. J. Schwinger: Phys. Rev. **82**, 914 (1951)
5. R.P. Feynman: *Statistical Mechanics* (Benjamin Inc., London 1972)
6. K. Huang: *Statistical Mechanics* (John Wiley & Sons, New York 1963)
7. H.J. Kreuzer: *Nonequilibrium Thermodynamics and its Statistical Foundations* (Clarendon, Oxford 1981)
8. P. Drude: Annalen der Physik **1**, 566 (1900)
9. P. Drude: Annalen der Physik **3**, 369 (1900)
10. M. Lundstrom: *Fundamentals of Carrier Transport*, 2nd edn. (Cambridge University Press, 1999)
11. R. Kubo: J. Phys. Soc. Japan **12**, 570 (1957)
12. A. Forghieri, R. Guerrieri, P. Ciampolini, A. Gnudan, M. Rudan, G. Baccarani: IEEE Trans. on Computer-Aided Design **7**, 407 (1988)
13. R. Stratton: Phys. Rev. **126**, 2002 (1962)
14. G. Arfken: *Mathematical Methods for Physicists*, 2nd edn. (Academic Press, New York 1970)
15. R. Feynman and A. Hibbs: *Qunatum Mechanics and Path Integrals* (Mc. Graw-Hill, New York 1965)
16. R.P. Feyman, R. Leighton, M. Sands: *The Feyman Lectures on Physics, Vol. III* (Addison-Wesley Publ. Comp. Inc., Reading MA 1963)
17. A. Messiah: *Quantum Mechanics*, 5th edn. (North Holland Publ. Comp. Amsterdam Oxford 1975)
18. P.A.M. Dirac: *The Principles of Quantum Mechanics*, 4th edn. (Clarendon Oxford University Press, Oxford 1958)
19. G. Baym: *Lectures on Quantum Mechanics*, W.A. Benjamin, Reading Massachusetts 1973)
20. E. Merzbacher: *Quantum Mechanics* (John Wiley & Sons, New York 1970)
21. L.D. Landau, E.M. Lifshitz: *Quantum Mechanics*, Pergamon Press Ltd., London (1958)
22. C. Cohen-Tannoudji, B. Diu, F. Laloë: *Quantum Mechanics, Vol. I,II* (John Wiley & Sons, New York 1977)
23. R.P. Feynman: Rev. Mod. Phys. **20**, 367 (1948)
24. P. .M. Morse, H. Feshbach: *Methods of Theoretical Physics: Vol. I and Vol. II* (Mc.Graw-Hill, New York 1953)
25. S. Flügge: *Practical Quantum Mechanics* (Springer Verlag, New York - Heidelberg - Berlin 1974)

26. F.F. Fang, W.E. Howard: Phys. Rev. Lett. **16**, 797 (1966)
27. I. Yanetka: phys. stat. sol. (b) **208**, 61 (1998)
28. B. Robertson: Am. J. Phys. **41**, 678 (1973)
29. A.L. Fetter, J.D. Walecka: *Quantum Theory of Many-Particle Systems* (McGraw-Hill Inc., New York 1971)
30. G.D. Mahan: *Many-particle physics* Plenum Press, N.Y. (1993)
31. E.T. Jaynes: Phys. Rev. **106**, 620 **106**, 620 (1957)
32. R. Kubo: Can. J. Phys. **34**, 1274 (1956)
33. H. Mori: Phys. Rev. **112**, 1829 (1958)
34. K.G. Wilson: Rev. Mod. Phys. **55**,583 (1983)
35. W. Schoenmaker: Nucl. Phys. **B285[FS19]**, 316 (1987)
36. R. Horsley, W. Schoenmaker: Nucl. Phys. **B280[FS18]**, 716 (1987)
37. R. Horsley, W. Schoenmaker: Nucl. Phys. B **B280[FS18]**, 735 (1987)
38. W. Schoenmaker: Phys. Lett. **B182**, 373 (1986)
39. W. Schoenmaker, M. Ichiyanagi: 'Covariant Derivation of Relativistic Hydrodynamics', report Institute for Theoretisal Physics, Catholic University of Leuven, Leuven (1987)
40. T. Matsubara, H. Matsuda: Prog. Theor. Phys. **16**, 416 (1956)
41. D.N. Zubarev: Fortschr. Phys. **18**, 125 (1970); **20**, 471 (1972)
42. D.N. Zubarev: *Non-Equilibrium Statistical Mechanics* (Plenum New York, 1974)
43. E.P. Wigner: Phys. Rev. **40**, 749 (1932)
44. W. Hänsch: *The Drift-Diffusion Equation And Its Applications in MOSFET Modeling* (Springer-Verlag Wien New-York, 1991)
45. A. Wettstein: *Quantum Effects in MOS Devices* Ph.D. Thesis, Swiss Federal Institute of Technology, Zurich (2000)
46. M.G. Ancona: Phys. Rev. **B42**, 1222 (1990)
47. Zhiping Yu, R.W. Dutton, D.W. Yergeau, M.G. Ancona: 'Macroscopic Quantum Transport Modeling' In: *Simulation of Semiconductor Processes and Devices, SISPAD 01, International Conference at Athens, Greece, September 5–7, 2001*, ed. by D. Tsoukalis, C. Tsamis (Springer Wien New-York, 2001) pp. 1–9
48. F.M. Peeters, J.T. Devreese: Phys. Rev. **B 23**, 1936 (1981)
49. X.L. Lei, C.S. Ting: Phys. Rev. **B30**, 4809 (1984)
50. X.L. Lei, C.S. Ting: Phys. Rev. **B 32**, 1112 (1985)
51. X.L. Lei, C.S. Birman, C.S. Ting: J. Appl. Phys. **58**, 2270 (1985)
52. W. Magnus, W. Schoenmaker: J. Math. Phys. **39**, 6715 (1998)
53. D.N. Zubarev: Sov. Phys. Usp. **3**, 320 (1960)
54. C. Kittel: *Introduction to Solid State Physics*, chapter5 (John-Wiley & Sons, New York, London 1976)
55. W. Magnus, W. Schoenmaker: J. Appl. Phys. **88**, 5833 (2000) *reprinted from Journal of Applied Physics, Vol. 88, pp. 5833–5842 Copyright 2000, with permission from American Institute of Physics*
56. J. Bardeen: Phys. Rev. Lett. **6**, 57 (1961)
57. H.P. Joosten, H.J.M.F. Noteborn, D. Lenstra: Thin Solid Films **184** (1990)
58. H.J.M.F. Noteborn, H.P. Joosten, D. Lenstra: Physica Scripta T **33** (1990)
59. H.J.M.F. Noteborn, H.P. Joosten, D. Lenstra, K. Kaski: *SPIE Vol. 1675, Quantum Well and Superlattice Physics IV*, 57 (1992)
60. J. Sune, P. Olivio, B. Ricco: J. Appl. Phys. **70**, 337 (1991)
61. W. Magnus, W. Schoenmaker: Proc. of the 29th European Solid-State Device Research Conference (ESSDERC'99), Leuven 13-15 September 1999, p. 248
62. G. Breit, E.P. Wigner: Phys. Rev. **49**, 519 (1936)
63. S.-H. Lo, D.A. Buchanan, Y. Taur, W. Wang: IEEE Elec. Dev. Lett. **18**, 209 (1997)

64. S.-H. Lo, D.A. Buchanan, Y. Taur: IBM J. Res. Develop. **43**, 327 (1999)
65. A. Schenk: Advanced Physical Models for Silicon Device Simulation, Ed. by Siegfried Selberherr, Springer-Verlag, Wien (1998)
66. W. Magnus, C. Sala, K. De Meyer: J. Appl. Phys. **63**, 2704 (1988)
67. W. Magnus, C. Sala, K. De Meyer: phys. stat. sol(b) **153**, K31 (1989)
68. C. Sala, W. Magnus, K. De Meyer: J. Appl.Phys. **69**, 7689 (1991)
69. B. Brar, G.D. Wilk, A.C. Seabaugh: Appl. Phys. Lett. **69**, 2728 (1996)
70. M. Depas, R.L.V. Meirhaeghe, W.H. Laflere, F. Cardon, Solid-State Electron. **37**, 433 (1994)
71. A. Wettstein, A. Schenk, W. Fichtner: Proceedings of the SISPAD'99 Conference, Kyoto, 1999, p. 243.
72. A.K. Ghatak, K. Thyagarajan, M. Shenoy: IEEE J. Quantum Electron., **24**, 1524 (1988)
73. Y. Nakajima, Y. Takahashi, S. Horiguchi, K. Iwadate, H. Namatsu, K. Kurihara, M. Tabe: Appl. Phys. Lett. **65**, 2833 (1994)
74. M. Ono, M. Saito, T. Yoshitomi, C. Fiegna, T. Ohguro, H. Iwai: IEEE Trans. Electron. Devices, **42**, 1822 (1995)
75. L. Guo, P.R. Krauss, S.Y. Chou: Appl. Phys. Lett. **71**, 1881 (1997)
76. S.N. Balaban, E.P. Pokatilov, V.M. Fomin, V.N. Gladilin, J.T. Devreese, W. Magnus, W. Schoenmaker, M. Van Rossum, B. Sorée: Solid-State Elec. **46**, 435 (2002) *reprinted from Solid-State Electronics, Vol. 46, pp. 435–444 Copyright 2002, with permission from Elsevier Science*
77. J.C.S. Woo, K.W. Terril, P.K. Vasudev: IEEE Trans. Elect. Dev. **ED-37**, 1999 (1990)
78. H.O. Joachim, Y. Yamaguchi, K. Ishikawa, Y. Inoue, T. Nashimuras: IEEE Trans. Electron. Devices, **40**, 1812 (1993)
79. S. Pidin, M. Koyanagis: Jpn. J. App. Phys. **37**, 1264 (1998)
80. T. Sekigawa, Y. Hayashi: Solid State Electronics, **27**, 827 (1984)
81. F.G. Pikus, K.K. Liharev: Appl. Phys. Lett. **71**, 3661 (1997)
82. E.P. Pokatilov, V.M. Fomin, S.N. Balaban, V.N. Gladilin, S.N. Klimin, J.T. Devreese, W. Magnus, W. Schoenmaker, N. Collaert, M. Van Rossum, K. De Meyer: J. Appl. Phys. **85**, 6625 (1999)
83. W. Frensley: Rev. Mod. Phys. **62**, 745 (1990).
84. H. Tsuchiya, M. Ogawa, T. Miyoshis: Jpn. J. Appl. Phys. **30**, 3853 (1991).
85. K.L. Jensen, A.K. Ganguly: J. Appl. Phys. **73**, 4409 (1993).
86. A. Isihara: *Statistical Physics* (State University of New York, Buffalo, Academic Press New York - London 1971).
87. A. Abramowitz, L.A. Stegun (Eds.): *Handbook of Mathematical Functions with Formulas, Graphs, and Mathematical Tables* (National Bureau of Standards, Washington, DC 1972).
88. *Semiconductors. Physics of II-VI and I-VII Compounds, Semimagnetic Semiconductors*, edited by K.H. Hellwege, Landolt-Börnstein, New Series, Group III, Vol. 17, (Springer, Berlin 1982).
89. M.A. Reed, W.P. Kirk: Nanostructure physics and fabrication, Proc. of the International Symposium, College Station, Texas, March 13–19, 1989, Academic Press Inc., N.Y.
90. A.O. Caldeira, A.J. Leggett: Phys. Rev. Lett. **46**, 211 (1981)
91. W. Magnus, W. Schoenmaker: Phys. Rev **B 47**, 1276 (1993) *reprinted from Physical Review Vol. B47, 1276. Copyright 1993 by the American Physical Society*
92. A.O. Caldeira, A.J. Leggett: Ann. Phys. **149**, 374 (1983)
93. K. Fujikawa, S. Iso, M. Sasaki, H. Suzuki: Phys. Rev. Lett. **68**, 1093 (1992)
94. C. Kittel: *Quantum Theory of Solids* (Wiley, New York, 1963)

95. N.N. Bogolyubov: *Studies in Statistical Mechanics Vol. 1* (North-Holland, Amsterdam, 1962)
96. I. Ojima, H. Hasegawa, M. Ichiyanagi: J. Stat. Phys. **50**, 633 (1987)
97. M. Ichiyanagi: J. Phys. Soc. Jpn **58**, pp. 2297–2315, 2727–2740 (1989).
98. R. Landauer: IBM J. Res. Dev. **1**, 223 (1957)
99. R. Landauer: Philos. Mag. **21**, 863 (1970)
100. R. Landauer: Phys. Lett. **85 A**, 91 (1981)
101. M. Büttiker: Phys. Rev. **B 64**, 3764 (1986)
102. E.N. Economou, C.M. Soukoulis: Phys. Rev. Lett. **46**, 618 (1981)
103. Y. Imry, R. Landauer: Rev. Mod. Phys. **71**, S306 (1999)
104. S.C. Zhang: Int. J. Mod. Phys. **B 6**, 25 (1992)
105. E.W. Fenton: Phys. Rev. **B 46**, 3754 (1992)
106. E.W. Fenton: Superlatt. Microstruct. **16**, 87 (1994)
107. H.A. Baranger, A.D. Stone: Phys. Rev. **B 40**, 8169 (1989)
108. W. Magnus, W. Schoenmaker: Phys. Rev. **B 61**, 10883 (2000) *reprinted from Physcial Review Vol. B61, 10883. Copyright 2000 by the American Physical Society*
109. Y. Aharonov, D. Bohm: Phys. Rev. **B 115**, 485 (1959)
110. J.J. Sakurai: *Advanced Quantum Mechanics* (Addison Wesley Publishing Company Inc., Massachusetts, 1976)
111. G. 't Hooft: Nucl. Phys. **B 153**, 141 (1979)
112. D.A. Wharam: T.J. Thornton, R. Newbury, M. Pepper, H. Ahmed, J.E.F. Frost, D.G. Hasko, D.C. Peacock, D.A. Ritchie, G.A.C. Jones: J. Phys. **C 21**, L209 (1988)
113. B.J. van Wees, H. van Houten, C.W.J. Beenakker, J.G. Williamson, L.P. Kouwenhoven, D. van der Marel, C.T. Foxon: Phys. Rev. Lett. **60**, 848 (1988)
114. A. Kawabata: J. Phys. Soc. Jpn. **58**, 372 (1989)
115. S. Datta: Electronic Transport in Mesoscopic Systems (Cambridge University Press, UK 1995)
116. A.D. Stone, A. Szafer: IBM J. Res. Dev. **32**, 384 (1988)
117. L. Wendler, V.M. Fomin: Phys. Rev. **B 51**, 17814 (1995)
118. R.B. Laughlin: Phys. Rev. **B 23**, 5632 (1981)
119. S. Kivelson, D.-H. Lee, S.-C. Zhang, Scientific American, March , **64** (1996)
120. T. Figielski, T. Wosinski: J. Appl. Phys. **85**, 1984 (1999)
121. A. van Oudenaarden, M.H. Devoret, Y.V. Nazarov, J.E. Mooij: Nature **391**, 768 (1998)
122. J.R. Barker: *Handbook on Semiconductors, Vol. 1*, North Holland Publishing Company, Amsterdam, 1982, pp. 617–631.
123. D.Y. Xing, M. Liu, C.S. Ting: Phys. Rev. **B37**, 10283 (1988)
124. W. Magnus, C. Sala, K.D. Meyer: *Proceedings of the 4th international conference on computational physics* (1992)
125. J. Bjorken, S. Drell: *Relativistic Quantum Mechanics* (Mc Graw-Hill Book Company, New York 1964)
126. G.J. Papadopoulos, J.T. Devreese: Phys. Rev. **D13**, 2227 (1976)
127. R.P. Feynman: Phys. Rev. **76**, 749 (1949)
128. P. Van Nieuwenhuizen: Phys. Rep. **68**, 189 (1981)
129. S. Tomonaga: Progr. Theor. Phys. **I, 27** (1946)
130. J. Schwinger: Phys. Rev. **74**, 1439 (1948)
131. R.P. Feynman: Phys. Rev. **80**, 440 (1950)
132. L.D. Faddeev, V.N. Popov, Phys. Letters **25B**, 29 (1967)
133. L.D. Faddeev: Theor. Math. Phys. **1**, 3 (1969)
134. W. Siegel: Phys. Lett. **151B**, 391 (1985)

Index

Springer Series in Solid-State Sciences

Editors: M. Cardona P. Fulde K. von Klitzing H.-J. Queisser

1 **Principles of Magnetic Resonance** 3rd Edition By C. P. Slichter
2 **Introduction to Solid-State Theory** By O. Madelung
3 **Dynamical Scattering of X-Rays in Crystals** By Z. G. Pinsker
4 **Inelastic Electron Tunneling Spectroscopy** Editor: T. Wolfram
5 **Fundamentals of Crystal Growth I** Macroscopic Equilibrium and Transport Concepts By F. E. Rosenberger
6 **Magnetic Flux Structures in Superconductors** 2nd Edition By R. P. Huebener
7 **Green's Functions in Quantum Physics** 2nd Edition By E. N. Economou
8 **Solitons and Condensed Matter Physics** Editors: A. R. Bishop and T. Schneider
9 **Photoferroelectrics** By V. M. Fridkin
10 **Phonon Dispersion Relations in Insulators** By H. Bilz and W. Kress
11 **Electron Transport in Compound Semiconductors** By B. R. Nag
12 **The Physics of Elementary Excitations** By S. Nakajima, Y. Toyozawa, and R. Abe
13 **The Physics of Selenium and Tellurium** Editors: E. Gerlach and P. Grosse
14 **Magnetic Bubble Technology** 2nd Edition By A. H. Eschenfelder
15 **Modern Crystallography I** Fundamentals of Crystals Symmetry, and Methods of Structural Crystallography 2nd Edition By B. K. Vainshtein
16 **Organic Molecular Crystals** Their Electronic States By E. A. Silinsh
17 **The Theory of Magnetism I** Statics and Dynamics By D. C. Mattis
18 **Relaxation of Elementary Excitations** Editors: R. Kubo and E. Hanamura
19 **Solitons** Mathematical Methods for Physicists By. G. Eilenberger
20 **Theory of Nonlinear Lattices** 2nd Edition By M. Toda
21 **Modern Crystallography II** Structure of Crystals 2nd Edition By B. K. Vainshtein, V. L. Indenbom, and V. M. Fridkin
22 **Point Defects in Semiconductors I** Theoretical Aspects By M. Lannoo and J. Bourgoin
23 **Physics in One Dimension** Editors: J. Bernasconi and T. Schneider
24 **Physics in High Magnetics Fields** Editors: S. Chikazumi and N. Miura
25 **Fundamental Physics of Amorphous Semiconductors** Editor: F. Yonezawa
26 **Elastic Media with Microstructure I** One-Dimensional Models By I. A. Kunin
27 **Superconductivity of Transition Metals** Their Alloys and Compounds By S. V. Vonsovsky, Yu. A. Izyumov, and E. Z. Kurmaev
28 **The Structure and Properties of Matter** Editor: T. Matsubara
29 **Electron Correlation and Magnetism in Narrow-Band Systems** Editor: T. Moriya
30 **Statistical Physics I** Equilibrium Statistical Mechanics 2nd Edition By M. Toda, R. Kubo, N. Saito
31 **Statistical Physics II** Nonequilibrium Statistical Mechanics 2nd Edition By R. Kubo, M. Toda, N. Hashitsume
32 **Quantum Theory of Magnetism** 2nd Edition By R. M. White
33 **Mixed Crystals** By A. I. Kitaigorodsky
34 **Phonons: Theory and Experiments I** Lattice Dynamics and Models of Interatomic Forces By P. Brüesch
35 **Point Defects in Semiconductors II** Experimental Aspects By J. Bourgoin and M. Lannoo
36 **Modern Crystallography III** Crystal Growth By A. A. Chernov
37 **Modern Chrystallography IV** Physical Properties of Crystals Editor: L. A. Shuvalov
38 **Physics of Intercalation Compounds** Editors: L. Pietronero and E. Tosatti
39 **Anderson Localization** Editors: Y. Nagaoka and H. Fukuyama
40 **Semiconductor Physics** An Introduction 6th Edition By K. Seeger
41 **The LMTO Method** Muffin-Tin Orbitals and Electronic Structure By H. L. Skriver
42 **Crystal Optics with Spatial Dispersion, and Excitons** 2nd Edition By V. M. Agranovich and V. L. Ginzburg
43 **Structure Analysis of Point Defects in Solids** An Introduction to Multiple Magnetic Resonance Spectroscopy By J.-M. Spaeth, J. R. Niklas, and R. H. Bartram
44 **Elastic Media with Microstructure II** Three-Dimensional Models By I. A. Kunin
45 **Electronic Properties of Doped Semiconductors** By B. I. Shklovskii and A. L. Efros
46 **Topological Disorder in Condensed Matter** Editors: F. Yonezawa and T. Ninomiya

Springer Series in Solid-State Sciences

Editors: M. Cardona P. Fulde K. von Klitzing H.-J. Queisser

47 **Statics and Dynamics of Nonlinear Systems**
Editors: G. Benedek, H. Bilz, and R. Zeyher

48 **Magnetic Phase Transitions**
Editors: M. Ausloos and R. J. Elliott

49 **Organic Molecular Aggregates**
Electronic Excitation and Interaction Processes
Editors: P. Reineker, H. Haken, and H. C. Wolf

50 **Multiple Diffraction of X-Rays in Crystals**
By Shih-Lin Chang

51 **Phonon Scattering in Condensed Matter**
Editors: W. Eisenmenger, K. Laßmann, and S. Döttinger

52 **Superconductivity in Magnetic and Exotic Materials** Editors: T. Matsubara and A. Kotani

53 **Two-Dimensional Systems, Heterostructures, and Superlattices**
Editors: G. Bauer, F. Kuchar, and H. Heinrich

54 **Magnetic Excitations and Fluctuations**
Editors: S. W. Lovesey, U. Balucani, F. Borsa, and V. Tognetti

55 **The Theory of Magnetism II** Thermodynamics and Statistical Mechanics By D. C. Mattis

56 **Spin Fluctuations in Itinerant Electron Magnetism** By T. Moriya

57 **Polycrystalline Semiconductors**
Physical Properties and Applications
Editor: G. Harbeke

58 **The Recursion Method and Its Applications**
Editors: D. G. Pettifor and D. L. Weaire

59 **Dynamical Processes and Ordering on Solid Surfaces** Editors: A. Yoshimori and M. Tsukada

60 **Excitonic Processes in Solids**
By M. Ueta, H. Kanzaki, K. Kobayashi, Y. Toyozawa, and E. Hanamura

61 **Localization, Interaction, and Transport Phenomena** Editors: B. Kramer, G. Bergmann, and Y. Bruynseraede

62 **Theory of Heavy Fermions and Valence Fluctuations** Editors: T. Kasuya and T. Saso

63 **Electronic Properties of Polymers and Related Compounds**
Editors: H. Kuzmany, M. Mehring, and S. Roth

64 **Symmetries in Physics** Group Theory Applied to Physical Problems 2nd Edition
By W. Ludwig and C. Falter

65 **Phonons: Theory and Experiments II**
Experiments and Interpretation of Experimental Results By P. Brüesch

66 **Phonons: Theory and Experiments III**
Phenomena Related to Phonons
By P. Brüesch

67 **Two-Dimensional Systems: Physics and New Devices**
Editors: G. Bauer, F. Kuchar, and H. Heinrich

68 **Phonon Scattering in Condensed Matter V**
Editors: A. C. Anderson and J. P. Wolfe

69 **Nonlinearity in Condensed Matter**
Editors: A. R. Bishop, D. K. Campbell, P. Kumar, and S. E. Trullinger

70 **From Hamiltonians to Phase Diagrams**
The Electronic and Statistical-Mechanical Theory of sp-Bonded Metals and Alloys By J. Hafner

71 **High Magnetic Fields in Semiconductor Physics**
Editor: G. Landwehr

72 **One-Dimensional Conductors**
By S. Kagoshima, H. Nagasawa, and T. Sambongi

73 **Quantum Solid-State Physics**
Editors: S. V. Vonsovsky and M. I. Katsnelson

74 **Quantum Monte Carlo Methods in Equilibrium and Nonequilibrium Systems** Editor: M. Suzuki

75 **Electronic Structure and Optical Properties of Semiconductors** 2nd Edition
By M. L. Cohen and J. R. Chelikowsky

76 **Electronic Properties of Conjugated Polymers**
Editors: H. Kuzmany, M. Mehring, and S. Roth

77 **Fermi Surface Effects**
Editors: J. Kondo and A. Yoshimori

78 **Group Theory and Its Applications in Physics**
2nd Edition
By T. Inui, Y. Tanabe, and Y. Onodera

79 **Elementary Excitations in Quantum Fluids**
Editors: K. Ohbayashi and M. Watabe

80 **Monte Carlo Simulation in Statistical Physics**
An Introduction 4th Edition
By K. Binder and D. W. Heermann

81 **Core-Level Spectroscopy in Condensed Systems**
Editors: J. Kanamori and A. Kotani

82 **Photoelectron Spectroscopy**
Principle and Applications 2nd Edition
By S. Hüfner

83 **Physics and Technology of Submicron Structures**
Editors: H. Heinrich, G. Bauer, and F. Kuchar

84 **Beyond the Crystalline State** An Emerging Perspective By G. Venkataraman, D. Sahoo, and V. Balakrishnan

85 **The Quantum Hall Effects**
Fractional and Integral 2nd Edition
By T. Chakraborty and P. Pietiläinen

86 **The Quantum Statistics of Dynamic Processes**
By E. Fick and G. Sauermann

87 **High Magnetic Fields in Semiconductor Physics II**
Transport and Optics Editor: G. Landwehr

88 **Organic Superconductors** 2nd Edition
By T. Ishiguro, K. Yamaji, and G. Saito

89 **Strong Correlation and Superconductivity**
Editors: H. Fukuyama, S. Maekawa, and A. P. Malozemoff

Springer Series in Solid-State Sciences

Editors: M. Cardona P. Fulde K. von Klitzing H.-J. Queisser

Managing Editor: H. K.V. Lotsch

90 **Earlier and Recent Aspects of Superconductivity**
Editors: J. G. Bednorz and K. A. Müller

91 **Electronic Properties of Conjugated Polymers III** Basic Models and Applications
Editors: H. Kuzmany, M. Mehring, and S. Roth

92 **Physics and Engineering Applications of Magnetism** Editors: Y. Ishikawa and N. Miura

93 **Quasicrystals** Editors: T. Fujiwara and T. Ogawa

94 **Electronic Conduction in Oxides** 2nd Edition
By N. Tsuda, K. Nasu, F. Atsushi, and K. Siratori

95 **Electronic Materials**
A New Era in Materials Science
Editors: J. R. Chelikowsky and A. Franciosi

96 **Electron Liquids** 2nd Edition By A. Isihara

97 **Localization and Confinement of Electrons in Semiconductors**
Editors: F. Kuchar, H. Heinrich, and G. Bauer

98 **Magnetism and the Electronic Structure of Crystals** By V.A. Gubanov, A.I. Liechtenstein, and A.V. Postnikov

99 **Electronic Properties of High-T_c Superconductors and Related Compounds**
Editors: H. Kuzmany, M. Mehring, and J. Fink

100 **Electron Correlations in Molecules and Solids** 3rd Edition By P. Fulde

101 **High Magnetic Fields in Semiconductor Physics III** Quantum Hall Effect, Transport and Optics By G. Landwehr

102 **Conjugated Conducting Polymers**
Editor: H. Kiess

103 **Molecular Dynamics Simulations**
Editor: F. Yonezawa

104 **Products of Random Matrices** in Statistical Physics By A. Crisanti, G. Paladin, and A. Vulpiani

105 **Self-Trapped Excitons**
2nd Edition By K. S. Song and R. T. Williams

106 **Physics of High-Temperature Superconductors**
Editors: S. Maekawa and M. Sato

107 **Electronic Properties of Polymers**
Orientation and Dimensionality of Conjugated Systems Editors: H. Kuzmany, M. Mehring, and S. Roth

108 **Site Symmetry in Crystals**
Theory and Applications 2nd Edition
By R. A. Evarestov and V. P. Smirnov

109 **Transport Phenomena in Mesoscopic Systems** Editors: H. Fukuyama and T. Ando

110 **Superlattices and Other Heterostructures**
Symmetry and Optical Phenomena 2nd Edition
By E. L. Ivchenko and G. E. Pikus

111 **Low-Dimensional Electronic Systems**
New Concepts
Editors: G. Bauer, F. Kuchar, and H. Heinrich

112 **Phonon Scattering in Condensed Matter VII**
Editors: M. Meissner and R. O. Pohl

113 **Electronic Properties of High-T_c Superconductors**
Editors: H. Kuzmany, M. Mehring, and J. Fink

114 **Interatomic Potential and Structural Stability**
Editors: K. Terakura and H. Akai

115 **Ultrafast Spectroscopy of Semiconductors and Semiconductor Nanostructures**
2nd Edition By J. Shah

116 **Electron Spectrum of Gapless Semiconductors**
By J. M. Tsidilkovski

117 **Electronic Properties of Fullerenes**
Editors: H. Kuzmany, J. Fink, M. Mehring, and S. Roth

118 **Correlation Effects in Low-Dimensional Electron Systems**
Editors: A. Okiji and N. Kawakami

119 **Spectroscopy of Mott Insulators and Correlated Metals**
Editors: A. Fujimori and Y. Tokura

120 **Optical Properties of III–V Semiconductors**
The Influence of Multi-Valley Band Structures
By H. Kalt

121 **Elementary Processes in Excitations and Reactions on Solid Surfaces**
Editors: A. Okiji, H. Kasai, and K. Makoshi

122 **Theory of Magnetism**
By K. Yosida

123 **Quantum Kinetics in Transport and Optics of Semiconductors**
By H. Haug and A.-P. Jauho

124 **Relaxations of Excited States and Photo-Induced Structural Phase Transitions**
Editor: K. Nasu

125 **Physics and Chemistry of Transition-Metal Oxides**
Editors: H. Fukuyama and N. Nagaosa